ASPECTS OF MATERIALS HANDLING

ASPECTS OF MATERIALS HANDLING

By

DR. K.C. ARORA
B.E., M.E., Ph.D.
Prof. Mech. Engg.
MITS, Gwalior
(Madhya Pradesh)

VIKAS V. SHINDE
B.E., M.Tech.
Lecturer,
Singhgad Institute of Technology
The Lonavala Campus
Kusgaon (Maharashtra)

UNIVERSITY SCIENCE PRESS

(An Imprint of Laxmi Publications Pvt. Ltd.)

BANGALORE • CHENNAI • COCHIN • GUWAHATI • HYDERABAD
JALANDHAR • KOLKATA • LUCKNOW • MUMBAI • RANCHI
NEW DELHI • BOSTON, USA

Published by :

UNIVERSITY SCIENCE PRESS

(*An Imprint of Laxmi Publications Pvt. Ltd.*)
113, Golden House, Daryaganj,
New Delhi-110002

Phone : 011-43 53 25 00
Fax : 011-43 53 25 28

www.laxmipublications.com
info@laxmipublications.com

Price : ₹ **350.00** *Only.*

First Edition : 2007
Reprint : 2012, 2013

OFFICES

✆ **Bangalore**	080-26 75 69 30	✆ **Chennai**	044-24 34 47 26
✆ **Cochin**	0484-237 70 04, 405 13 03	✆ **Guwahati**	0361-251 36 69, 251 38 81
✆ **Hyderabad**	040-24 65 23 33	✆ **Jalandhar**	0181-222 12 72
✆ **Kolkata**	033-22 27 43 84	✆ **Lucknow**	0522-220 99 16
✆ **Mumbai**	022-24 91 54 15, 24 92 78 69	✆ **Ranchi**	0651-220 44 64

UAM-9654-350-ASPECT OF MATERIAL HAND-ARO

Typeset at : ABRO Enterprises, Delhi

C—6344/013/02

Printed at : Ajit Printers, Delhi.

CONTENTS

PREFACE

Materials handling engineering, simply, consists of determining "how a product should be moved from one place to another, within the shortest allowable period of time, for the least cost and with the least amount of manual effort".

This publication will give you an idea of the different materials handling equipments and their uses. Also in this book all related aspects such as plant layout, selection, maintenance and design is discussed up to the possible extent keeping in view scope of the book. As far as possible new developments related to the basic equipments are also added. New thing to this book is that it contains the Indian standards used for material handling and the list of manufacturers of different material handling equipments from India.

We hope that this publication will help to guide you to the best possible solution to the many materials handling problems, which you may encounter. It is extremely difficult to put in to writing the many years of problem solving experienced by "old-timers" in different industries, no two solutions are identical.

—Authors

1 Introduction to Materials Handling

Every modern and ancient wonder has one common factor between them *i.e.,* the time, which had taken to complete these wonders. Take example of any wonder in the world whether it is wonderful Taj Mahal, heavy Pyramids or Robust Eiffel Tower. They all had taken years to complete. Now the question may be raised why had taken so much years? Because if at present if we wish to construct all these wonders they may take few months. So where the difference arises. Answer is very simple:- "Materials Handling."

Materials handling is defined as movement of material of any form (raw, finished, packaged, solid, liquid, gas, light and heavy) from one location to another location either in a restricted path by manual or mechanical aids. The movement may be horizontal, vertical or may be combination of both.

1.1 HISTORY OF MATERIALS HANDLING

Materials handling have very old and vast history. If we look into history this concept is thousands of years old. In ancient times materials handling is not a concept but it is only formal work which is needed in any activity. Also point wasn't highlighted as it is today.

In older days materials handled generally using animals such as Elephants, Horses, Bulls, Axes, Yaks etc. But now-a-days we see very few such activities. Also human-being is the most important material transporter. History suggests that in ancient days kings used to keep slaves, these slaves are used to transport material from one place to another. He used his muscular power to transport.

A tonnes of materials from one place to another. Also materials are rolled over, or slide down, glided by different means such as wooden blocks, taper land and ice. Also some times river flow is used to transport the material before invention of wheels. In middle age history, wheels were invented. At this time the job of material handling became easier than earlier. All means such as human beings, animals are now used to coupled with carts and they are used to transport the material.

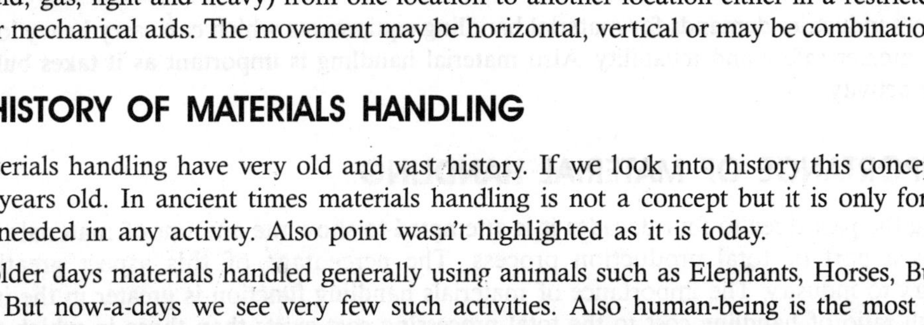

Figure 1.1. *Animal carts used for moving marbles from Rajasthan to build up Taj.*

As we turn towards modern history materials handling gained lots of importance. Modernization of industries and civilizations needs material handling equipment which can handle heavy loads with high speed and great reliability and safety.

In May, 1913 operations were started on the first progressive assembly line at the Ford Motor Company, within a short time, number of other production assembly had been set upon various types of conveyors.

1.2 NEED OF MATERIALS HANDLING

Need comes from daily activities, past experiences, future predictions. Every activity we do need movement from one place to another place whether it is eating food, catching balls, travelling. Each activity needs means of moving the material. Hence there is a need of material handling and hence material handling equipments.

Although in daily activities we use very negligible amount of material which is to be handled as compared to industrial material which is heavy and bulky, this heavy and bulky materials cannot be handled efficiently. That's why we needs material handling equipments.

Any type of industry whether small or big, production plant or process plant, agriculture industry or service sector in which such application there is only one common thing *i.e.,* movement of material, material handling. Thus material handling directly affects the outcome of the plant. Hence material handling is important. Due to this reason the equipments, which are used, are also important.

Today's industries demands for material handling equipments which can carry heavy loads with high speed, greater safety and reliability. Also material handling is important as it takes bulk of the time of any activity.

1.3 IMPORTANCE OF MATERIAL HANDLING

During the past three to four decades it is attempted to show the amount of materials handling provided as a part of total production process. The percentage of this aspect greatly varies from industry to industry. The importance of materials handling function is greater in the industries where a high ratio of handling cost to the total processing cost exists than those in which this ratio is smaller.

Material handling does not involve only handling of materials but it also involves packaging and warehousing. The following sentences will give you some idea:

"Materials handling generates something like 80% of indirect labour costs in the plant."

"Materials handling may be responsible for one-third to one-half of the total cost of production in some plants."

"An uninterrupted flow of materials through a plant can increase the productivity of existing men and equipment by 15%."

1.4 ADVANTAGES OF MATERIALS HANDLING

There are so many advantages of materials handling systems, some of important advantages are listed below :

(*a*) Reduction in material handling and indirect labour cost.

(*b*) Improvement in the productivity.

(*c*) Better use of floor space and facilities.

(*d*) Reduced cost of transportation

(*e*) Less fatigue to the workers

(*f*) Improved work environment.
(*g*) Reduction in bottlenecks.
(*h*) Greater flexibility.
(*i*) Improvement in the safety and reduction in the accidents.
(*j*) Improved material flow *i.e.*, less backtracking and reduction in inventory.

1.5 DISADVANTAGES OF MATERIAL HANDLING

Although material handling systems offers so many advantages but there are few disadvantages also which are listed below :

(*a*) Initial cost of the equipment in the most of cases is high.
(*b*) Skilled labour or operator is required to operate almost all types of material handling systems.
(*c*) If not operated well there are concerns of horrible accidents and hazards.
(*d*) Regular maintenance is required.
(*e*) Inventory carrying cost is high in some cases.
(*f*) Installation and operating cost is high.

1.6 FUNCTIONS INVOLVED IN MATERIALS HANDLING

(*a*) Surveys and Analysis
(*b*) Engineering Drawings
(*c*) Testing and Experimental work
(*d*) Installation
(*e*) Maintenance
(*f*) Assembling and packaging.

1.7 TERMS AND DEFINITIONS USED IN MATERIAL HANDLING

Analysis of Handling Economies

An analysis of materials handling economies requires an investigation of cost factors that enter into both the present and the proposed methods. The savings expected by the new equipment should be sufficient to cover equipment maintenance, interest, depreciation, obsolescence, taxes and other fixed charges. The amount of savings show the maximum that can be spent on the new equipment.Usually industries feel that such type of handling devices must pay for itself within a relatively short period, say 2 years or so.

In material handling some terminology is used. Before understanding the material handling as a subject it is necessary to understand these terms as these terms will appear now and then in this book.

Transport

It is a movement of material from one production centre to another production centre or from one storage area to other storage area. The purpose of transport is to locate the product for another operation or storage.

Transfer

The movement of material from container to machine or vice versa.

Materials

Almost all the materials are handled in a bulk or packaged form.

Bulk Materials

Any material that are loose and handled in quantities without being contained in bags, boxes, barrels and the like the material handling equipment supplied with temporary container.

Packaged Materials

Materials contained in convenient sized packages, such as bags, boxes, crates, cartoons, cans, barrels or other type of containers that are handled as individual pieces by material handling system.

Unit Load

A certain number of packaged units mounted together on skids, pallets, and platform or skid box for movement as a single unit. The unit load may be made of certain number of individual pieces attached to a pallet or contained in a box.

Logistics

Integrated control and physical transfer of material (raw) and semi finished and finished products from a supplier to consumer via raw materials store, processing and finished products store. Also called as Material Administration.

Charge to Pallet Ratio

It is the proportion of goods, in a given time and quantity that loaded on pallets.

UNSOLVED QUESTIONS

1. What is material handling?
2. Write the history of material handling.
3. Why we need material handling equipments?
4. Explain importance of material handling.
5. Explain relative advantages and disadvantages of the material handling.
6. Define following terms :

   Logistics, Bulk Material, Packaged Material, Unit Load.

❑❑❑❑

2 Material Handling Equipments

INTRODUCTION

Starting from the time, the raw material enters the factory gate and goes out of the factory gate in the form of finished products, it is handled at all stages in between, no matter it is in the stores or on the shop floor. It has been estimated that average handling cost is roughly 20 to 60% of the total production cost and the ratio between the weight of handled material to produce an item and the weight of the finished item may vary between 40 to 50. A component may be handled even 50 times or more before it changes to finished product. It thus becomes clear that the cost of production of an item can be lowered considerably by making a saving in the material handling cost.

Material handling involves the movements of materials, manually or mechanically in batches or one item at a time within the plant. The movement may be horizontal, vertical or the combination of horizontal and vertical.

Material handling emphasizes upon the need of installing efficient and safe methods and equipments for material handling. It has been found that 35 to 40% of the plant accidents are the results of bad methods of materials handling.

There are so many ways by which material handling equipments are classified. In this chapter we will discuss classification of material handling equipments.

2.1 CLASSIFICATION OF MATERIAL HANDLING EQUIPMENTS

The main groups of materials handling equipments classified according to

I. By their design features
II. By nature and type of work
III. By working area
IV. By movement or motion

2.1.1 By their Design Features

I. **Hoisting equipment** is a group of machines with lifting gear intended for moving loads mainly in batches.

II. **Conveying equipment** is a group of machines which may have no lifting gear and which move loads in a continuous flow.

III. **Surface and overhead equipment** is a group of machines which also may not be provided with lifting gear and which usually handle loads in batches.

Each of these groups of machines is distinguished by a number of special features and by their specific spheres of application. The differences in the design of these groups also depend on the nature of the load they handle, the direction of working movement and the nature of the handling process.

2.1.2 Nature and Type of Work

Material handling equipment (MHE) is used for the movement and storage within a facility or at a site. MHE can also be classified into the following five major categories:

I. **Transport Equipment:** Equipment used to move material from one location to another (*e.g.*, between workplaces, between a loading dock and a storage area, etc.). Material can also be transported manually using no equipment.

II. **Positioning Equipment:** Equipment used to handle material at a single location so that it is in the correct position for subsequent handling, machining, transport or storage. Unlike transport equipment, positioning equipment is usually for handling at a single workplace. Material can also be positioned manually using no equipment.

III. **Unit Load Formation Equipment:** Equipment used to restrict materials so that they maintain their integrity when handled a single load during transport and for storage. If materials are self-restraining (*e.g.,* a single part or interlocking parts), then they can be formed into a unit load with no equipment.

IV. **Storage Equipment:** Equipment used for holding or buffering materials over a period of time. Some storage equipment may include the transport of materials (*e.g.,* the S/R machines of an AS/RS, or storage carousels). If materials are block stacked directly on the floor, then no storage equipment is required.

V. **Identification and Control Equipment:** Equipment used to collect and communicate the information that is used to co-ordinate the flow of materials within a facility and between a facility and its suppliers and customers. The identification of materials and associated control can be performed manually with no specialized equipment.

I. Transport Equipment

A. Conveyors

1. Chute Conveyor
2. Wheel Conveyor
3. Roller Conveyor
4. Chain Conveyor
5. Slat Conveyor
6. Flat belt Conveyor
7. Magnetic belt Conveyor
8. Troughed belt Conveyor
9. Bucket Conveyor
10. Vibrating Conveyor
11. Screw Conveyor
12. Pneumatic Conveyor
13. Vertical Conveyor
14. Cart-on-track Conveyor
15. Tow Conveyor
16. Trolley Conveyor
17. Power-and-free Conveyor
18. Monorail
19. Sortation Conveyor
20. Hydraulic Conveyor

B. Cranes

1. Jib Crane
2. Bridge Crane
3. Gantry Crane
4. Stacker Crane
5. Fixed Crane
6. Mobile Crane
7. Floating Crane
8. Wall Crane

9. Revolving Crane
10. Tower Crane
11. Crawler Crane
12. Locomotive Crane

C. Industrial Trucks

1. Hand Truck
2. Pallet Jack
3. Walkover Stacker
4. Pallet Truck
5. Platform Truck
6. Counterbalanced lift truck
7. Narrow-aisle straddle truck
8. Narrow-aisle reach truck
9. Turret Truck
10. Order Picker
11. Side loader
12. Tractor-trailer
13. Personnel and burden carrier
14. Automatic guided vehicle (AGV)

II. Positioning Equipment

1. Manual (no equipment)
2. Lift/tilt/turn table
3. Dock leveler
4. Ball transfer table
5. Rotary index table
6. Parts feeder
7. Air film device
8. Hoist
9. Balancer
10. Manipulator
11. Industrial robot

III. Unit Load Formation Equipment

1. Self-restraining (no equipment)
2. Pallets
3. Skids
4. Slip sheets
5. Tote pans
6. Pallet boxes/skid boxes
7. Bins/baskets/racks
8. Cartoons
9. Bags
10. Bulk load containers
11. Crates
12. Intermodal containers
13. Strapping/tape/glue
14. Shrink-wrap/stretch-wrap
15. Palletizers

IV. Storage Equipment

1. Block stacking (no equipment)
2. Selective pallet rack
3. Drive-through rack
4. Drive-in rack
5. Flow-through rack
6. Push-back rack
7. Sliding rack
8. Cantilever rack
9. Stacking Frame
10. Shelves/bins/drawers
11. Storage Carousel
12. Automatic Storage/Retrieval
13. Split case order picking system
14. Mezzanine

V. Identification and Control Equipment

1. Manual (no equipment)
2. Bar codes
3. Radio frequency (RF) tag
4. Magnetic stripe
5. Machine vision
6. Portable data terminal
7. Electronic data Interchange (EDI)/Internet

List of material handling equipments is so large and unlimited. Students may get confused. Students are advised to note only major classification of above equipments. In preceeding chapters efforts has been made to explain everything listed above.

2.1.3 By Working Area

Unrestricted

These equipments can move in any direction. These include all types of trucks and tractors and carts.

Area Restricted

They work in the restricted area *i.e.,* their work area is constant. These include overhead travelling cranes, gantry cranes and stacker cranes.

Line Restricted

Such type of equipments generally works as a continuous transportation device traveling on a specified path. These include almost all types of conveyors, railways, lifts and elevators and automatically guided vehicles etc.

Position Restricted

They are fixed units and works in specified area from its position. These includes jib cranes and other industrial robots.

Auxiliary Equipments

These equipments helps in improving the effectiveness of handling, these includes load carriers, warehouse, goods reception and dispatch equipments ,assembling and securing equipments.

2.1.4 By Movement or Motion

1. Vertical motion (lifting and lowering devices)
2. Horizontal motion (transportation devices)
3. Combination motion (lifting and lowering plus transportation)

Although all material handling equipments are important but in industries most of the work is done by cranes, conveyors and trucks because of their variety, capacity to handle heavy and bulky loads and suitability to such kind of operation. In the next chapter we will discuss these three giants in materials handling and in subsequent chapters, we will discuss other equipments.

UNSOLVED QUESTION

1. Classify material handling equipments.

❑❑❑❑

3 Principles of Materials Handling Equipments

INTRODUCTION

These are qualitative statements that are used to guide the analysis of and the decision regarding a handling problem. The purpose of the principles is to prevent serious mistakes resulting from oversights or incomplete examination of a problem. The principle does not give solution rather they point out important consideration in the solution. Principles generally derived from past experiences of engineers as well as from the man working on such equipments. These are nothing but the accepted guidelines for MH Equipment, so as to achieve safe and efficient operations. Principles derived from past experiences can be used as guidelines to operate such equipments.

3.1 PRINCIPLES AND THEIR USES

The experience of an expert in the field of materials handling management is difficult to transfer among people. Fortunately, however, some of its experience has accumulated over the years in the form of principles.

Principles, then, must be applied carefully, in the light of what we like to call common sense, for they represent tremendeous condensations of extensive and complicated experience. Principles, however, have important uses in the study of materials handling problems:

(*a*) Taken literally, they may be used as general guides to action, with appropriate caution.

(*b*) Their best use is perhaps as clues or suggestions as to what possibilities should be looked for and considered in a problem. They may be used, in fact, as rough hypothesis to be tested in connection with particular instances.

(*c*) Many problems are best treated by the first application of principles before undertaking analysis. Principles suggest the achievement of the obvious, quick, "big pay off" attacks on a problem.

(*d*) Principles may help, and indeed may be the only resort, in those problems where analysis is too complex, too costly, or beyond the analyst's ability.

3.2 PRELIMINARY CONSIDERATIONS

(*a*) Handling contributes to the value of material which is attributable to its position in time and space but not to the value attributable to its form.

(*b*) For every kg of finished product many kgs of material must be handled and stored.

(*c*) Materials handling is responsible for a large portion of a firm's indirect labour charges, a large portion of industrial accidents and a large portion of product damage.

(*d*) A steady, uninterrupted flow to each production operation can by itself significantly increase productivity.

(*e*) Money spent to move materials cannot be recovered.

(*f*) A useful objective to work toward is the "thin, swiftly flowing stream of production".

3.3 PRINCIPLES OF MATERIAL HANDLING EQUIPMENTS

These are categorized in two types major and minor principles :

Major Principles

(*i*) Reduction in time.

(*ii*) Reduction in Handling.

(*iii*) Equipment Design.

Minor Principles

(*a*) Planning Principle.

(*b*) Operations Principle.

Now, we will consider and understand all these principles one by one.

3.3.1 Major Principles

(*i*) Reduction in Time

As everybody knows Time is Money. Any time lost during production will not help to any organization to achieve their targets. So every effort should be made to save the time. To do this it is very much necessary to know where the idle time (non-productive) is. In any industry time is consumed principally in three things. *i.e.,*

(*a*) Waiting time.

(*b*) Loading and Un-loading time.

(*c*) Travel time.

(*a*) **Waiting time** is a time lost in waiting such as waiting of raw material, waiting for set-ups, waiting for labours, waiting for equipments. Such types of losses may arise due to bad scheduling, bad organization of labour force. This time may be due to in plant congestions as well as external to plant due to traffic jams (truck waiting due to traffic jam). Waiting time may not be related directly to MHE but it should be reduced as much as possible although it cannot be eliminated.

(*b*) **Loading and unloading time**

This time is directly related to MHE. Loading and unloading directly involves the material. Handling such putting the raw material to machine for processing (*i.e.,* loading) or putting the raw material to the platform from loaded truck. (*i.e.,* unloading) as shown in Fig. 3.1.

The loading and unloading can be done by two ways *i.e.,* manually or using specialized equipments. Manual process takes more time but it is cost effective but specialized equipment is quick but may not be cost effective. In general larger the unit loaded or unloaded greater the reduction that can be made in loading time. So the equipment used should reduce loading and unloading time.

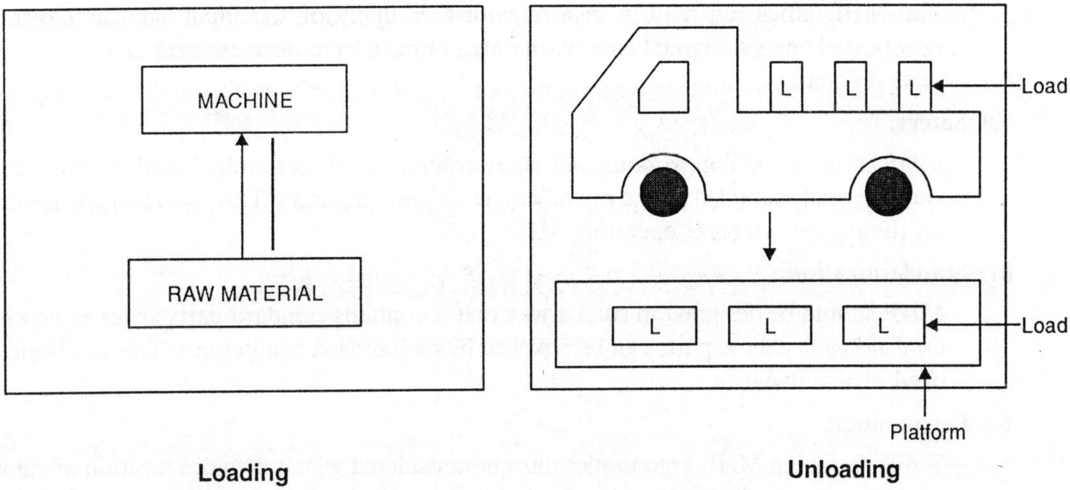

Figure 3.1. *Loading and Unloading Methods.*

(*c*) **Travel time**

– It is a time lost for traveling from one point to another point in plant or external to plant. This is essentially related to speed of equipment and distance to be travelled.

– Travel time can be minimized using high speeds and short distances.

– Hence aim is to minimize travel time.

(*ii*) Reduction in Handling

– No matter what is to be handled it should be handled in such a way that it should minimize handling itself.

– In this aim is to reduce the handling of material itself. By doing this other principles will be automatically fulfilled.

– Reduction in handling can be done by using following means:

(*a*) Unit load concept.

(*b*) Proper Containerization.

(*c*) Proper methods and Process changes.

(*d*) Proper selection of MHE.

(*e*) Minimizing backtracking.

(*f*) Layout improvement.

(*iii*) Equipment Design

– This principle is related with the equipment design. It is basically the important considerations which affect the selection and working of MHE.

Following should be considered for MHE:

(*a*) **Total weight** or load of MHE which includes dead weight and live load. For good MHE dead weight should be low but it should not be much low so as to affect strength and rigidity of MHE.

(*b*) **Efficiency:**

– The MHE should be efficient enough to ensure it's usability for it's purpose.

- The MHE efficiency may be related with the supply of the input such as electrical, energy, fuel energy, manual energy (mainly muscular), etc. And amount of work done to raise the load.

(*c*) **Safety:**

- MHE should be safe in working. All safety aspects as well as standard working procedures were generally supplied by the manufactures to the companies. This aspect largely depends on the person which is operating MHE.

(*d*) **Standardization:**

- MHE should be designed in such a way that it contains standard parts as far as possible. So if necessary these parts can be replaced by its standard counterpart. This is a basically need of maintenance.

(*e*) **Ergonomics:**

- While designing MHE ergonomics must be considered which includes location of cabins, levers, knobs etc. Sometimes local ergonomics must be considered.

(*f*) **Working:**

- MHE should be easy in working. It should be easier to operate. For safety working range should be specified over it. It should be marked on the ground beneath.

(*g*) **Speed, Size and Capacity:**

- MHE should be handle material quickly *i.e.,* MHE should transport material from one location in a plant to another location with high speed.
- Size of MHE or space requirement for working of MHE is important. It should not accommodate large working place.
- Capacity Of MHE should be large with versatility. MHE capacity is specified as amount of material which can be handled per operation say one tonne, 10 tonne, 100 tonne etc.
- Capacity can also be stated as number of operations per hour.
- It also may be defined as amount of material discharged per hour.
- Versatility means MHE should be capable of performing more than one operation on it. It can be used for multioperations.

(*h*) **Obsolescence:**

- It occurs when it is possible to buy improved equipment so much superior to the present equipments in terms of efficiency, working, flexibility etc. and it is economical to discard present one with new one.

(*i*) **Flexibility and Adoptability:**

- Flexibility comes into picture when we need equipment to perform different operations and adaptability comes into picture we need equipment to perform at different locations and environments. Such as different soils, rocks, atmospheres, nill region etc.

3.3.2 Minor Principles

Systems Operation Principles

(*a*) Move materials as swiftly as possible and establish operating controls to keep movement going.

(*b*) Move as close as possible to the next operation, and preposition materials if possible.

(*c*) Keep a backlog of material ahead of each operation.

(*d*) Relieve production operators of materials handling tasks.

(*e*) Reduce the terminal time or turn-around time of power units.

(*f*) Work toward a balance between service to the production system and utilization of the materials handling equipment.

(*g*) Develop effective and rapid communication between production and materials handling.

System Design Principles

(*a*) Eliminate handling where possible.

(*b*) Keep the patterns of flow simple, moving in straight lines whenever possible.

(*c*) Use equipment that sets a uniform pace for production.

(*d*) Keep the rate at which materials flow as high as possible, consistent with other factors.

(*e*) Palletize materials and use unit loads, working always toward the largest practical handling unit.

(*f*) Mechanize handling problems and keep mechanization modern through an economic replacement policy.

(*g*) Consider the movement not only of materials but of men, machines and tools.

(*h*) Develop a policy of preventive maintenance for materials handling equipment.

(*i*) Make safety a major consideration in the design of materials handling systems.

(*j*) Select versatile equipment, noting the possibilities for the application of accessories.

(*k*) Use gravity whenever possible.

(*l*) Standardize equipment and methods. Make the system self-programming and self-controlling where possible.

(*m*) Where possible, combine handling and production operations, including sorting, inventorying, packaging, inspection and assembly.

UNSOLVED QUESTIONS

1. Explain what is the use of principles.
2. Explain major principles of materials handling.
3. Explain minor principles of materials handling.
4. Explain different time factors which are taken into account in material handling.
5. Explain how reduction in handling can be done.
6. Explain the principles related to the equipment design.
7. Explain operation principles.

□□□□

4 Selection of Material Handling Equipments

INTRODUCTION

As discussed in previous chapters so many varieties of material handling equipments are available. Question here is how to select one particular equipment? At first instant it seems to be very tedious job but it is not, if anybody has the knowledge about the factors involved in the selection of material handling. These are only the important considerations while selecting the material handling equipments not the guidelines.

Because of the large amount of money involved in equipment selection, the material handling specialist must take a caution and conservative approach in this matter. A piece of equipment can be a vital element of the handling process in a plant, or it can be disruptive to plant operations and the focal point of labour problems and other disturbances for equipment systems, it may be necessary to employ a consultant to obtain the best possible solution to the handling problem.

As labour costs continue to increase, the justification for using specialized and even general purpose equipment in specific appreciations becomes less difficult primarily because the cost per hour of labour is often higher than the cost per hour of the equipment to be selected. However, with labour costs escalating, it also becomes easier to justify operator less automation, in which the operator has been completely or almost completely eliminated.

There are several beneficial aspects such as:

(*i*) High productivity of the operation or process.

(*ii*) Good working climate with little cause for labour unrest.

(*iii*) High rate of return on the investment.

(*iv*) Labour and equipment down time are minimized.

(*v*) Maintenance cost may be relatively low with maintenance easily accomplished.

While selecting MHE the material handling specialist should first obtain data concerning to –

(*i*) The volume of material to the handled.

(*ii*) How far the material must be moved.

(*iii*) The way the material is packed, or not packed.

(*iv*) Where and how it is to be stored.

(*v*) Whether or not the volumes to be handled vary from day to day or during the day, week, or month etc.

4.1 FACTORS IN SELECTION OF MATERIAL HANDLING EQUIPMENTS

So many types of Material Handling Equipments are available commercially in the market. One has to choose MHE, best fit for his purpose. As MHEs are in wide range, the selection becomes more difficult. Here are some important factors which are useful in selection of MHE.

1. Plant Layout.
2. Product/Process.
3. Building Construction.
4. Type of Organization or Industry.
5. Factors Related to Equipment.
6. Money Factors.

Above stated are important factors in selection of MHE. Now we will discuss them in detail.

4.1.1 Plant Layout

It is defined as systematic arrangement of machines, facilities and amenities in a industry so as to achieve maximum production.

'Plant Layout is such a systematic and effective functional arrangement of different departments, machines, equipments and services of an organization that will facilitate the processing of the proposed or undertaken product in the most efficient and economical manner in the minimum possible time.'

'A technique of locating different machines and plant services within the factory so that the greatest possible output of high quality at the lowest possible total cost can be available.'

'It is the arrangement of machines within a factory, so that each operation is performed at the point of greatest convenience.'

4.1.1.1 Types of Plant Layout

Plant layouts fall into following categories.

A. Product Layout.
B. Process Layout.
C. Fixed Position Layout.
D. Just in Time Layout.
E. Group Layout.

A. Product Layout

In product layout machines are arranged according to process to be carried out on the product. In its simplest form product layout is made for a specific product such as tyres, automobiles etc. Raw material enters from one side and leaves from other side (*i.e.,* finished material). Figure 4.1 shows an arrangement of this type. The operations are performed in sequence. All parts, sub-assemblies etc., are started at right time to ready at the required time are kept moving until the finished product is available at the end of the assembly line.

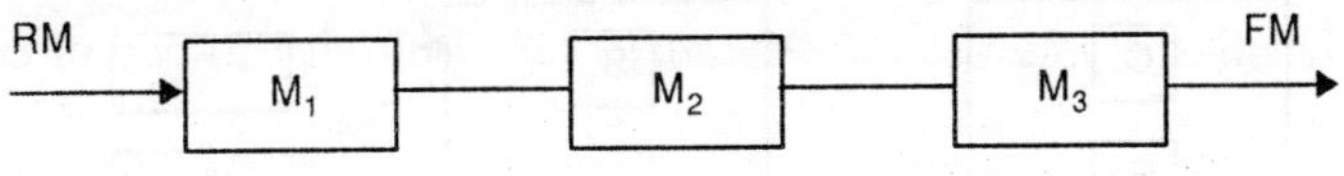

Figure 4.1

Advantages of Product Layout

1. It lowers the overall manufacturing time as product under it proceeds rapidly from operation to operation with lessen handling.
2. Less space is needed for placing machines.
3. There is minimum of handling or transportation.
4. There is less work-in-process.
5. Better utilization of machines and labour.
6. It minimizes counting, inspecting and clerical work etc.
7. There is smooth flow of materials.
8. Promises for supply of products to consumers are more reliable.
9. Use of gravity and power conveyors for material handling reduce the necessity of aisles.
10. Production control is simplified greatly.

Disadvantages of Product Layout

1. When the model or type of product changes, the layout or machinery also requires a change.
2. All machines in the arrangement cannot be used to their maximum capacity.
3. The manufacturing cost rises with a fall in the volume of production.
4. If one or more lines are running light, there is great machine idleness.
5. Specialized and strict supervision is needed.
6. If any one of the machines in the shop breaks down the other machines have to remain idle till that machine again gets ready to commence operation.
7. For expansion purpose, it is not possible to add more machines.
8. Each worker performs job at a particular machine, so he is not skilled for other machines.

B. Process Layout

In process layout machines are arranged according to the process which can be carried out such as drilling, milling, reaming etc.

Machines are grouped in certain numbers according to their utility in industry.

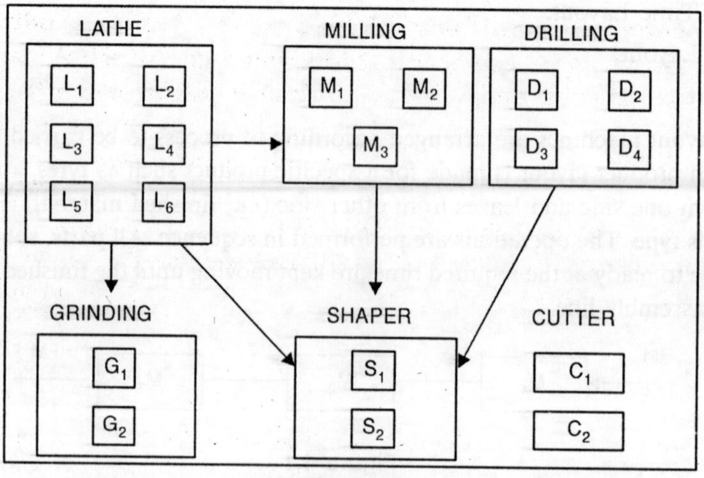

Figure 4.2

In this similar equipment and similar machines are grouped together. Thus it is very useful where low volume of production is required. As shown in the figure there is no definite flow of material.

Advantages of Process Layout

1. Similar jobs are manufactured on similar machines, hence supervision is simple.
2. Less machines are required.
3. This offers areas of incentive for each worker to raise the level of performance.
4. There will be better control on precision or complicated processes.
5. The layout is flexible, hence it can easily be managed to change in the rate of production, in raw materials used.
6. Capital outlay for machines is minimum.
7. New workers have better training facilities on the job. Workers know, how to run various machines in the group and also to set up work.
8. Foremen become specialized in the performance of the job and know all about the equipments and their operations.

Disadvantages of Process Layout

1. Handling and back-tracking of materials is too much.
2. It makes necessary to plan and supervise the work of each department, each worker and machine. This makes the production control more difficult and more costly.
3. Generally more floor area is required.
4. Total production cycle time is more, due to long distances and waiting.
5. Routing and scheduling is more difficult, as the work does not flow through any definite mechanical channels.
6. Inspection of work for each operation is necessary as the material passes to the next department. This results in more inspection and causes delays.

C. Fixed Position Layout

It is layout in which all the processes carried out on a product at one place *i.e.,* product is stationery and all the machines come towards the product. Such layout is used in ship construction, aeroplane, rocket, missile, turbine, and generator industries. This type is known as 'Static Product Layout.' M_1, M_2, M_3, M_4, M_5, M_6 are machines. P is product.

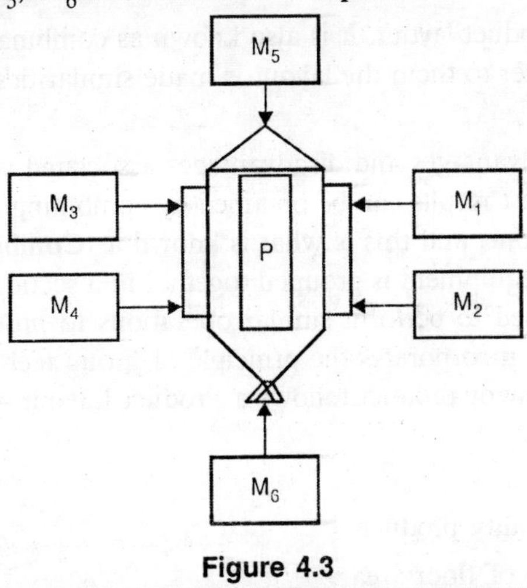

Figure 4.3

Advantages of Fixed Position Layout

1. Capital investment is minimum.
2. Continuity of operations ensured.
3. Less total production cost.
4. Less material movement.

Disadvantages of Fixed Position Layout

1. Machines and tools etc. take more time to reach at the work place.
2. Highly skilled workers are required.
3. Complicated jigs and fixtures may be required in fixing jobs and tools etc.

D. JIT Layout

It is just in time layout. It is an extension synchronization of all machines. It is a layout in which there is no inventory. The layout of zero inventory. The material is not stored as it is received. It is directly put on the machine to get processed. This is a Japanese concept.

Now it is clear from study of all layouts that selection of MHE depends on plant layout. In product layout MHE has to work in a line but in process layout MHE has to work in all the area. *i.e.,* in product layout choose conveyors, but as in process layout area is fixed so MHE has to be chosen such that it can work in all area such as mobile trucks, mobile cranes. In product layout-working area is fixed. Hence fixed MHE with fixed area of working can be used. Important criteria of selection MHE also the space between machines, distance to be travelled, speed with which MHE should move all such factors depends on plant layout. Hence plant layout is major factor in selection of MHE.

E. Group Layout (Combination Layout) Cellular Layout

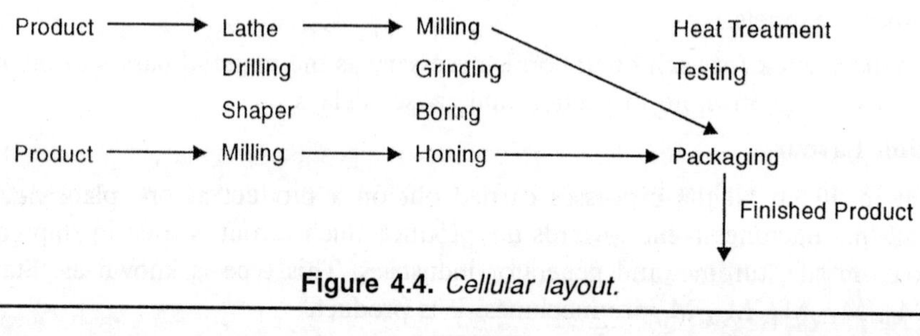

Figure 4.4. *Cellular layout.*

It is again extension of product layout. It is also known as combination or mixed layout. In this similar items are grouped together to them the layout is made similarities such as shape, size, process to be carried out etc.

Because of the inherent advantages and disadvantages associated with both the Process Layout as well as Product Layout the best results can be obtained by combining the advantageous features of both these layouts into a single one, and this is what is known as **Combination Layout.** In this type of layout a set of machinery or equipment is grouped together in a section, so that each set (or group) of machines or equipment is used to perform similar operations to produce family of components. Since this type of process layout incorporates the principle of group technology, this layout is known as Group layout as well. The flow of product follows a Product Layout sequence. Hence, it is known as combination (or group) layout.

4.1.1.2 Goals of Good Layout

1. To produce better quality products.
2. Maximum utilization of floor area.

3. To produce internal transport from one operation to the next as much as possible.
4. Lighting and ventilating of areas.
5. Lower cost of scrap and waste.
6. Fewer accidents.
7. Minimizing production delays.
8. Space for future expansion.
9. Safety of equipment and personnel.
10. Better working conditions for both executive and operative employees.
11. Avoidance of unnecessary changes.
12. Savings of costs.
13. Easy supervision.
14. Neatness.
15. Proper production control.
16. To estimate waste effort and speeding of production.

4.1.1.3 Principles of Plant Layout

1. **Principle of over-all integration.** According to this principle, the best layout is one which integrates the men, materials, machinery, supporting activities and any other such factors that results in the best compromise.
2. **Principle of minimum distance.** According to this principle, other things being equal, the best layout is one in which men and materials have to move the minimum required distance between operations.
3. **Principle of flow.** According to this principle, other things being equal, the best layout is one which arranges the work area for each operation or process in the same order or sequence that forms, treats or assembles the materials.
4. **Principle of cubic space.** According to this, the best layout is one in which all the available space both vertical or horizontal is most economically and effectively used.
5. **Principle of satisfaction and safety.** According to this, other things being equal, the best layout is one which makes work satisfying, pleasant and safer for workers.
6. **Principle of flexibility.** It means the best layout is one which can be adopted and re-arranged at a minimum cost and least inconvenience.

4.1.1.4 Aspects Related Plant Layout

There are two important aspects of plant layout. While selecting the MHE either existing layout is to be improved or new plant is designed after selecting the MHE.

(*a*) Revising the Plant Layout

A typical study for revising the plant layout includes the following steps.

(*i*) The determination of the objectives to be attained by the survey.

(*ii*) The compilation of the layout drawings of the existing plant.

(*iii*) The undertaking of a cost study as a guide to layout analysis.

(*iv*) The analysis of all the data on the present layout and new additions.

(*a*) Flow diagram analysis.

(*b*) Flow process chart analysis.

(*c*) Operational analysis.

(*d*) Overall analysis of the material handling.

(*e*) Analysis of service activities and auxiliary departments.

(*v*) The development of an improved and revised layout that attains

(*a*) An improved flow diagram.

(*b*) An improved process chart.

(*c*) A final detailed drawing of the revised layout.

(*b*) Plant Layout Design

Various procedural steps involved in new plant layout design have been listed and described below:

(*a*) Accumulate basic data.

(*b*) Analyze and coordinate basic data.

(*c*) Decide the equipment and machinery required.

(*d*) Select the material handling system.

(*e*) Sketch plan of the plot for making factory building

(*f*) Determine a general flow pattern.

(*g*) Design the individual work stations.

(*h*) Assemble the individual layouts into the total layout.

(*i*) Calculate the storage space required.

(*j*) Make flow diagrams for work stations and allocate them to areas on plot plan.

(*k*) Plan and locate service areas.

(*l*) Make master layout.

(*m*) Check final layout.

(*n*) Get official approval of the final layout, and

(*o*) Install the approved layout.

(a) ***Accumulate the Basic Data***, such as

- Volume and rate of production.
- Product specification and bill of materials.
- Process sheets indicating tools, equipments, the method and the product which will be manufactured.
- Flow process charts.
- Standard time to complete each operation.

(b) ***Analyze and Coordinate Basic Data*** in order to find,

- The work force size and type.
- Number of work stations required.
- Type of equipment required.
- Storage and other space requirements.

Assembly chart and operation process chart help coordinating basic data.

(c) ***Decide the equipment and machinery required***

Number of equipments required to meet a particular production target can be calculated by knowing:

(*i*) Number of articles or products to be produced.

(*ii*) Capacity of each equipment.

(*iii*) Time in which the order is completed, etc.

(*d*) ***Select the material handling system*** for moving raw material, semi finished goods and final products. The type of material handling equipment to be selected depends upon:

- Material/product to be moved.
- Container in which it will be moved.
- Length of movement.
- Frequency of movement.
- Speed of movement etc.

(*e*) ***Sketch plan of the plot*** to mark building outline, roads, storage and service areas, etc.

- The plan orientation should utilize maximum, the natural heat, light and other weather conditions.

(*f*) ***Determine a general flow pattern***

- Machinery may be laid as per production process requirements and plant building be erected around the same.
- The flow pattern (movement from store to finished product) of materials should be such that the distance involved is least between the store and the shipping department through the production centers. There should be minimum back tracking and bottle-necks.
- Flow pattern (can be I, L, U, O, S, flow) may be analyzed using operation process charts or travel charts.
- Based upon the process or product requirements, one may adopt:

  (*i*) A process layout.

  (*ii*) A product layout.

  (*iii*) A combination layout.

- Plant layout should be flexible so that it can accommodate changes in the product or product diversification.

(*g*) ***Design of the individual work stations.*** Each work station should be laid for achieving optimum

- Performance of operations.
- Materials and space utilization.
- Safety and comfort of employees.

(*h*) ***Assemble the individual work station layout*** into total layout in accordance with the general flow pattern and the building facilities.

(*i*) ***Calculate the storage space required***

- Stores accommodate raw materials, in-process goods and finished products.
- Storage size increases with the plant size.
- Storage space can be calculated by knowing :

  (*i*) Volume of each store item.

  (*ii*) Number of items to be kept in stores.

  (*iii*) The time (date-wise) each item may be kept in the stores.

(*j*) ***Make flow diagrams for work stations and allocate them to areas on plot plan.***

Flow diagram is a diagram or modes substantially to scale which shows the location of specific activities carried out and the routes followed by workers, material or equipment in their execution.

(*k*) ***Plan and locate service areas,*** such as

- Offices.
- Toilet and wash rooms.
- Tool rooms.
- Rest and lunch rooms.
- Cafeterias.
- Dispensary.
- Power generation areas.
- Parking areas, etc.

(*l*) ***Make master layout*** by using templates and models.

(*m*) ***Check final layout*** and associate in this act, all the persons concerned, *e.g.,* those from plant, production, personal, safety and other departments.

Final plant layout is checked as regards the following aspects :

(*i*) Safe and economical material handling.

(*ii*) Adequate production and production control.

(*iii*) Plant building and it's surroundings.

(*iv*) Product design.

(*v*) Service areas.

(*vi*) Employee safety and comfort, etc.

(*n*) ***Get official approval of the final layout***

- After the final layout has been checked, it is got officially approved and signed by the team which checked the final layout.
- A final layout accompanies:

(*i*) Product drawings.

(*ii*) Bill of material (parts list).

(*iii*) Assembly and operation process charts,

(*iv*) Manpower requirements,

(*v*) Equipment requirements, and

(*vi*) Estimated expenditures and revenues, etc.

(*o*) ***Install the approved layout***

- Make detailed plans for installing production, service and other centers.
- Install equipment, machinery, workbenches, offices etc.

4.1.1.5 Factors Influencing Selection of Layout

These are:

1. **Type of Industry.** Industries are generally classified according to their process of manufacture. The process manufacture can be classified into four categories.

(*a*) **Synthetic Process.** When two or more materials are mixed to get a product, the process is known as synthetic process. The example of such a process is to produce the cement by mixing limestone and clay.

(*b*) **Analytical Process.** This is opposite of synthetic process. It is the breaking up of a material into several parts. The refining of petroleum is the example of this case.

(*c*) **Conditioning Process.** In the conditioning process, the form of raw materials is changed into desired product as in jute industry.

(*d*) **Extractive Process.** In this type, by applying heat, desired product is extracted from the original raw material, *e.g.,* manufacture of aluminum from bauxite.

2. **Type of Products.** Type of product means whether the product is heavy or light, large or small, liquid or solid. It is also a consideration of plant layout.

3. **Volume of Production.** According to this point, while plant layout is being done, it should be kept in mind that, what volume is required to be produced. In this case, it should be seen that whether :

(*a*) Job Production

(*b*) Batch Production or

(*c*) Mass Production is being adopted.

4. **Influence of Processes.** Last, but not least factor to consider, is the process through which the material passes in the concern. Much care should be given to the material handling problem, position of store room and tool room...

Table 4.1: Relative comparison of Process Layout and Product Layout

Criteria	Process Layout	Product Layout
1. Flexibility of changes in layout	Better	Too Less
2. Capital investment required	Less	More
3. Product quality	Better	Poorer
4. Utilization of available machinery	Better	Poorer
5. Monotony in working	Less	More
6. Flexibility in work allotment to workmen and machinery	More	Less
7. Space required for same amount of production	More	Less
8. Ease and effectiveness in production, planning and control	More difficult	Better and easier
9. In-process inventory required	Large	Small
10. Time taken in completion of production	More	Less
11. Material handling movements and costs	More	Less
12. Ease of automation in material handling	Not applicable	Full automation possible

4.1.1.6 Advantages of Good Layout

A good layout has got the so many important advantages:

1. It reduces time of transportation to minimum.
2. It increases safety and minimizes accidents. It makes repairs and maintenance easy.
3. It reduces production delays to large extent. It eliminates waste efforts and thus speeds of production increases.
4. Back-tracking is reduced.
5. It reduces set-up and total operation time.
6. Time and motion study can be easily and accurately performed.
7. Better utilization of manpower.
8. Improvements in methods of production can be made.

4.1.2 Product/Material

- In selection of MHE this is important factor as the MHE is going to handle the product (*i.e.,* material).
- Material properties such as physical and chemical properties are important consideration in selection of MHE.
- Physical properties such as size, shape, weight, density, hardness, fragileness, brittleness.
- Chemical properties such as toxicity, flammability, volatility.
- In material handling materials are broadly classified as packaged material and bulk material. Packaged material is a material which is packed in containers such as bins, bags, boxes etc. Bulk material is a material which is loose in nature such as powders, coal, sand, stones, bullets, medicines, tablets etc. Hence product or material has important bearing on MHE selection. The MHE used for handling packaged material cannot be used to handle the bulk material unless they are packaged in containers.

4.1.3 Building Construction

It is walls and construction of building which is actually takes the load of material handling equipments. If these are not capable to handle the load of MHE then how could it will be installed?

Floor area is also important in some MHEs. As far as possible floor area must flat and smooth for easy efficient handling.

4.1.4 Type of Organization or Industry

Selection of any MHE will largely depends on the organization and the management. This is due to the capability of management to invest in the material handling equipment. Variety of MHEs are available it can be selected according to organization.

Selection also depends on the production of the industry such as batch or mass production.

4.1.5 Factors Related to Equipment

As discussed earlier variety of equipments are available. Each equipment has definite configuration and capacity. All the aspects related to the equipment must be considered.

These are area of working, speed, flexibility, versatility, efficiency, must be considered.

4.1.6 Money Factors

Everything depends on the money and this is a very important factor. This factor includes cost such as cost of equipment; installation, operating and maintenance are the major one.

4.1.7 Material Handling Costs

Correct determination of MH Cost is necessary because of the following reasons :

(*i*) To identify all the material handling costs to exercise effective control so as to minimise the handling costs.

(*ii*) To help in improving methods.

(*iii*) To help in company costs of present method and improved method.

(*iv*) To help in budgeting.

Handling activities can be classified as:

(*i*) Specialised MH activities—Unloading, Receiving, Storage, Issuing, Material distribution, Interdepartmental handling, Loading.

(*ii*) Mixed MH activities—Separation is not possible. These activities are done by production people.

In process handling, transferring the material from one work place to another, material handling at workplace, in process storage, maintenance, tool room, tool crib.

(*iii*) Packaging and packing.

(*iv*) Distribution activities—loading by vendor, transportation to the enterprise plant, from dock/ yard to plant, transportation to warehouse, shipping operation, transportation from the plants/warehouse.

Elements of MH Costs

(1) **Direct Costs:**

(*a*) Equipment related direct costs—Capital investment costs—Equipment cost—Transportation, transmit insurance. Inspection expenses, installation charges providing facilities for the equipment—power, fucl.

Fixed costs—Depreciation interest on investment, taxes, insurance.

Variable charges—Operating persons, fuel, power, lubricants, maintenance.

(*b*) Manpower related direct costs—Salary, overtime, bonus, fringe benefit.

(2) Indirect costs—Related to equipment—Related to the occupied space, inventry downtime —Related to the effectiveness of the management follow up, relayout, training of personnel, overtime.

(3) Indeterminense costs—Changes in overhead equipments, inventry control, production control, quality control, improved workflow.

UNSOLVED QUESTIONS

1. Write down important factors in selection of MHE.
2. Explain how product is important in selection of MHE.

3. Explain with neat sketch good plant layout minimizes material handling.
4. Explain with neat sketches different types of plant layouts.
5. What is meant by product layout? Write down advantages and disadvantages.
6. What is meant by process layout? Write down advantages and disadvantages.
7. What is meant by combination layout? Write down advantages and disadvantages.
8. What is meant by fixed layout? Write down advantages and disadvantages.
9. What is meant by group layout? Write down advantages and disadvantages.
10. What is meant by JIT layout? Write down advantages and disadvantages.
11. Differentiate between product and process layout.
12. What are the advantages of good plant layout?

5 Materials Handling and Role of Management

INTRODUCTION

Whenever someone think about setting up a plant he has to answer questions such as what type of the product, land ,finance, labour and machinery. These all decisions are important and should be taken after considerable thought. After selecting product and land we need machinery. Talking about machinery it may be light or heavy equipments, processing equipments or handling equipments. He has to select this machinery according to his need and here the role of management starts and continues until the management of handling ruins if happened so.

In an important sense, many of the decisions made in the design of the production system will dictate the answers to material handling problems and the questions involved in the design can only be answered when considered together with the problems of the materials handling function.

5.1 ROLE OF MANAGEMENT IN MATERIALS HANDLING

It starts as soon as someone thinks about setting up a factory or plant. Following are the important areas of material handling in which role of management is important

1. Selection policy.
2. Purchasing policy
3. Installation policy
4. Space utility / layout design
5. Operation policy
6. Maintenance policy
7. Investment policies

Above factors are exhaustively discussed in separate chapters except purchasing and installation since they are beyond scope of this book.

Besides above listed areas, there are some areas of material handling in which management plays part in some way or other such as

1. Analysis of material handling problems
2. Collection of material handling data.
3. Provision of the funds.

4. Training and up gradation of labour and personnel.
5. Keeping up maintenance policies.
6. Setting up safety standards.
7. Modifications in existing system.
8. Provision of funds for future revisions.

5.2 MATERIAL HANDLING MANAGEMENT PROBLEMS

To organize and present some of the methods of analysis which are appropriate for making recommendations to management concerns the design, operation and control of material handling systems. Following statement shows the importance of the material handling.

"Materials handling generates something like 80% of indirect labour costs in the plant."

"Materials handling may be responsible for one-third to one-half of the total cost of production in some plants."

'An uninterrupted flow of materials through a plant can increase the productivity of existing men and equipment by 15%.'

5.3 APPROACHES TO MATERIALS HANDLING PROBLEMS

In any actual instance to assist for management decisions, a complex mixture of experience and analysis is likely to be used.

5.3.1 Analysis

Analysis has begun to make inroads into the difficult problems of managing the material handling functions of production systems. The problems of managing material handling systems are complicated by the fact that these systems are found within the context of larger and more complicated production systems. The areas where analysis seems to have some usefulness includes decisions as to selecting, maintaining, routing (the specification of the flow the process to be followed in producing a particular manufacturing lot), dispatching (deals with the release of resources, along with the necessary instruction for the production function), scheduling (relates to time table of production function) and controlling (Progress reporting, Data processing, Expediting corrective action replan) materials handling equipment as a part of manufacturing organizations. These problems are, of course, intimately related to the problem of plant layout and design, organizational design, and, in fact, the design of the production system as a whole.

Analysis falls more in the usual traditions of science and its applications in engineering. It is rather explicit process involving such steps as:

1. A careful statement of the management problem in terms which permit measurement.
2. The formulation of a mathematical model or abstraction of the problem which hopefully will permit the analyst to predict the consequences of various management policies and perhaps even suggest something of the relative value of these policies.
3. The collection of data which must be used for testing the model to see if it satisfactorily describes the situation at hand, suggesting appropriate modifications of the model and finally forming the basis for the predictions and evaluations upon which the analyst's advice to management will be based.

5.3.2 The Uses of Analysis

There are many instances in which analytical expressions or mathematical models have been used with great skill to represent situations faced by managers and to predict the consequences of various policies. We can be clearer about the benefits of analysis as we develop specific instances of it; however, one might note here the possibility of results at two levels:

1. Exact quantitative solutions to a very few problems.
2. An increased general understanding of a larger number of problems, leading to the formulation of principles which can be only inexactly applied to complex decision-making problems.

It is in the latter area that many of the more useful products of analysis are found. Much is to be learned from seeking answers to questions such as:

1. What is the general nature of the relationships among the variables in a problem? If management does something, what will go up?
2. How sensitive is a recommendation to errors of measurement or changes in the parameters of the problem?
3. If one does not or cannot adopt the optimal solution, how mush will profits decrease or costs increase as a result?
4. Does analysis suggest any non obvious relations among the variables which might not be dependably revealed through managerial experience?
5. Does analysis permit one to consider more details and more variables than one conveniently handle unassisted?

Failure to appreciate these considerations has led to some disappointments and sometimes to the identification of the analytical approach as theoretical rather than practical.

5.4 THE CRITERION PROBLEM

A persistent and most difficult problem which arises in all attempts at analysis is the problem of selecting and developing a suitable criterion or measure of effectiveness. One must explore the questions:

1. What is the management's objective in general and in each particular problem?
2. What is it we wish to maximize or minimize?
3. How can we avoid the selection of a measure of effectiveness which, while seemingly suitable from a materials handling point of view, actually operates to the detriment of the firm as a whole?
    1. Distance moved.
    2. Number of handlings.
    3. Amount of crisscrossing in a flow plan.
    4. Investment in equipment.
    5. Indirect labour costs.
    6. Product of weight times distance moved.
    7. Number of trips.
    8. Number of pieces times distance moved.

9. Volume times distance moved.
10. Load factor (actual load divided by capacity).
11. Ratio of pay load to dead weight.
12. Equipment utilization.
13. Service to protection (production delays attributable to materials handling).
14. Amount of in-process inventory.

Such measures as these do not themselves express the objectives of a firm or of a manager, but they do have the important attribute of being measurable in many situations.

In the examples of analysis which follow, we have used a variety of measures of performance. These measures must ultimately be considered as examples rather than prescriptions for any given materials handling problem.

5.5 COLLECTING THE DATA

It seems clear enough that a successful piece of analysis must meet three requirements:

1. The mathematical model developed must be rich enough in reality to lead to improved policy decisions, yet it must not be so complicated that deductive reasoning which is beyond the talents of the analyst.
2. The data required must be reasonably possible to obtain. A model containing unmeasurable variables is of little use.
3. Finally, the conclusion derived must be acceptable to management. No matter how reasonable the results, there is little to call success until management accepts them and begins to act upon them.

The problem of data collection is thus an intimate part of the analytic and quantitative approach to materials handling problems and is done by data sheet such as below:

Operation Number	Operation Description	Manufacturing Operation	Handling Operation	Move	Inspection	Storage	Distance Feet	Handling Equipment	Remarks
		○	◎	→	□	△			
		○	◎	→	□	△			
		○	◎	→	□	△			
		○	◎	→	□	△			
		○	◎	→	□	△			
		○	◎	→	□	△			
		○	◎	→	□	△			
		○	◎	→	□	△			
		○	◎	→	□	△			
		○	◎	→	□	△			
		○	◎	→	□	△			

5.6 A CASE STUDY ON MATERIAL HANDLING PROBLEMS AND ROLE OF MANAGEMENT

Materials handling system–Problems of Design and Maintenance

Introduction

Jiwaji cotton Mills, Gwalior had their own thermal power house requiring typical system of coal handling. Being a composite mill it was to handle inputs like cotton-yarn, dyes and pigments, maintenance spare parts and finished cloth. The Gwalior Rayons, being predominantly a weaving and processing unit, has a system which partly resembles that of Jiwaji cotton Mills. The steel foundry of Central India Machinery Mfg. Co. Gwl. Ltd. was a system for handling inputs like sand, steel scrap, oil dolomite, sand additive, and for handling finished castings in various departments, mould etc. Textile Machinery Mfg. Division of M/s Central India Machinery Manufacturing Company Limited was manufacturing weaving machines, warping machines, filter presses, winding machines etc. It was having its own grey iron foundry, machine shop and other manufacturing facilities.

Problems of Layout and Systems Engineering

Let us first study the system of material handling in grey iron foundry. The inputs of the foundry are brought in by trucks and bullock carts. Sand is exclusively brought in trucks where as pig iron is brought in trucks or bullock carts from Railway Station to factory, the distance being half kilometer. These were kept in the open and taken to respective units like - cupola or sand mixers manually in wheel trolleys etc. Pig iron, hard coke, lime stone etc. once brought to the Cupola are handled by powered mechanized charging system. All these materials such as sand cupola inputs, once fed to sand mill and cupola, are handled by mechanized system such as belt conveyors, metal handlers and gravity roller conveyors etc. The finish castings are handled manually on wheel barrows or trucks or electric trolleys from knock out to Fettling Shop and then to the castings godown. The heavy castings are directly fed to the machining centre and are not routed through casting godown.

The problems with the material handling system of this grey iron foundry when analyzed in detail were a very good example of what we call total systems engineering. Each item can not be handled without its relation to other item. In design improvement over the existing system has to be done in such a manner that it does not affect day to day production. One of the factors in foundry sand handling system is bucket elevator which has been avoided. The troubles of bucket elevators, as observed in many other places are (a) faulty design (b) overload (c) Longer time required for repairs. The design of bucket elevator has since been improved considerably. It is noticed that the rate of return sand from knocked in short time, resulting sand fed to the bucket elevator per minute is more than what it was designed for. Therefore, either the bucket elevator of bigger size should be provided so that the foundry castings can not be knocked out in short time.

The "U" shaped gravity roller system to carry molding boxes from moldings machines for closing and pouring and them returning back to the molding machine has its own advantages and disadvantages as compared continuous pallet conveyor. The speed of pallet conveyor is fixed and therefore it can not take higher production achieved by improvements somewhere else. In gravity roller system this is not the case. We have been benefited by this system because production had been more than doubled on individual pair of machines. The disadvantages is that the molten metal has to be carried to different places where as in pallet conveyor system pouring can be done at one place and knocked out at one place. This is another example of thoughtful system engineering as applied to material handling.

The layout of the foundry plays a very important part on material handling system to be adopted considering right from the area where input like sand, coke, pig iron, wood, lime stone are stored are located in such a manner that if mechanization is designed and implemented now it would be very costly. This should have been thought when layout was originally made. Many times we are forced to buy complete rake load of pig iron instead of at the rate at which we would like to have with regard to our consumption. Storing the large amount of coke and pig iron requires extra material handling cost because of distance of the inputs having increased. Therefore, this is an example of giving more thoughts at the time of making layout of equipments and building while designing material handling systems.

All the castings of smaller size were kept in racks and then issued to various machining centers, subsequently transferred from one machine to other machine, inspection point and finished parts stores. This system has lot of potential for improvement, *viz.,* instead of keeping material in racks the same can be kept in stackable containers, and container as such can be moved from one place to another and stacked in Finished Part Godown.

The assembly shop had rail system where the machine under assembly was moved from one station of assembly to the next station of assembly.

Dispatch after packing was by means of mobile crane. This was proving very costly and is under study for improvement which again pertains to material handling system engineering and as such to layout of shop.

Handling of long bars and sheets is also not proper because of improper layout.

Problems of Components/Elements of Material Handling Equipments

Many a time we find that elements of mechanism of material handling equipments have not been properly designed with reference to their shape and material selection. Take for example, reduction gear boxes manufactured in our country. Gears do not have proper sealing systems to stop oil leakage from inside of the gear box and also observed that the shaft rubbing against the oil seal wear out earlier that the worm wheel gear profile and when any attempt is made to build up material at the worn out surface, it causes additional problems of poor surface, finish and bending of the shaft. This is due to improper hardness of the shafts. Alternative could be adopted by the designer by providing a hard end and ground sleeves over the shafts at such place so that the sleeves when worn out can be replaced along with the oil seal. The quality of rubber used and oil seals is also not good.

The designs needs standardization from ISI in physical dimension of the gear boxes at least for the sizes manufactured and used in bulk quantity in our country. This causes additional strain to the user when he has to replace one brand of gear box with another in case of non-availability. Other elements of the drive systems should be standardized for certain rating where which is used in large quantity in our country.

Rubber tyred wheels of the material handling equipments also used standardization with reference to its chemical composition or other properties like shore hardness wear resistance with reference to their carrying capacity. Oil holes for pouring and checking of oil level should have covers of such design that the covers could be replaced easily and the best method is spring back to its required position so as to prevent entrance of foreign particles due to negligence of lubrication person. It is surprising to note that there is no Indian standard giving specifications of lubricants (in line with similar ones to DIN) for various types of duties. This results unnecessary flexibility taken by oil companies in recommending their lubricants and also unnecessary varieties which at least is not reasonable specially when oil companies in our country are now under Govt. Corporation. Indian standards Institution can play a very important part in standardization of such materials and also of other components like driving chain, sprockets with reference to group of ratings required by designer for various duties. This will also reduce unnecessary varieties and monopoly with subsequent higher price for the components of material handling equipments.

Problems of Design

This problem can be divided into two parts: -

(*a*) For simple material handling systems which can be fabricated by the user himself.

(*b*) Complicated or advanced systems which are made available only by the organization which are elaborately geared for this job.

Problems related to (a) only. Designs of simple material handling equipments/elements should be standardized and made available by the Institution of Engineers in the form of a hand book so that user engineers can take help from it for quick design and fabrication. Some times teaching institutions take objections to this method of hand book because, they prefer graduate engineers to do this by first principle. This is a very healthy way but this takes lot of time and will give variety of designs with subsequent deviations from a fresh engineer graduate is not even able to design a sample jib crane normally used in shops for day half tonne capacity and four meter turning radius. This has been given as an example because it has been experienced frequently. Standardized design of gravity roller pallets, winches, pulley system etc. which are very frequently required, is a necessity its felt by the Industry. In case an Institution still insists that such type of designs should not be made available in hand books then graduate engineer should be able to design the same quickly, and a student coming out from the college should be prepared for the same.

Problems of Maintenancc

These are partly related to the problems of components/elements and also problems of in adequate designs. A proper management's attitude has to build up for the ability required by the maintenance person. Many a times management gives less importance to standards of filters etc. to repair these equipments because majority of such equipment are considered by them to be of rough type. It is not to be forgotten that the very act of material handling equipment in use is rough and subject to poor working condition, it needs better skilled man to repair and maintain these equipments. Just to give an example, fitter to dismantle the gear box of material handling equipment and to replace the bearing seals etc. require to gear box of machine tools. The maintenance persons are also heavy inventory of spare parts. The ray of hope as seen by some of the enlighten industries is in having a senior material handling engineer so that such difficulties are minimized.

Problems of Funds

Invariably the project engineer and manager overlooks the necessity of allocation of sufficient funds for providing efficient material handling systems. By the time, the need of proving such system is felt, funds are already exhausted in purchasing machine tools. The simple exercise of providing stackable container from casting godown to finished part stores through various machining centers and equipment to handle these containers need such financial requirements that either project is self sufficient or alternatives are implemented which are grossly in adequate. It is also observed that proper production planning control systems can reduce requirements of containers to a great extent and therefore this point is to be looked in to while deciding for funds.

UNSOLVED QUESTIONS

1. Explain the role of management in material handling.
2. Explain various problems faced by management in material handling.

6 Economics of Material Handling

INTRODUCTION

Money matters in the every field so as in the material handling. So many cost factors are considered in the material handling.

Survey made by organization like national productivity council and the Textile and Allied Industries Research Organization have revealed that material handling costs in Indian Industry range from 22 to 75 per cent in producing a finished product. Hence, cost of handling should be reduced as possible.

6.1 COST CONSIDERATIONS

- **Initial cost of equipment**

  This is a cost of equipment and varies from model to model. One can purchase the equipment according to his standards to suit his need.

- **Cost of installation**

  This is a cost of installation. Generally it is done by manufacturer. Some times while installing the equipment rearrangement and alterations to present equipment and building has to be done. This responsibility and decision mostly depends on the installer but the management can assist in the decision making.

- **Operating cost**

  This is cost which is incurred during operation of the equipment. This includes

  1. Cost of power.
  2. Cost of labour to operate.
  3. Cost of supervision.
  4. Cost of training

- **Cost of maintenance, repairs, supplies etc.**

  As a need maintenance must not be overviewed. It can be done by the industry persons or may be given contract to specialized agencies which are involved in this field. If maintenance is done by industry he has to carry inventories and must train the persons involved in this job. This will be added to the cost of maintaining the equipment. If maintenance contract is given everything will be done by agency, this will reduce the cost and will be done by experts.

- **Rate of depreciation**

  This is cost and starts as the equipment is installed and starts operating. This is important in case, there is need of replacing the equipment with new one. This should be as low as possible.

- **Rate of obsolescence**

  This is important in case of need of replacing the equipment with new equipment which is technologically superior. This should be as low as possible.

- **Probable salvage value**

  This is a cost comes into picture when equipment has worked for its useful life and is to be scrapped. This should be as high as possible.

- **Cost of any necessary auxillary equipment** (such as charging equipment for truck batteries etc.) this is necessary to enhance the productivity and versatility of equipment.

- **Taxes and insurance**

  This is to be borne by the purchaser and depends on the norms of the region or country.

- **Interest on investment**

  If it is not possible for company to buy the equipment in cash then it has to be taken on loan in that case this cost comes into picture.

- **License fees**

  These are related to the equipments which are operated only by licence such as trucks that may operate on highways.

- **Rent of space**

  This is cost of space which is taken into account if equipment is hired or space is to be hired. For example garage rent for trucks.

  In short material handling cost is given as

  Material handling cost = cost of handling + cost of transportation + cost of packaging + cost of space + cost of handling equipment including operation, maintenance and depreciation etc.

6.2 HOW TO REDUCE THE COSTS ?

As discussed material handling involves so many costs, efforts should be made to reduce the costs. Reduction in total cost of material handling can be achieved by either reducing material handling or by improved handling procedure or both. The objective of reduction in cost of production through improved material handling can be achieved by

- Reducing material handling labour.
- Material handling work should not be assigned to skilled or semi skilled labour.
- Reducing indirect labour expense on activities connected with storage inspection quality control, repair, tool room, shipping etc,.
- Reducing damage of material during handling.
- Better utilization of space.
- Reducing in process storage.
- Increasing productivity.
- Reducing expenditure on packaging and other protective devices.
- Decreasing inventory.

6.3 COST ANALYSIS

Cost can also be reduced by cost analysis.The cost of **handling** is something that must be calculated within each plant. The following aspects when analysed present a figure that will serve as a practical index of the handling efficiency of the plant.

(*a*) Cost of handling department.

(*b*) Individual accounts chargeable to handling.

(*c*) Handling performed as part of other operations.

(*d*) Total shown as percentage of factory cost.

In order to develop these standard loads the "Handling Load Analysis Chart" may be used.

Manufacturing cost – cost of raw materials – purchased parts cost – Assembles cost = Handling cost.

(*a*) Cost of Handling Departments

Some of the functions often setup within the accounting framework—receiving shipping, storage (raw material, in process and finished), tool rooms, packing, crating and internal transport.

(*b*) Individual Accounts Chargeable to Handling

Equipment costs, maintenance cost for handling devices, the wages of truck operators, handlers, checkers and other fillers.

(*c*) Handling Performed as Part of the Operation

Machining time, handling, setting or positioning time.

Internal handling cost = labour bill of the dept. x handling time/total time.

(*d*) Total Cost of Handling Shown as Percentage of Factory Costs

The cost of handling for the factory = cost of handling depts. + individual accounts chargeable to handling + handling performed as part of other operation.

6.4 THE COST ACCOUNTANT AND MATERIALS HANDLING

To facilitate cost accounting following should be done

(*a*) Adequate cost accounting system is in operation.

(*b*) Provide account numbers and classification which will provide the required cost data.

(*c*) Show the importance of the information gathered.

6.5 COST INDICES

These are useful to calculate the cost effectiveness of given material handling equipment. For this proper accounting is necessary.

(*a*) Labour productivity $= \dfrac{\text{Expense of handling personnel}}{\text{Volume of material moved}}$

(*b*) Equipment utilization factor $= \dfrac{\text{Operating hours}}{\text{Available hours}}$

(c) Equipment utilization factor $= \dfrac{\text{Output or production per hour}}{\text{Capacity of equipment per hours}}$

(d) Frequency of breakdown $= \dfrac{\text{Number of hours in breakdown}}{\text{Number of breakdown}}$

(e) Frequency of breakdown $= \dfrac{\text{Number of breakdowns}}{\text{Available machine hours}}$

UNSOLVED QUESTIONS

1. Explain why cost of materials handling is to be reduced.
2. Explain various cost factors involved in the materials handling.
3. Explain why cost analysis is necessary and how it is done.
4. Write down various cost indices which can be used in materials handling.

□□□□

7 Cranes Used for Material Handling

INTRODUCTION

Need of the present day is for the equipment to handle heavy loads with fast speed, reliability, safety, economy etc. Due to variety of requirements, equipments of different types and sizes have been developed and manufactured. Application of material handling device is prime consideration in designing new plants and for modifying existing plants. These devices increase output, improve quality, speed up the deliveries and therefore, decrease the cost of production. Utility of these devices have further been increased due to increase in labour costs and problems related to labour management.

Cranes are one of the most important and versatile equipment used in the industries along with conveyors and trucks. They are available in many varieties.

Cranes have wide application in construction projects, industries and in shipping etc. These are used for lifting the loads (may be construction materials, loose materials, packages, containers, finished and semi-finished products in industries etc.) and placing them at desired place. For this purpose the cranes have three motions in general, namely hoisting, derricking and slewing. The cranes are generally electrically operated; diesel operated, or may have diesel-electric drive.

7.1 DEFINITION

Cranes are a combination of separate hoisting mechanisms with a frame for lifting or a combination of lifting and moving loads which can be truly suspended on or secured to them.

They are very much useful to pick up a load at one point and be able to deposit it at another point within a restricted area. They are used for loads that are heavy for other types of equipment.

7.2 CLASSIFICATION OF CRANES USED FOR MATERIAL HANDLING

➢ The cranes may be classified according to their construction

1. **Revolving or Rotary Cranes**
    (*a*) Jib crane
    (*b*) Tower crane
    (*c*) Cranes with rotary turntable

(*d*) Derrick

(*e*) Wall

2. Mobile Cranes

(*a*) Truck mounted

(*b*) Rail mounted

(*c*) Floating cranes

(*d*) Sky cranes

(*e*) Crawler

3. Bridge Cranes.

(*a*) EOT(Electrically operated overhead traveling crane)

(*b*) Gantry crane

(*c*) Cable cranes

- According to their movability
  1. Stationary cranes
  2. Mobile cranes
- According to the use
  Indoor and outdoor
- According to the load to be lifted
  1. Light duty
  2. Medium duty
  3. Heavy duty

7.3 ADVANTAGES OF CRANES

- For intermittent lifting jobs.
- An advantage possessed by a cranes, one with a long boom, is its ability to reach out from considerable distance from its base. Also, it can reach up to a considerable height.
- An important advantage of a crane equipped with a rotating boom is the high degree of flexibility providing in picking up and placing roads. Such a crane can go into comparatively narrow aisle with a load and then, with a rotating boom, swing the load in to a small space with little difficulty.

LIMITATIONS OF CRANES

- They do not operate continuously.
- Their loads are not controlled.
- They swing about and do not follow smooth vertical path as with lift and elevators.
- Their range of activity is quite limited both as to area covered and height of elevating.
- They are usually confined to relatively low lifting jobs.

7.4 IMPORTANT PARTS OF CRANES

Following table shows important parts of the cranes which are common for all cranes.

Table 7.1: Important Parts of the Cranes

Serial No.	Name of the Part	Use or Function
1.	Supporting frame or structure	These are generally walls or girders. Walls made up of cement. Girders are made up of different sections or angles. Girders generally support hoisting trolley. These take the dynamic and static loads.
2.	Hoisting mechanism	This is a heart of any crane. They are manually operated or power operated. They consist of hoisting element such as wires ropes or chains, motor, gear drive, pulley, sprocket, drums etc.
3.	Load lifting attachments	These are the attachments used to lift the load. It generally uses crane hook. These differs according to the load to be lifted. Sometimes grab buckets or special attachments are used suiting to the load to be lifted.
4.	Controlling equipments	These include those equipments which are essential for safe working of the crane. These include brakes, limit switches, arresting gears, relays, stops and sensors.
5.	Drives	These are essential to lift the load. Cranes are driven by electrical drive, gear drive or diesel generator.
6.	Crab or cabin	This is a place from where the operator can operate the crane. The location of the crab is very important in the crane operation.

7.5 BASIC MOTIONS OF THE CRANES

1. **Hoisting Motion**

   This is a motion specifically useful for the lifting and lowering the load.

2. **Radical Movement of Load**

   In this movement load travels towards the fixed frame or towards the centre of crane.

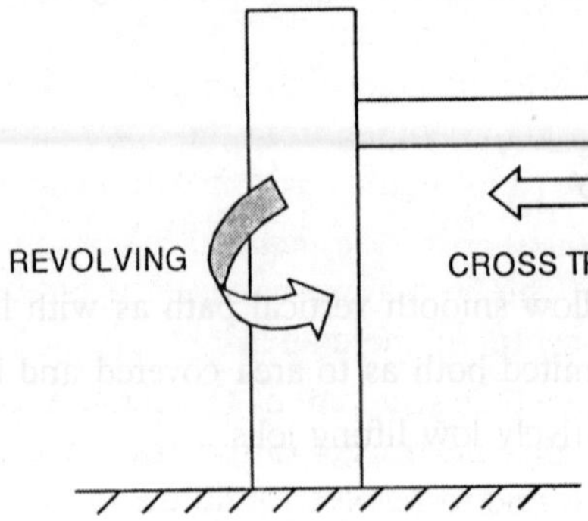

Figure 7.1. *Basic motions of crane.*

3. **Revolving Motion**

   In this motion the load can revolve about a vertical axis in 360 degrees.

4. **Traveling the Crane with Load**

   This is motion in which whole crane can travel with load.

7.6 REVOLVING OR ROTARY CRANES

As the name suggests a revolving crane swings the load through 360 degree. Revolving cranes are of different varieties. They can have basically 3 or 4 degrees of freedom. These are following:

(*a*) Jib crane

(*b*) Tower crane

(*c*) Cranes with rotary turntable

Now we will discuss these cranes one by one.

7.6.1 Jib Cranes

These cranes are characterized by the jib which is nothing but a long cantilever. Such cranes consist essentially of an inclined member, called the jib, supported by a rope or cable or some type of structural member attached to a vertical mast or frame. Load is usually suspended from the outer end of this inclined mast. The outreach of the jib may be fixed or variable. The crane as a whole may be either fixed or movable. Various sub-classifications of these cranes are possible.

Depending on the use, jib cranes may be classified into a number of varieties, primarily on the basis of their mountings. These are:

A. Floor mounted jib cranes

B. Wall-mounted jib crane

C. Roof mounted cranes

D. Vehicle mounted

7.6.1.1 Basic Motions of a Jib Crane

Jib cranes have three degrees of freedom: vertical, radical, and rotary. However, they can not reach into corners.

In general, a jib crane may have the following motions:

(*i*) Hoisting motion

(*ii*) Derricking or luffing motion

(*iii*) Slewing motion

(*iv*) Long travel motion

(*i*) Hoisting motion is used to lift or lower the load. This is usually achieved by steel wire ropes being affixed to a crane hook or a grab hanging from the outer end of the jib. The rope is applied through some sheaving arrangement and controlled and operated by a winch system.

(*ii*) Derricking or luffing motion is imparted to the inclined member or the jib to move in a vertical plane so that the angle of the jib may be changed in order to bring the load line nearer to or further off from the centre of the crane.

(*iii*) Slewing motion is imparted to the whole structure of the crane including the jib, so that it can turn about a central pivot shaft with respect to the non-revolving parts. This motion enables the crane to shift the load line to revolve round the crane.

(*iv*) Finally, the long travel motion may be required when the whole crane structure has to be shifted to a distant placed along a rail track or along a road.

A. Floor Mounted Jib Cranes

These are the cranes in which mast is attached to the ground. In other words they are some how attached to the ground.

(*a*) Fixed column jib crane

(*b*) Rotating column jib crane

(*c*) Stationary crane with a turntable

(*d*) Wharf crane

(*e*) Tower cranes

Every jib crane has a cantilever may be horizontal or inclined. This cantilever is nothing but a I-channel. This I-channel carries hoist trolley which travels on these I channel popularly known as guide rails. The jib is attached to a vertical column known as mast. Jib is supported by a tension member one end of which is attached to the jib and other end to a mast.

(*a*) Fixed Column Jib Crane

In this type mast is stationery but jib may be stationery or rotary. These two types of designs are shown below.

The difference is quite visible in figure 7.2 the load is raised by hoisting trolley and can be moved only along the jib but it cannot be rotated.

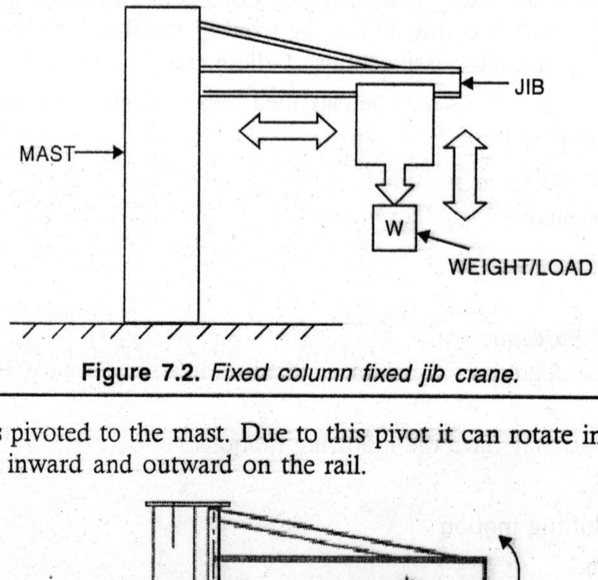

Figure 7.2. *Fixed column fixed jib crane.*

In figure 7.3 jib is pivoted to the mast. Due to this pivot it can rotate in a horizontal plane. As well as hoist can move inward and outward on the rail.

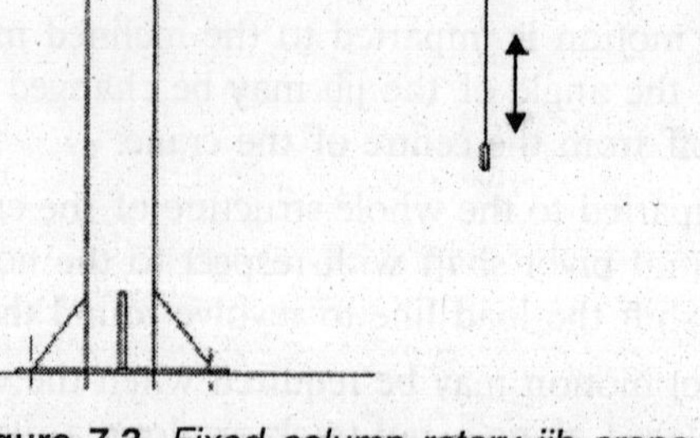

Figure 7.3. *Fixed column rotary jib crane.*

(b) Rotating Column Jib Crane

In this type of crane everything is same as that of above jib crane only exception of rotating mast. This mast can rotate through 360 degrees with whole jib and trolley. As seen from the figure it is clear that the column is mounted on the floor and it is pivoted at the bottom and top at the roof to provide support. These are generally hand operated but hoist trolley used may be simple (block and tackle) or may be electrical hoist.

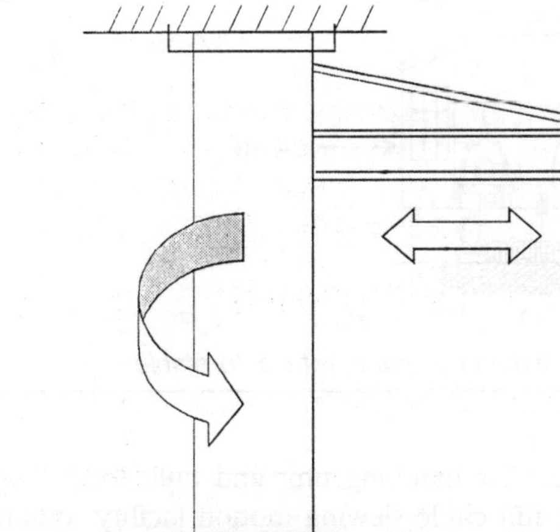

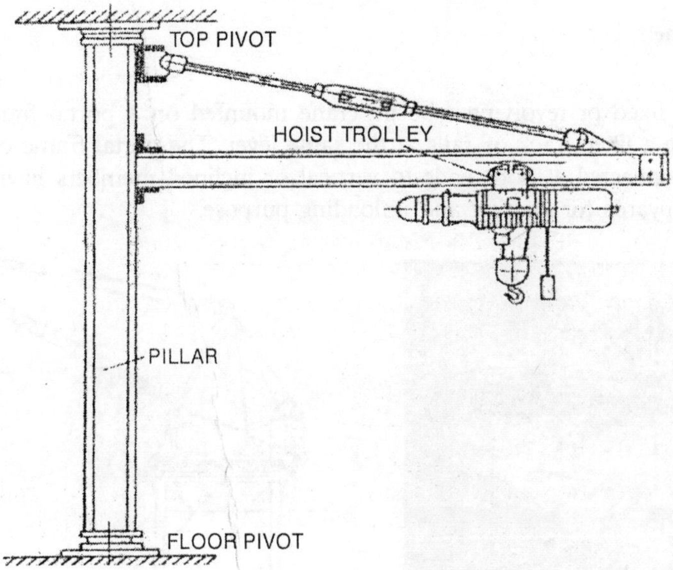

Figure 7.4. *Rotating column rotary jib crane.*

(c) Stationary Crane with a Turntable

These are the cranes in which the jib is inclined is made up of trusses it is popularly known as *boom*. This crane is mounted on the turntable and can rotate at 360 degrees about vertical axis. Turntable carries hoist and winches to pull the load. Operator's cabin is provided in front side so as to control the hoisting motion. This is electrically operated and commonly used in outdoors.

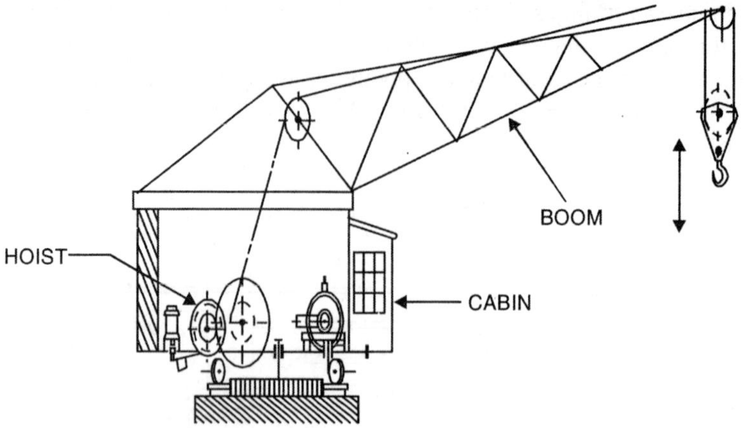

Figure 7.5. *Stationary crane with a turntable*

(*d*) *Wharf Crane*

These are used in shipyard and port for handling unit and bulk load. These are usually self propelled balanced level luffing type with full circle slewing motion facility. Wharf cranes may be of different types, depending on the type of structure on which it is mounted. The choice of structure for mounting depends on site condition. The principal types of wharf cranes are (*i*) high pedestal, (*ii*) full portal (*iii*) semi-portal.

(*i*) High Pedestal Crane

(*ii*) Full Portal Crane

Portal crane is a fixed or revolving type jib crane mounted on a portal frame in location or arranged to travel along a fixed track of rails at the same level. The portal frame consists essentially of horizontal girders connected at both ends to vertical or inclined members having equal lengths. These are useful in shipyards for loading and unloading purpose.

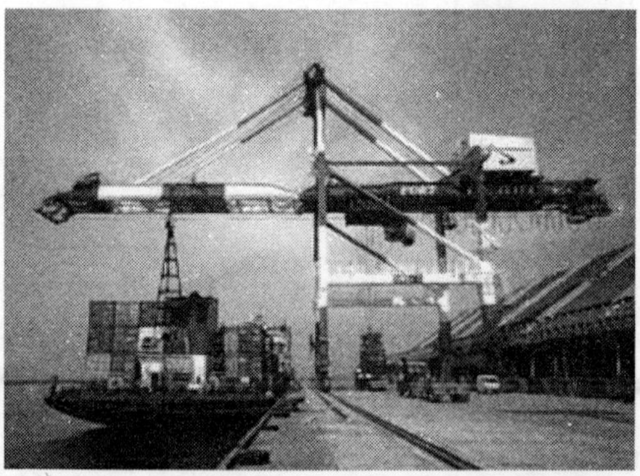

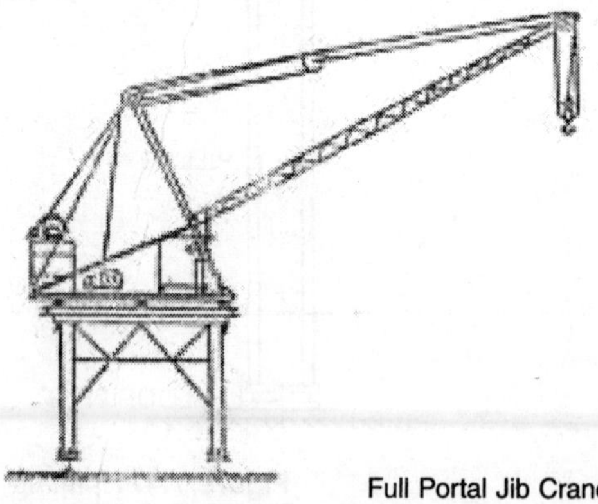

Full Portal Jib Cranes

Figure 7.6. *Portal cranes.*

(*iii*) Semi-Portal Crane

Semi-portal crane is a fixed or revolving type jib crane mounted on a semi-portal frame fixed in location or arranged to travel along a fixed track or rails at different levels. The semi-portal frame essentially consists of horizontal girders connected at both ends to vertical or inclined members which constitute a shorter side and a longer side. The shorter members may consist only of the trolley running along the elevated rail. These are useful in shipyards for loading and unloading purpose.

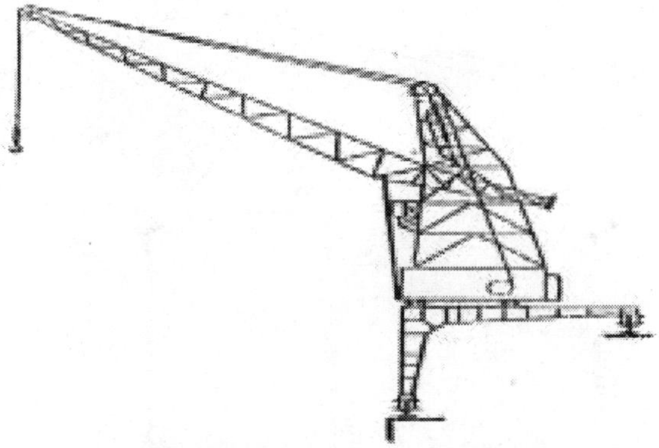

Figure 7.7. *Semi-portal crane.*

(*e*) *Tower Cranes*

Tower cranes are a common fixture at any major construction site. They often rise several meters into the air, and can reach out just as far. The construction crew uses the tower crane to lift steel, concrete, large tools like acetylene torches and generators, and a wide variety of other building materials.

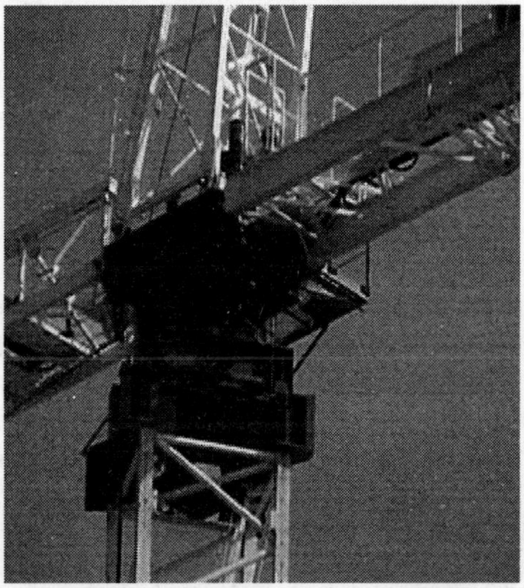

Figure 7.8. *Turntable of tower crane.*

Tower Crane

A typical tower crane has the following specifications:

- **Maximum unsupported height** – 80 meters

  The crane can have a total height much greater than 80 m if it is tied into the building as the building rises around the crane.
- **Maximum reach** – 70 meters
- **Maximum lifting power** – 19.8 tonnes, 300 tonne-meters
- **Counterweights** – 20 tonnes

The maximum load that the crane can lift is 18 tonnes, but the crane cannot lift that much weight if the load is positioned at the end of the jib. The closer the load is positioned to the mast, the more weight the crane can lift safely. The **300 tonne-meter** rating tells

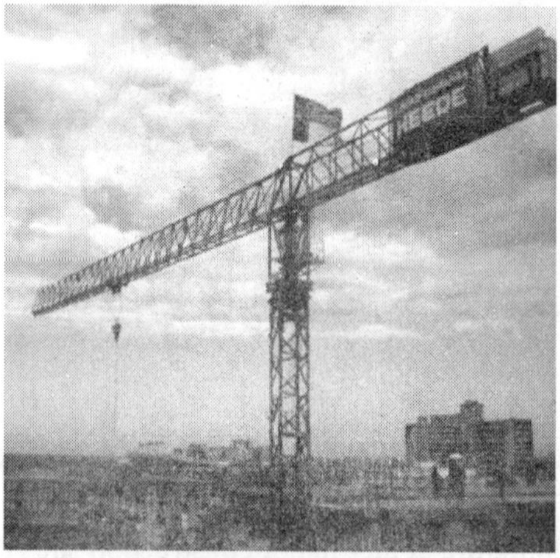

Figure 7.9. *Tower crane.*

you the relationship. For example, if the operator positions the load 30 meters from the mast, the crane can lift a maximum of 10.1 tonnes.

The crane uses two **limit switches** to make sure that the operator does not overload the crane: The **maximum load** switch monitors the pull on the cable and makes sure that the load does not exceed capacity. The **load moment** switch makes sure that the operator does not exceed the tonne-meter rating of the crane as the load moves out on the jib. A **cat head assembly** in the slewing unit can measure the amount of collapse in the jib and sense when an overload condition occurs.

Tower Crane Stability

The first element of the tower crane's stability is a large **concrete pad** that the construction company pours several weeks before the crane arrives. This pad typically measures 10 × 10 × 1.3 meters and weighs 182,000 kg—these are the pad measurements for the crane shown here. Large **anchor bolts** embedded deep into this pad support the base of the crane.

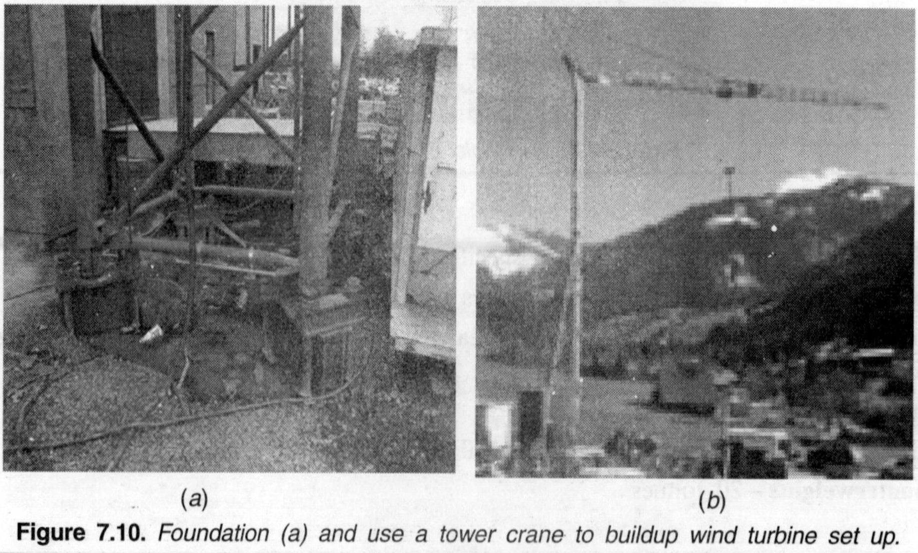

(*a*) (*b*)

Figure 7.10. *Foundation (a) and use a tower crane to buildup wind turbine set up.*

So these cranes are essentially bolted to the ground to ensure their stability. Generally these cranes were assembled at specified location for the work and disassembled at the site itself.

B. Wall-Mounted Jib Crane

The jib is provided pivots at the one end on the wall. Another pivot is attached at the end of tension member. The whole jib can turn or rotate about these pivots. This can rotate up to 150 degree in practical. The walls need to be strengthened to take a load of cranes. Advantage of these cranes is that they leave the floor space open.

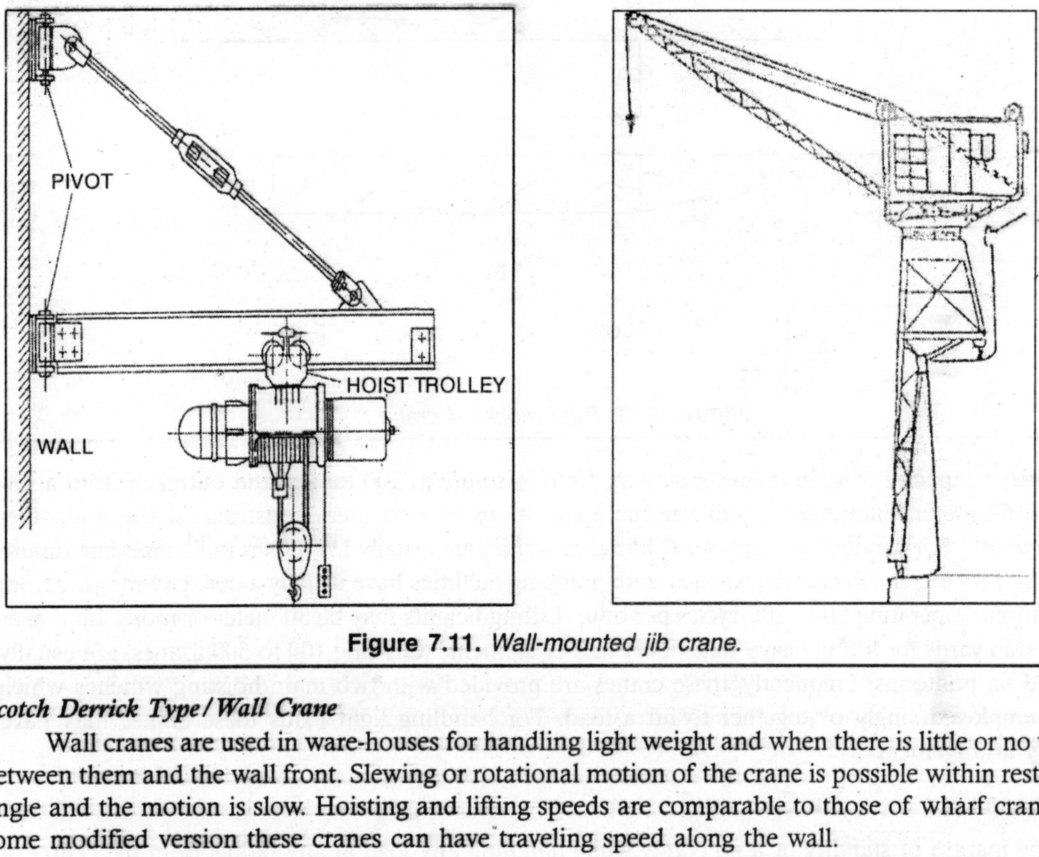

Figure 7.11. *Wall-mounted jib crane.*

Scotch Derrick Type / Wall Crane

Wall cranes are used in ware-houses for handling light weight and when there is little or no wharf between them and the wall front. Slewing or rotational motion of the crane is possible within restricted angle and the motion is slow. Hoisting and lifting speeds are comparable to those of wharf cranes. In some modified version these cranes can have traveling speed along the wall.

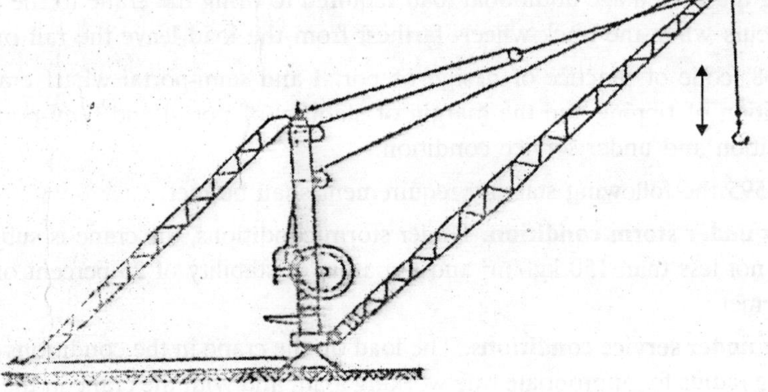

Figure 7.12. *Scotch derrick crane.*

C. Roof Mounted Cranes

These are used when there is less overhead space or material has to be picked and placed within restricted area. They are mounted on the roof it is exactly opposite to the floor mounted crane with rotary turntable. In this turntable is attached to the roof and jib hang from this turntable. It works similar to the floor mounted crane with rotary turntable except that the operator controls this crane from the ground instead sitting in the cabin. It is electrically operated. The roof must have high strength to sustain the weight of crane acting in downward direction.

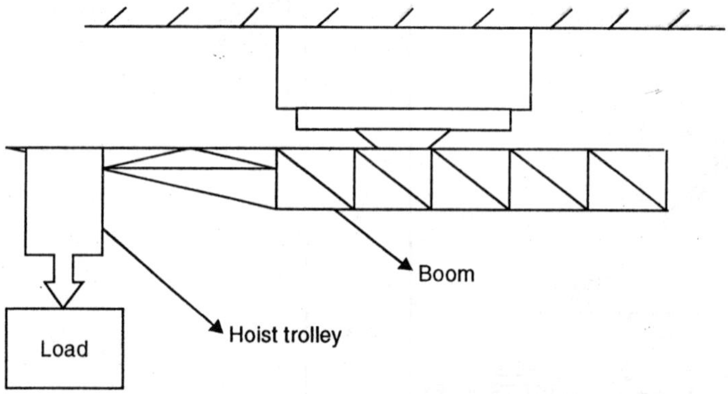

Figure 7.13. *Roof mounted crane.*

Lifting capacity of such cranes may vary from ½ tonne to 200 tonnes and outreach from a few meter to 50 meter. Such cranes find various applications in port area, construction site and other outdoor works. For handling general cargo, lifting capacities are usually 1½ tonnes to 5 tonnes maximum outreach of 30 meter. Jib cranes provided with grabbing facilities have usually a capacity ranging from 3 to 20 tonnes operating 50 to 100 cycles per hour. Lifting heights may be 30 meter or more. Jib cranes used in ship yards for lifting heavy machinery and equipment, weighing 100 to 300 tonnes, are usually mounted on pontoons. Frequently, these cranes are provided with two main hoisting winches which can be employed singly or together to lift a load. For handling light loads these cranes may have auxiliary arrangement.

7.6.2 Stability of a Jib Crane

The margin of stability of a jib crane, while handling any load at any radius from the centre of the crane and working on a track on firm level ground and with the jib at right angles to the track, is defined as being the percentage additional load required to bring the crane to the point of tipping. This condition occurs when the track wheels farthest from the load leave the rail or the road.

IS: 4594-1968 [code of practice of design of portal and semi-portal wharf cranes] provides in clause 4 the condition of tipping and the margin of stability of portal and semi-portal wharf cranes under storm condition and under service condition.

As per IS: 1595 the following stability requirement shall be met:

(*a*) **Stability under storm condition.** Under storm conditions, the crane is subjected to a wind force of not less than 150 kgf/m^2 and a margin of stability of 25 percent of this wind force shall be met.

(*b*) **Stability under service conditions.** The load on the crane in the conditions of tipping at the operating radius for appropriate safe working load, and with the crane in the position of least stability shall not be less than safe working load +50 percent of the same, without imposing undue stresses on centre pin or the centre column; and the safe working load +75 percent of the same, without causing any one of the track wheels leaving the track, and subject to a wind force of 25 kgf/m^2 acting at the same time.

The stability requirements are to be fulfilled with the jib set along, and at right angles to the track. The latter position creates a critical condition of loading and is consequently, relied on in stability calculations; for, in most traveling cranes the gauge is shorter than the wheel base.

Stability under service condition and that of the crane proper (without pay load) are determined with the crane in an inclined position, the angle being at least 3 degree for jib cranes (except the rail road type) and 1 degree for high pedestal cranes. The stability of rail road cranes operating without outriggers should be ascertained considering the relatives elevation of track rail on curves.

The stability under service condition is always determined with the jib set at the maximum reach and treating the wind effect and terrain gradient as factors increasing the chance of overturning.

Hosting Drum

For portal and semi portal wharf cranes, the diameter of drum measured at the bottom of a groove shall be not less than the appropriate value specified in Table 3 of IS: 4594-1968. It is given according to the construction of the rope and the classification of crane mechanism. The specified values are presented in table 7.2.

Table 7.2: Hoisting Drum Diameters

Rope Construction	Tensile Breaking Strengths of Wire kgf/mm^2	Minimum Diameter of Drum for Mechanism	
		Class 2 and 3	Class 4
6 × 19	160 to 175	24d	27d
	175 to 190	28.5d	30d
6 × 24	160 to 175	21d	30d
	175 to 190	24d	28d
6 × 37	160 to 175	19d	22d
	175 to 190	21d	24d
18 × 7	175 to 190	27d	30d
34 × 7	175 to 190	21d	27d

D = Diameter of rope in mm.

Rope drums shall be machine-grooved and the contour at the bottom of the groove shall be circular over an angle of approximately 120°. The radius of the groove shall be larger than the radius of the rope by not less than the appropriate amount given in Table 4 of IS: 4594–1968.

The depth of the groove shall be not less than 0.35 times the diameter of the rope.

The grooves of the drum shall be so pitched that the clearance between adjacent turns of the rope shall be not less than:

(*a*) 1.5 mm for ropes upto and including 12 mm diameter.

(*b*) 2.5 mm for ropes over 14 mm and including 29 mm diameter.

(*c*) 3.0 mm for ropes over 29 mm diameter.

Drums shall generally be of such length that there will be not more than one layer of rope on the drum when the rope is in its fully wound position and there will be in addition not fewer than two dead turns at each anchored end and one spare groove at the other end. The drum shall be flanged at both ends and the flanges shall project above the rope a distance of twice the diameter of rope.

Drums shall be made of cast iron, cast steel of mild steel conforming to the following Indian Standards:

(*a*) Gray Cast Iron : Grade 25 of IS: 210-1962

(*b*) Cast Steel : Grade 2 of IS: 1030-1963

(*c*) Mild Steel : IS: 226-1962

7.7 MOBILE CRANES

(*a*) Truck mounted

(*b*) Crawler mounted

(*c*) Rail mounted locomotive

(*d*) Floating cranes

(*e*) Sky cranes

Mobile crane (power driven) includes all type of traveling jib cranes, such as truck mounted, crawler mounted, and locomotive crane on rails.Mobile cranes are hoisting devices mounted on rubber tyred trucks, rollers or railroad wheels. Most of these cranes are full-rotating. Their mountings make them a distinctive group.

- Truck, crawler and locomotive cranes have now been highly developed for industrial use and they all have many possibilities of services to any plant or terminal that has heavy handling to do.
- The choice of crane, whether a truck, caterpillar or locomotive crane is a matter of determining the equipment which will perform the work most economically.
- Where greater flexibility of movement is needed, where clearance between buildings is sight, where there is free and dependable ground and where it is necessary to "walk close to the hob", a crawler or truck crane should be considered.
- A valuable characteristic of truck, crawler and locomotive cranes is the different ways in which a standard model can be put to use by the mere change of attachments, such as electromagnets or buckets for handling of bulk materials.
- Performances factors in which various cranes differ include lifting capacity, radius of operation, hoisting speed, speed of travel and speed of rotation of the boom.
- Lifting capacity of models usually purchased for industrial use ranges from 5 to 30 tonnes.

The radius of operation depends, of course, on the length of the boom. Auto-lift models with booms up to 18 m long are standard. Boom up to 30 m long are used on special work.

The group of trackless cranes is made up mainly of pillar cranes mounted on trucks, automobiles or tractors and intended for traveling on dirt, gravel and asphalt roads.

(*a*) Truck Mounted Crane

They are referred to frequently as yard or shop cranes because of their most general applications.

Truck crane is a crane mounted on a motorized vehicle (body fitted with large pneumatic tyres) for movement over streets, fields, lots and so on. It is mounted on chasis of standard design, on heavy duty chasis (when the capacity is up to 7.5 tonne), or on special purpose chasis with tyred wheels.

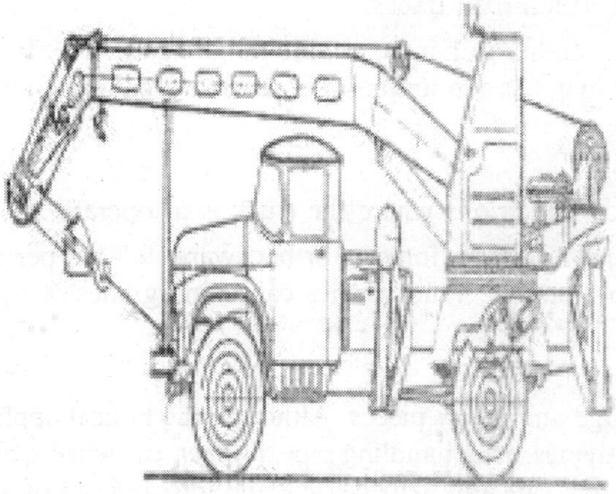

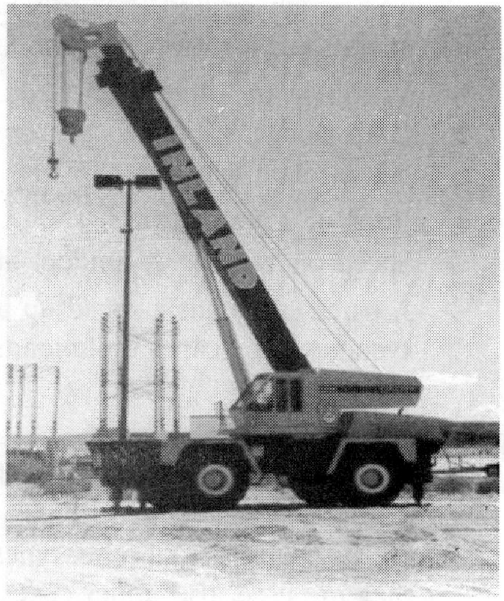

Figure 7.14. *Hydraulically operated truck crane.*

Types of Truck Cranes

Truck cranes are classed into general purpose cranes adopted to operate with the hook only, dual purpose cranes using both the hook and grab bucket and multipurpose cranes which can operate with any load-handling or earth moving attachment.

Two types of cranes are mounted on trucks. Those that have no traverse movement can lift and handle their loads in one plane only. Other varieties can be revolved about a vertical axis.

Machines of this category have three basic operating components: a revolving superstructure, mounting and front end equipment.

- The revolving superstructure includes the rotating frame (bed plate or turntable) which supports the power unit and the boom hinge bracket for attaching the boom to the frame. The boom hoist controls the angle of the boom. It may be driven by any type of gearing. The boom may be lowered by gravity or controlled by power. A braking mechanism is required regardless of the boom drive type.
- The machines may be powered by an internal combustion engine or an electric motor.
- A high gantry provides a wider angle between the crane boom and the supporting cables to reduce stresses in both when the crane boom is unusually long.
- The swing (slew) mechanism controls the revolving function of the machine usually through reversing clutches and bevel gears or some other type of drive.
- The travel mechanism on the revolving superstructure may be driven through clutches used for other functions, by means of a declutching device or through an independent mechanism.
- The truck mounting consists of a heavy framed, rubber-tyred carrier with two or more axles and having the general characteristics of a heavy duty motor truck on which is mounted a revolving superstructure.

Advantages of Truck Cranes

- A truck mounted crane can maneuver in any direction desired. This flexibility increases its efficiency (flexibility principle).

➢ Flexibility is provided by motor truck wheels fitted with tyres that permit them to operate any where they are needed without any restrictions as to path of travel, except on very soft and very rough ground. For this reason they have supplanted the locomotive cranes for some operations, for the latter are confined to rail-road tracks.

➢ Various crane attachments add to their ability to handle a wide range of articles effectively. Grabs, lings, specialized carriers and magnets are among the accessory equipments which makes them versatile handlers.

Specifications of Truck Cranes

➢ Speed of travel is dependent largely on conditions where the truck is to operate.

➢ A truck crane has a speed up to around 32 km/hr forward or backward. 9.5 km per hour is average. They can handle loads up to 20 tonne. Truck cranes can climb grades of up to 30 percent.

Applications of Truck Cranes

➢ Truck cranes are used for handling large and heavy pieces. Although the typical application of mobile cranes is in outdoor yard activities, as in handling pipe, lumber, car wheels, etc. and out of gondolas and other types of freight cars and flatbed truck and into and out of storage, they are used also under cover. For example marine warehouses, large assembly plants and similar locations are ideal for crane truck activities.

(*b*) Crawler Cranes

A special type of truck crane is so-called crawler crane mounted on caterpillar treads of the same basic type as used on tanks and half track military vehicles.

Crawler mounting consists of two parallel crawler belts with a series of tread shoes or links, encompassing rollers and drive tumbles supported by axles and a base frame which houses the propelling mechanism. This latter is driven and controlled from the revolving superstructure. Final drive to the crawler tread belts may be by chain or gears. A caterpillar (crawler) mounted can maneuver in any direction desired. This flexibility increases its efficiency (flexibility principle).

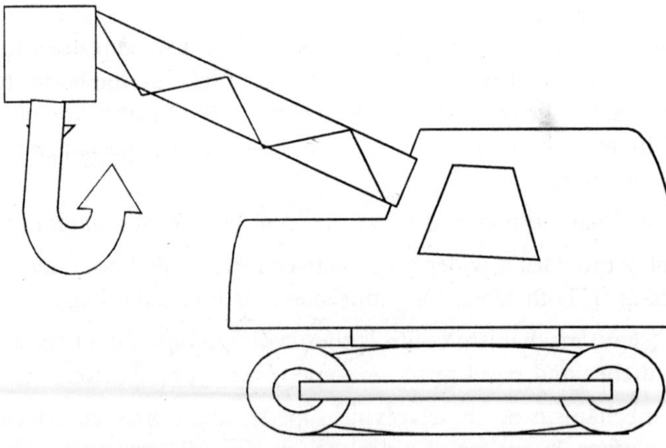

Figure 7.15. *Block diagram of crawler crane.*

Specifications of Crawler Crane

The crawler crane can be obtained in capacities ranging from 7 to 5 tonne. Usually the boom rotates through a full circle. The length of the boom varies, depending upon the type of work to be done. Crawler cranes have a travel speed of about 10 km per hour. The average crawler crane sufficient drawbar pull for moving railroad wagons. A crawler crane is capable of hoisting a 4 tonne load at a 3.2 m reach.

Applications of Crawler Crane

The most effective use of the crawler crane is in yards, fields, forests or other places where movement must be made over soft or rough surfaces. It is specifically used when soil under crane is loosely bound.

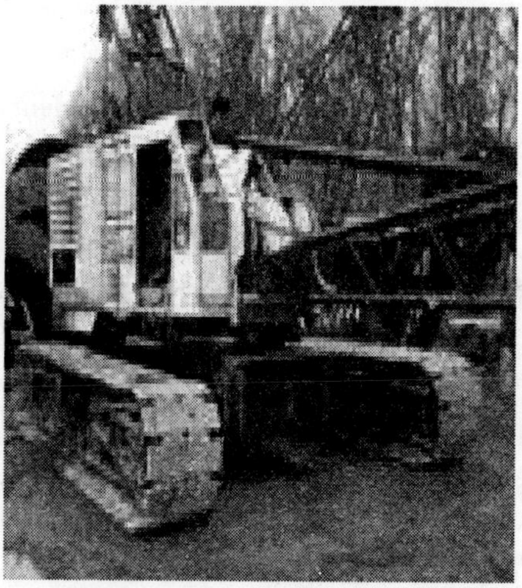

Figure 7.16. *Crawler in use.*

(c) Locomotive Mounted Crane or Railroad Crane

- It is a general purpose railroad crane equipped with replaceable handling attachments providing a wide field of application. It carries a power unit which is a combination of a diesel generator and a number of electric motors.
- Railroads cranes are mounted on trucks meeting railroad requirements and supported by axis of varying number. They have also an extra winch and a gooseneck type of jib enabling the handling of bulky metal structures.

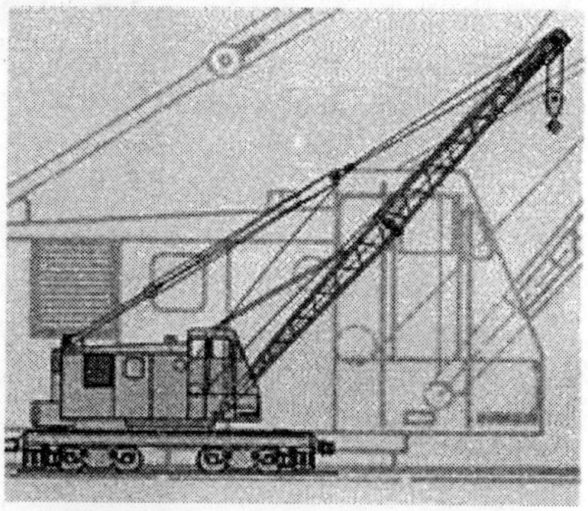

Figure 7.17. *Actual railroad crane.*

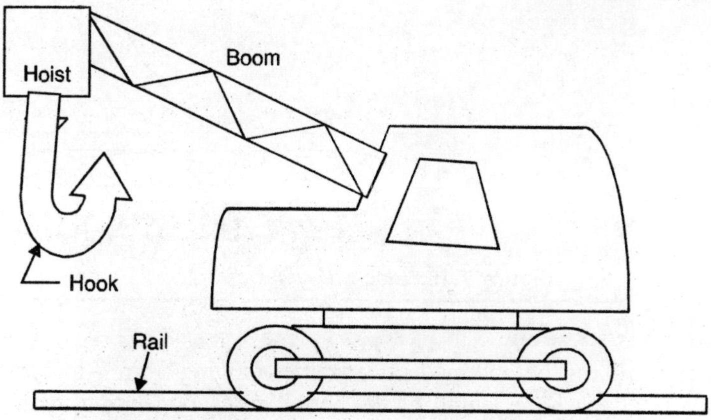

Figure 7.18. *Block diagram of railroad crane.*

Advantages of Locomotive Crane

Where free ground space is limited, the locomotive type is at its best.

Limitation of Locomotive Crane

A limitation of the locomotive crane is its confined movement in the direction of the track on which it operates.

Applications of Locomotive Crane

Such types of crane were used to handle the containers, accidental relief cranes for rail accidents, in plant handling.

(*d*) Floating Cranes

A floating crane is one mounted on barge, lighter or self-propelled towed pontoons. They are used for salvage, along shore and off shore duties.

Applications of Floating Cranes

They are used for handling cargo, but more especially for dredging operations, in which case they are equipped with buckets.

Advantages of Floating Cranes

They are compared with heavy duty cranes on shore, floating cranes have the advantage of flexibility, for they can be towed to ships or terminals any where about the port. An important advantage is that, while the heavy lifts are being loaded or unloaded, other cargo can be loaded or unloaded through other hatches. Otherwise the ship would have to be shifted to another point such as a ship yard, where a heavy duty shore crane could be used. Heavy lifts are handled into and out of vessels more economically and rapidly by heavy-duty floating cranes than by any other means.

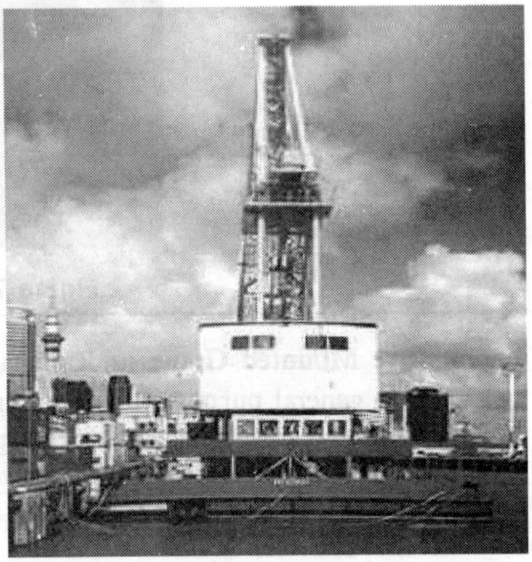

Figure 7.19. *Floating crane at rest near port.*

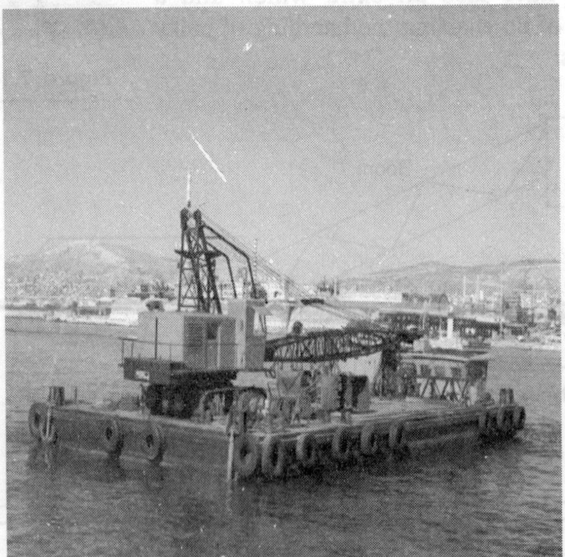

Figure 7.20. *Floating cranes in use at ports.*

Fully equipped floating cranes, with a lifting capacity ranging from 30 to 300 tonnes, offer varied solutions. They may be used to complete high standard harbour projects or to carry out loading, unloading, transhipping assignments, heavy load transportation and lifting or replacing anchors. To ensure immediate intervention, wherever that may be needed (marine accidents, vessels, groundings etc.), all vessels are available on a 24-hour basis.

***(e)* Sky Cranes**

Helicopters have many commercial uses. One of the most unusual is serving as an aerial crane. In this capacity, helicopters carry loads slung beneath their fuselages on cables. The first helicopters were used in this capacity in the 1950s, although it was not until the 1960s that they were heavily used for this purpose in the construction industry. Many different types of helicopters are used as aerial cranes. In the 1960s, the Sikorsky S-58, known as the H-34 in military service was used in this capacity, and many S-58s are still in use today carrying medium-size loads.

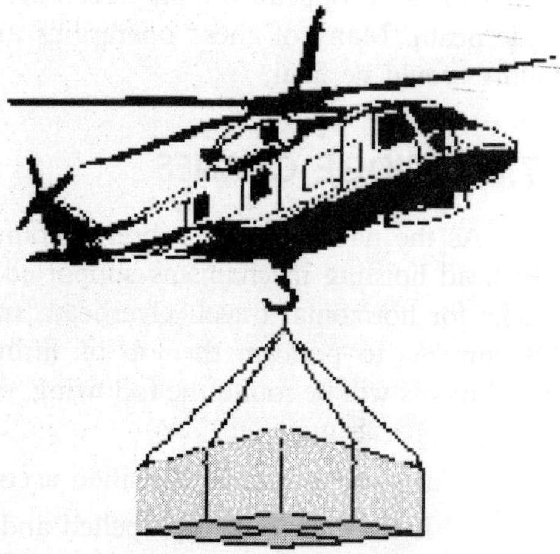

Figure 7.21. *Block diagram of helicopter crane.*

Crane helicopters are employed on an ever increasing scale as a means of moving load and doing variety of other jobs in regions which are difficult of access. Capable of descending vertically offers a safe and convenient way of load slinging. The use of helicopters has proved to be economically attractive providing for a 35 percent saving as compared with the other hoisting and haulage equipment.

Helicopter skycranes offer several advantages over conventional cranes. First, they can reach areas that may be hundreds of kms from the nearest roads or on inhospitable terrain. For instance, they are often used to erect communications towers on top of mountains or electrical transmission towers far from cities and towns. Large skycranes are often used in the lumber business to lift large trees out of terrain that is too rugged for a ground-based vehicle.

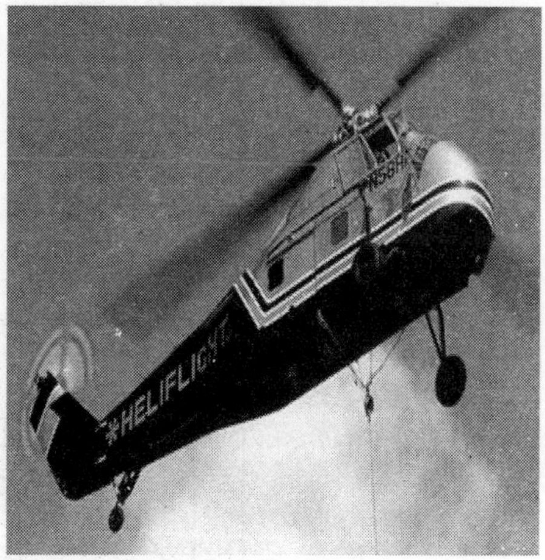

Figure 7.22. *Helicopter crane in use.*

They can also reach locations that a conventional crane cannot easily reach. For instance, sky cranes are often used to replace equipment such as transmission towers and large air conditioning units at the top of tall buildings. While they are not practical for the construction of the buildings themselves, sky cranes are highly useful when a single piece of equipment, such a large pump, needs to be brought in or removed. In addition, some large factories have equipment mounted on their roofs far from the edge. A conventional crane could not reach over the building far enough to place or remove this equipment, but a sky crane has no problem doing it.

The sky crane is exactly as its name implies-an aerial crane. It has a cockpit pod at the front of a long boom equipped with landing gear mounted on outriggers. This provides a large open area under which the carried load is connected to the fuselage.

During many lifting operations, pilots often move their helicopters very little. They usually lift off, while the load is attached, and immediately—but slowly—move over to where they will lower the load. If anything goes wrong, such as an engine warning light coming on, they are under orders to immediately drop their load so as not to risk the aircraft. For this reason, the area underneath the flight path has to be empty. If the sky crane is lifting an air conditioning unit to the middle of the roof of a large factory, for instance, the area inside the factory under the flight path has to be empty. A piece of machinery dropped from the helicopter would likely crash straight through the roof and kill anyone underneath. Many of these operations are also conducted at such low altitude that a total engine failure would be fatal.

7.8 BRIDGE CRANES

As the name implies a bridge crane resembles a bridge that spans the work area. They have overhead hoisting mechanisms supported by shear or pier legs which may be fixed or mounted on tracks for horizontal travel. Overhead, traveling bridge cranes can be fitted with a wide variety of attachments to perform the job of lifting and transporting materials. Among the more common attachments will be found the following: grabs, hooks, magnets, vacuum lifting pads, counter balanced c-hooks, special grips etc.

Bridge cranes can be classified according to the work have to perform for example:

- Light duty hand—propelled and very light weight, up to 5 tonne capacity.
- Medium duty—for general use in factories and warehouses from 5 to 20 tonnes capacity.
- Heavy duty—for foundry work and large magnet work, 20 to 50 tonnes capacity.
- Extra heavy duty—dock side and steel mill use for loads over 50 tonnes capacities.

Current is supplied to the trolley and hoist motors by way of cross crane wiring made of round or angle bars. Important bridge cranes are

A. EOT Crane

B. Gantry Crane

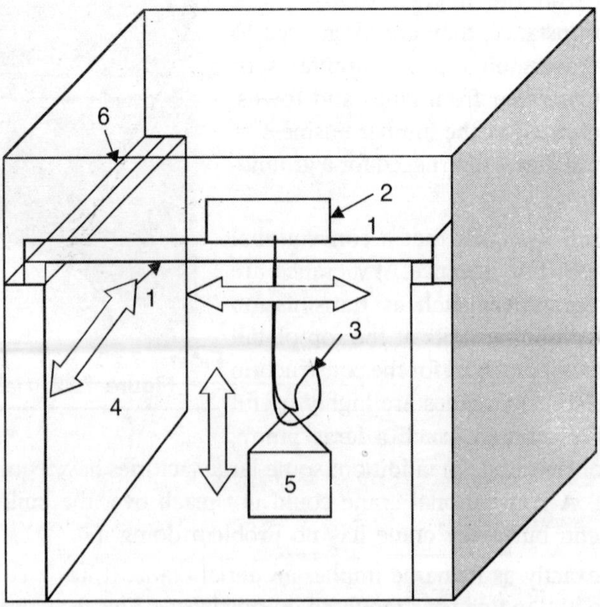

Figure 7.23. *Bridge crane motion.*

Important parts and motions of the bridge crane :

1. Transverse Girder, 2. Hoist Trolley, 3. Crane Hook, 4. Side Wall, 5. Load, 6. Longitudinal Rail.

A. Electric Overhead Traveling (EOT) Crane

The most adaptable and the most widely used type of power driven crane for indoor service is undoubtedly the three motion E.O.T. Crane. It serves a large area of floor space within its own traveling restrictions than any other permanent type hoisting arrangement. The three motion of such crane are the hoisting motion, the long travel motion and the cross travel motion. Each of the motions is provided by electric motors. A E.O.T. consists two parts—The bridge and the crab.

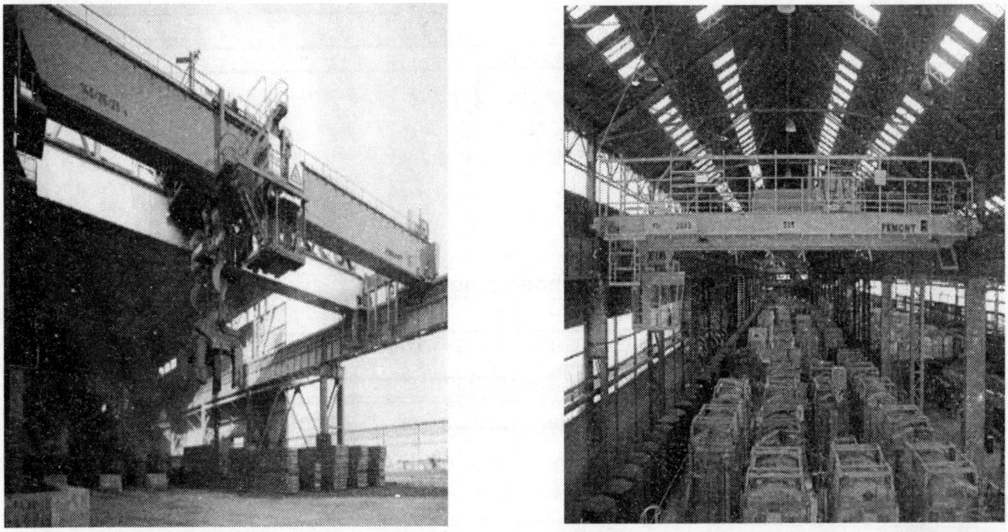

Figure 7.24. *Industrial EOT.*

EOT Cranes Classification

(*a*) According to the number of girders

Single girder and double girder

(*b*) According to trolley attachment

Under slung or over slung trolley

(*c*) Remote control EOT cranes

Remote control operation is offered with Infra-red Remote Controls/Radio Frequency remote Controls.

Infrared Remote Range-up to 50 meters.

Radio frequency Remote Range-up to 500 meters.

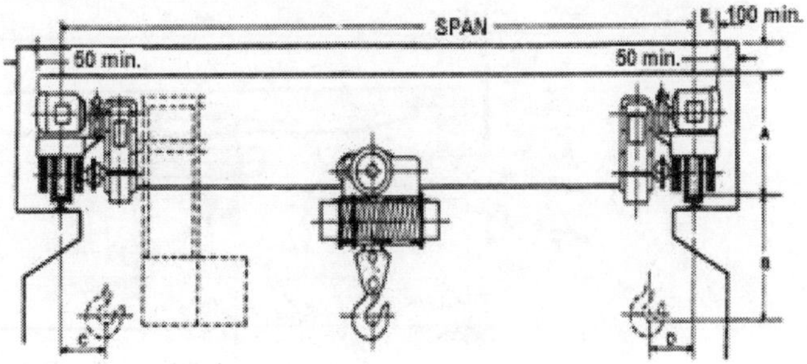

Figure 7.25. *EOT specifications.*

(*a*) Single Girder Overhead Cranes

Capacity Range

0.5 tonne to 20 tonnes

Span Range

3 meters to 30 meters

Description

Single girder dual motor cranes provide greater floor coverage than other type cranes. The close crane end approaches coupled with a close trolley approach hoist give you more working floor coverage. This can reduce space requirements in new buildings, or expand your usable space in existing buildings.

- Adjustable torque magnetic disc/shoe brake for smooth operation.
- Forged steel crane wheels for long life.
- Top quality hoists matched to the duty required.
- Electric hoist conforms to IS : 3938.

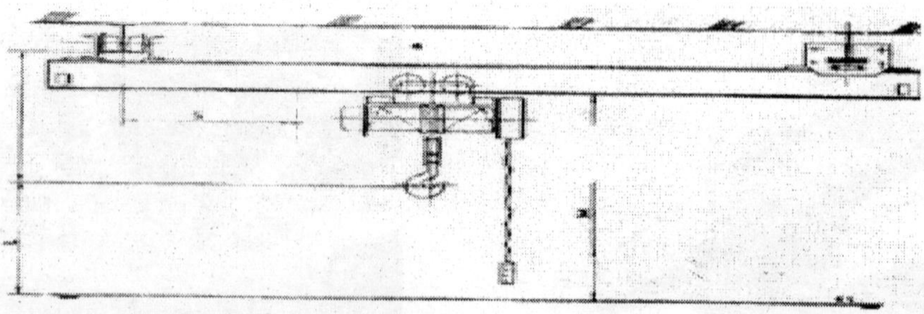

Underslung

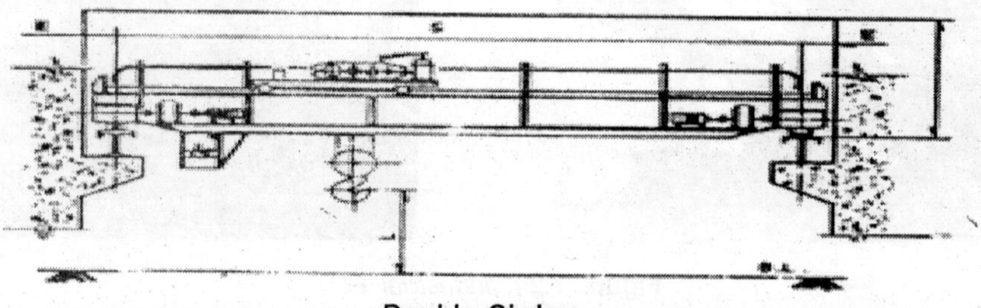

Double Girder

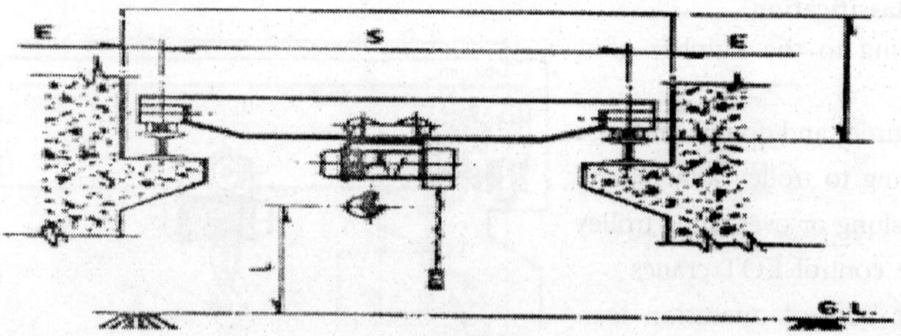

Single Girder

Figure 7.26. *Types of EOT cranes.*

(*b*) Double Girder Overhead Cranes

Capacity Range

1.0 tonne to 150 tonnes

Span Range

3 meters to 36 meters

Types

(*i*) **Class I :** Light duty cranes for repair shops, light assembly operations, service buildings, light warehouses, where service requirements are infrequent.

(*ii*) **Class II :** Medium duty cranes for machine shops and general industrial use both indoor and outdoor application.

(*iii*) **Class III :** Extra heavy duty cranes for steel mills and other heavy machine shops, steel warehouses, heavy fabrication shops, paper mills etc. with magnet and bucket operation for scrap yards, fertilizer plants and induction furnace.

(*iv*) **Class IV :** Extra heavy duty cranes for steel mills and other heavy engineering purposes.

Features

(*i*) **Structure :** Bridge girder, end carriages, crab frame manufactured from box plate design/ Rolled Sections. If required 100% radiography of all but welding joints of bridge and end carriages can be provided. The material of construction conforms to IS:2062, and the bridge is designed for minimum vibration.

(*ii*) **Wheel Assemblies :** Trolley and crane wheels are straight treaded double flanged forged steel C55Mn75, heat treated to achieve a special sorbatized structure upto a depth of 15-20 mm from the surface and gradually diminishing towards the centre retaining the core properties. These wheels are supported on antifriction ball/roller bearings, in special L-type bearing blocks. Wheels also available of gear and pinion type.

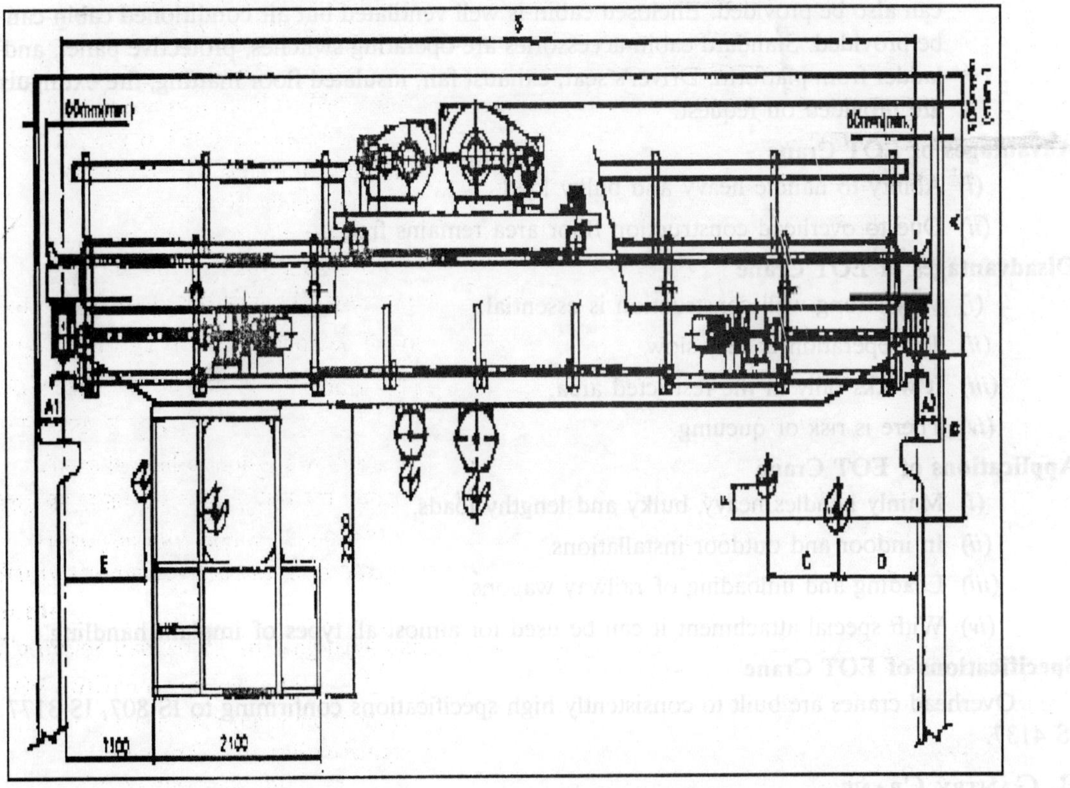

Figure. 7.27. *EOT details.*

(*iii*) **Gearing :** The gear reducers are constructed either in double, triple and quadruple reduction using rugged wide faced helical and spur gears that are precision cut from special alloy steel forgings. And to protect the gear train from shock loads, all gears and pinions are supported between two antifriction ball/roller bearings. All gear units are totally enclosed with oil bath splash lubrication.

Micro drive special gear boxes—The special micro drive provides fine inching movements of hoist, trolley or crane at the rate of 5% to 20% of the main speed. The micro drive is achieved by a separate gear box, with unique sun and planet system. Gearing system fitted on the input side of main gear unit and pony motor with separate brake unit. It is a positive drive and avoids the use of clutches etc.

(*iv*) **Cranes Drives :** A.C. control packages to meet electrical requirements.

(*v*) **Variable Frequency Drives and Remote Controls :** Variable Frequency Drive in hoisting, cross travel, long travel for stepless speed control can be implemented. also cranes with radio remote control cranes for application in inacessible and hazardrous areas.

(*vi*) **Brakes :** Failsafe magnetic or hydraulic brakes of 150% torque or more are fitted on shaft extensions. For heavy duty cranes a second brake can be mounted to ensure failsafe operation.

(*vii*) **Controls :** Pendant control from floor, master control switches, master controllers in the driver's cabin can be provided.

(*viii*) **Cabin :** Open or enclosed type driver's cabin available. The cabins can be provided either at one end of the bridge girder or center of bridge girder. Moving type cabin fitted with crab can also be provided. Enclosed cabin is well ventilated but air conditioned cabin can also be provided. Standard cabin accessories are operating switches, protective panel, and safe ladder from platform. Driver's seat, exhaust fan, insulated floor matting, fire extinguishers are provided on request.

Advantages of EOT Crane

(*i*) Ability to handle heavy and bulky load.

(*ii*) Due to overhead construction floor area remains free.

Disadvantages of EOT Crane

(*i*) Very strong wall construction is essential.

(*ii*) The operation is very slow.

(*iii*) It works only in the restricted area.

(*iv*) There is risk of queuing.

Applications of EOT Crane

(*i*) Mainly handles heavy, bulky and lengthy loads.

(*ii*) In indoor and outdoor installations.

(*iii*) Loading and unloading of railway wagons.

(*iv*) With special attachment it can be used for almost all types of implant handling.

Specifications of EOT Crane

Overhead cranes are built to consistently high specifications confirming to IS 807, IS 3177 and IS 4137.

B. Gantry Crane

The gantry or goliath crane spans a work area in a manner similar to the bridge crane, however, it is generally floor supported rather than overhead supported on one or both ends of the spanning section. The support can either be fixed in position or travel on runways.

Gantry crane means a powered crane consisting of one or more bridge beams supported at each end by leg mounted on end carriages and that is capable of traveling on supporting surfaces or deck levels, whether fixed or not, and that has a crab that has one or more hoisting units that are able to travel across the bridge beam or beams, but does not include a crane has 3 or fewer powered operations and that is controlled from a location remote to a permanent cabin or control station on the crane.

Cranes are widely used to transport heavy loads and hazardous materials in shipyards, factories, nuclear installations and high-building construction. They can be classified into two categories based on their configurations: gantry cranes and rotary cranes. Gantry cranes are commonly used in factories. This type of cranes incorporates a trolley, which translates in a horizontal plane. The payload is attached to the trolley by a cable, whose length can be varied by a one-dimensional pendulum with one-degree–of-freedom sway. There is another version of these cranes, which can move also horizontally but in two perpendicular directions. The analysis is almost the same for all of them because the two-direction motions could be divided into uncoupled one-direction motions.

Applications of Gantry Crane

Gantries serve their most useful purpose in industry for shipping and receiving.

Gantry cranes are employed to handle unit loads at storehouses, transportation field, heavy industries, unloading towers, docksides, powerhouses, marine terminals, steel mills, fabrication plants, railroad yards and cargo terminals.

The heavy capacity gantries are usually required on hydroelectric installations. They are used primarily for installation of heavy equipment and subsequent intermittent maintenance.

A rather specialized type of gantry is one that lifts the gates of dams.

Specifications of Gantry Crane

Gantry cranes commonly feature a capacity in excess of 200 tonne, a span of 48 m and a hoisting height of 18 m.

Limitations of Gantry Crane

Due to the center of gravity of the crane being at a higher level than the track level, these cranes can not travel fast. Rail tracks on the workshop floor must not project above the floor level so that other vehicles can travel unhindered.

Advantages of Gantry Crane

The advantage of the legged structure is that it is traveling on the floor and does not require specially made over head tracks like the overhead traveling crane. The gantry crane offers advantages for many difficult applications.

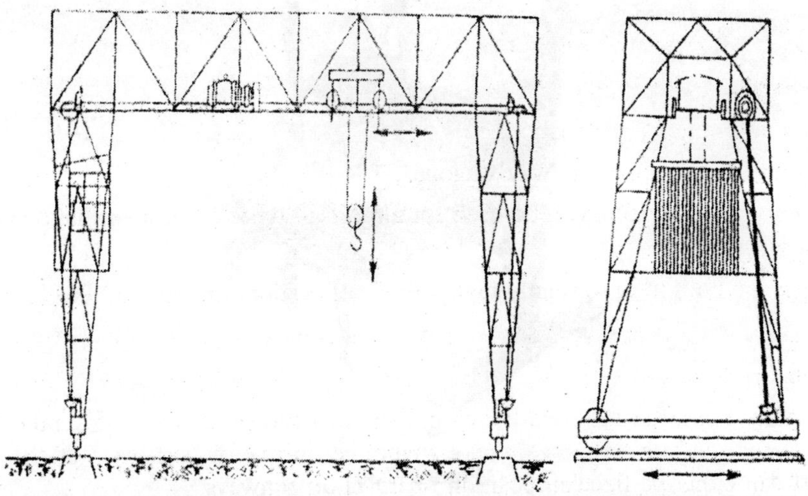

Figure 7.28. *Gantry crane used for container handling.*

Types of Gantry Cranes

***(i)* Semi-Gantry Cranes**

A Semi-gantry crane has one leg shorter than other in which one pair of legs travel over a track placed on a low foundation while the other pair receives support from an elevated track laid on a trestle or track girder attached to the building.

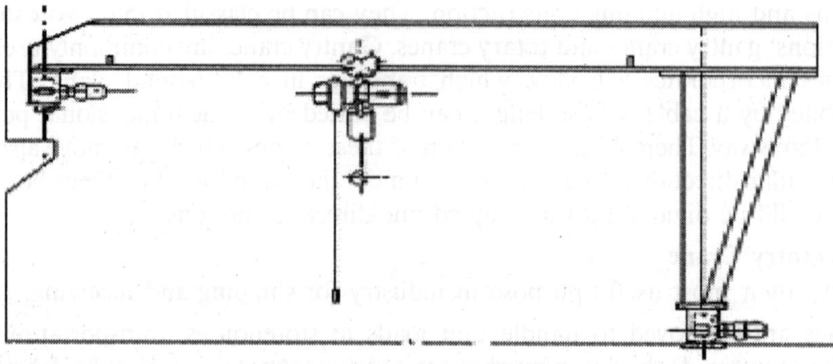

Figure 7.29

- **Adjustable :** The height and clear span (distance between braces) are adjustable. Both options are available or combined for maximum flexibility.
- **Navi Gantry :** An option that provides steering of powered gantry cranes eliminating any form of tracks. This feature allows the gantry crane to service many areas of the plant.
- **Track Mounted :** Casters of the gantry are equipped to run either on inverted angle, 15 and 20 kg. rail or a system that tracks along a floor mounted angle rail.
- **Utility :** High-beam design, beam bolted to support legs without bracing. Quick assembly service crane.
- **Capacity :** Up to 15 tonnes.

***(ii)* Portable Gantry Crane**

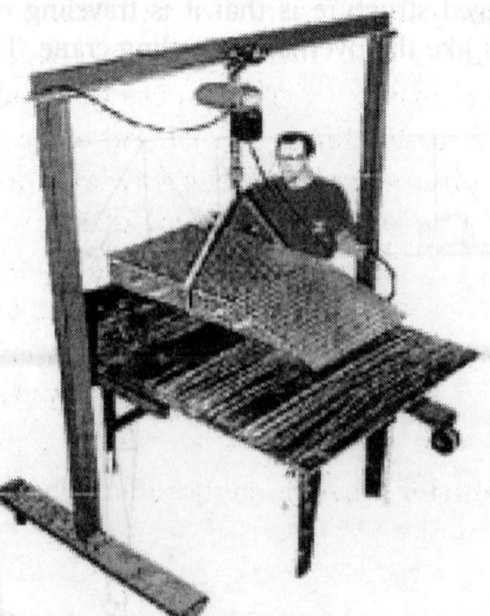

Figure 7.30

Designed to maximizes material handling requirements in light-duty lifting applications. Lightweight design allows for easy mobility. Includes four polyurethane-on-steel swivel casters. Welded steel construction. The straddle width and base width is 170 cm. Powder coat finish. Hoist and trolley not include.

***(iii)* Fixed and Adjustable Steel Gantry Cranes**

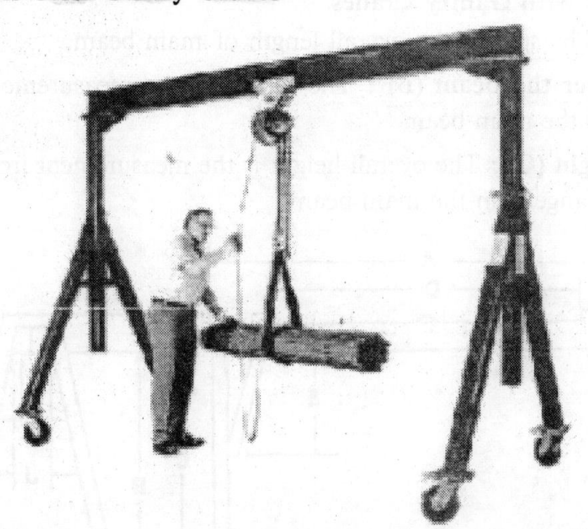

Figure 7.31. *Fixed and adjustable steel gantry crane.*

Industrial steel gantry cranes are designed for transporting and positioning materials. Solid steel construction will provide years of service. Choose from a variety of sizes. Large 20 cm diameter locking phenolic casters with roller bearings will facilitate easy mobility from one area to another. More economical and flexible than permanent cranes. Features quick setup design. Order optional lever ratchet for easy one-person height adjustment (not for fixed models). Height is adjustable in 15 cm increments on adjustable-height models. Blue finish. Hoist and trolley sold separately.

Fixed Gantry Cranes Specifications

Semi-steel heavy-duty swivel casters with roller bearings are standard equipment. Two casters are supplied with brakes. Gantries are easily disassembled and assembled for relocation. Contact factory for special heights, spans and options. Our fixed Height Gantry cranes are ideal for use

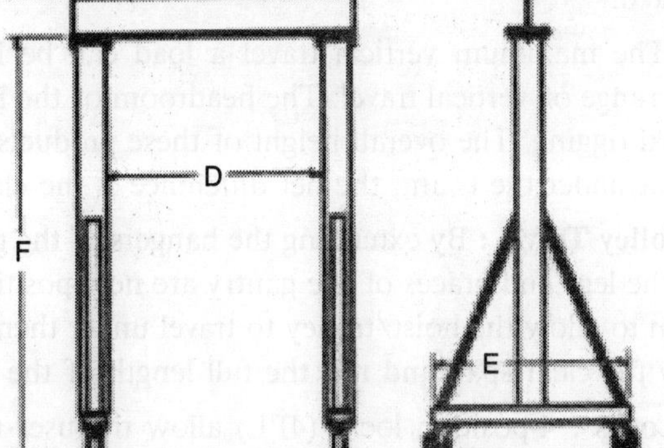

Figure 7.32. *Gantry specifications.*

in aircraft, automotive and marine repair shops. They also work well in warehouses, machine shops and loading areas. Gantry cranes are supplied with four roller-bearing steel wheels for easy maneuverability. The structural steel legs are welded together at the factory. Gantry crane requires assembly before use.

Terminology Associated with Gantry Cranes

(*i*) **Span (A) :** The span is the overall length of main beam.

(*ii*) **Height Under the Beam (B) :** The H.U.B is the measurement from the floor to the underside of the main beam.

(*iii*) **Overall Height (C) :** The overall height is the measurement from the floor to the top of the gantry hangers on the main beam.

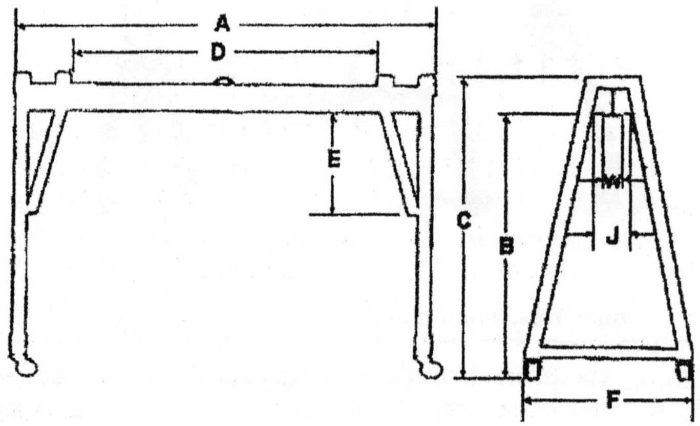

Figure 7.33

(*iv*) **Clear Span (D) :** The clear span is the unobstructed work area between the braces.

(*v*) **Brace Length (E) :** Attached midway down each leg bolted to the hangers.

(*vi*) **Beam Flange Width (W) :** The width measurement of the beam flange.

(*vii*) **A Frame Width (F) :** The "A" frame width is the overall measurement of the "A" frame at the base.

(*viii*) **Hangers :** Four (4) individual pieces welded to the top of the beam to which the legs and braces are bolted.

(*ix*) **Clear Lift :** The maximum vertical travel a load can be lifted. A number of factors determine the range of vertical travel. The headroom of the hoist/trolley, depth of lifting beam or special rigging. The overall height of these products must be deducted from the available height under the beam, the net difference is the clear lift available.

(*x*) **Full Beam Trolley Travel :** By extending the hangers of the gantry to fit the width of the hoist/trolley, the legs and braces of the gantry are now positioned far enough away from the main beam to allow the hoist/trolley to travel under them. Now the hoist/trolley are not limited by the clear span and run the full length of the beam.

(*xi*) **Directional Locks :** 4-position locks (4PL) allow the user to lock the direction of the wheel in a north, south, east or west position. For power driven units, only 2 (4PL) or 1 set, would be required if necessary.

(*xii*) **Port-A-Cart System :** With minor changes, Motivation Industrial Equipment can take the spreaders and a caster of our gantry crane and transforms them into a transport cart.

(*xiii*) **Jack Kit :** Used on adjustable height gantries only, 2 are required for most gantry and 4 (1 for each leg) are required for high capacity gantries. A screw jack allows height adjustment without the use of an overhead device or forklift.

Gantry Construction

Pre casting the box girder segments speeds up the construction of the bridge structure in two ways:

1. It permits to be fabricated elsewhere (precast) while the substructure (*i.e.*, piles, footings, pier column and pier segments) is being built.
2. It allows rapid assembly of the segments when building the cantilevers since the delays in waiting for newly cast concrete to harden are eliminated.

The decision to precast obviously requires a method for transporting the segments in place to assemble them on the structure. Often times, when bridges are constructed over water, the segments can be shipped to the site on barges and lifted into place by equipment positioned at the end of the cantilevers. Most of the river is not deep enough to readily permit barging without "dredging" or lowering a portion of the river bed.

The massive size and length of the gantry is not so much for lifting the segments but to allow it to move from pier to pier as it completed one section of the bridge and moves on to the next. It normally sits on supports near the center and the rear of the gantry while lifting segments so build a cantilever. With the aid of temporary props, the gantry can be raised so the supports can be moved to the next pier. The gantry then follows by sliding along the top of the supports.

Advantages of Gantry Cranes are:

- Can be used where overhead runways are not practical.
- A much greater bridge weight than an equivalent overhead crane.

Because of these design limitations, Gantry Cranes are generally used for moving heavy loads, often outdoors.

Gantry Cranes – Uses in Various Plants—Power Plant, Steel Plant, Air Craft Manufacturing Plant.

ISO Standards of Gantry Cranes

ISO TR 16880:2004

Collects the design and manufacturing requirements and recommendations for bridge and gantry cranes in ISO and IEC International Standards.

ISO 4301-5:1991

Cranes – Classification – Gantry cranes and Overhead traveling cranes. Establishes the cranes based on the number of operating cycles to be carried out during the expected life of the appliance and its mechanisms, and a load spectrum factor, which represents the nominal state of loading.

C. Cable Cranes

The taut-line cableway crane consists of two towers suspended between which a track cable of the locked construction is serving as a bridge structure. A carriage with a handling attachment travels on the track cable with the aid of a haul rope and a winch. The haul rope is reeved through sheaves provided at both towers and passes over a drum of the winch, forming a closed loop attached to which is the carriage.

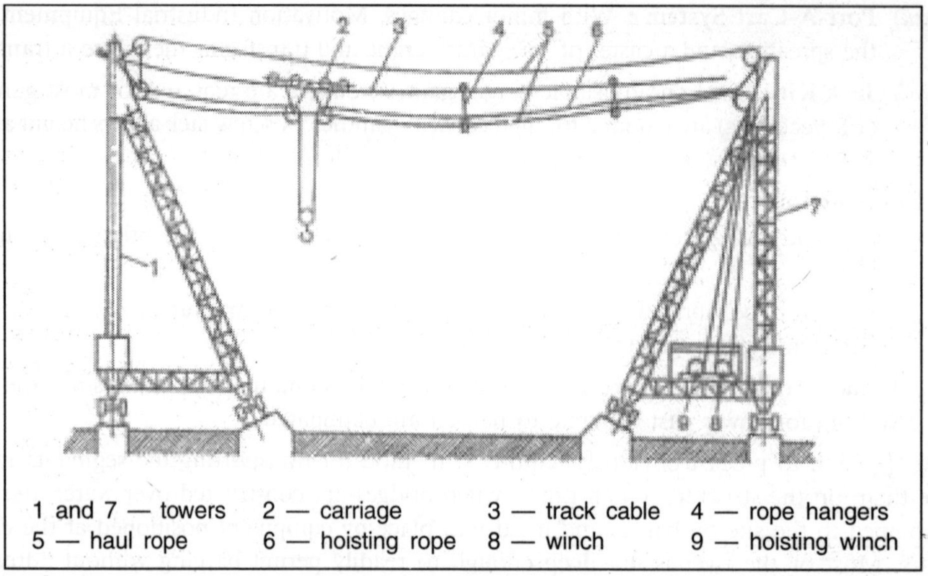

Figure 7.34. *Taut-line cableway crane.*

Types of Cable Crane

Taut-line cableway towers can be fixed structures or traveling ones moving over parallel tracks.

Specifications of Cable Crane

Spans range between 1,000 and 4,000 m. But 10,000 m. spans are also met with. Standard capacity is 1 to 25 tonne while cableways rated at 150-tonne capacity are also in use. The load hoisting speed is 1.5 to 0.9 m per second depending on the height and the conveying speed varies between 8 to 12 m/sec. Towers are spotted into the working position at a rate of 6 to 25 m/min.

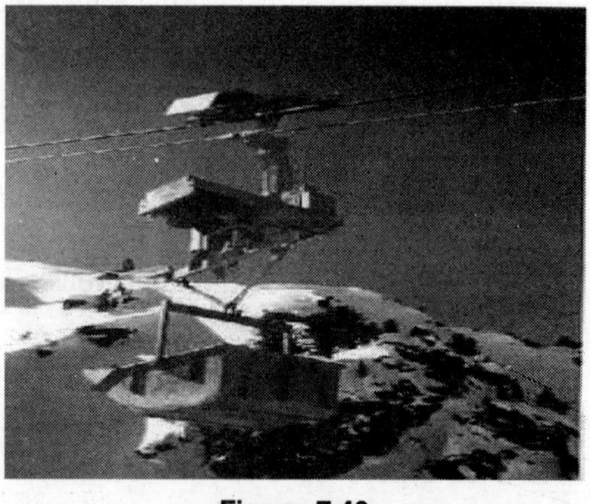

Figure 7.40

Applications of Cable Crane

Storage and production areas of considerable extent are served by taut-line cableways. They also find application at building sites.

UNSOLVED QUESTIONS

1. Explain the importance of the cranes.
2. Classify different types of the cranes.
3. Explain with neat sketch basic motions of the cranes.
4. What is meant by jib cranes? Classify.
5. Explain with neat sketch different types of jib cranes.

6. Explain stability of jib crane.
7. Explain stability of tower crane.
8. Explain the basic motions of the bridge cranes and classify them.
9. Explain the utility of EOT crane.
10. Classify EOT cranes.
11. Explain with neat sketch working of gantry cranes.
12. Classify gantry cranes.
13. Write short note on cable crane.
14. What is meant mobile cranes? Explain their utility.
15. Classify mobile cranes.
16. Write short note on truck mounted cranes.
17. Write short note on rail mounted cranes.
18. Write short note on floating cranes.
19. Write short note on sky crane.

❑❑❑❑

8 Conveyors

INTRODUCTION

- Mechanical conveyors are used for transporting most bulk solids (sand, ash, grains etc.) handling.
- They are used not only in transportation of bulk solids from one location to another location but also for feeding, discharging, metering and proportioning such materials to and from bulk solids storage silos and other solids handling and processing equipments.
- Conveyor is a device which is used for continuous transportation.

(A) It can be used for horizontal or inclined conveying. Also vertical conveying as an elevator.

(B) They may be fixed or portable

(*a*) Portable Conveyors.

(*b*) Straight Conveyors.

(*c*) Circular Conveyors.

- Conveyors which are driven with the help of power known as traction conveyors or those without the power known as tractionless conveyors as material moves due to gravity.

(C) Conveyors can also be classified in following ways :

- Type of product being handled : *unit* load or *bulk* load.
- Location of the conveyor : *overhead, on-floor,* or *in-floor.*
- Whether or not loads can accumulate on the conveyor.

- There are so many types of conveyors. Some important are :

(*a*) Belt Conveyors.

(*b*) Screw Conveyors/Elevators.

(*c*) Bucket Conveyors/Elevators.

(*d*) Chain Conveyors.

(*e*) Roller Conveyors.

(*f*) Pneumatic Conveyors.

(*g*) Slat Conveyors.

(*h*) Vibrating Conveyors.

(*i*) Overhead Conveyors.

(*j*) Sliding friction gravity Conveyor also known as chutes.

– Besides this there are other conveyors which are used for specific applications such as

(*a*) Can Conveyor.

(*b*) Bottle Conveyor.

(*c*) Wire or cable Conveyor.

(*d*) Drag Conveyor.

(*e*) Monorail.

(*f*) Trolley Conveyor.

(*g*) Sortation Conveyor.

(*h*) Push through Conveyor.

(*i*) Indexing Conveyor.

Among all these conveyors one may choose it suitable to him. Here are some points, which are considered for selection/choice of conveyor. Choice of conveyor largely depends on

(*a*) Specified Capacity (m^3/hr).

(*b*) Conveying distance (m).

(*c*) Configuration (horizontal, inclined or vertical).

(*d*) Bulk solids and individual particle properties such has bulk density, particle size (maximum) abrasiveness, flowability, toxicity corrosiveness, explosibility and temperature.

Advantages of Conveyors

(*a*) It is a continuous operation.

(*b*) Variety of materials can be handled.

(*c*) Variety of configurations can be achieved.

(*d*) Quick transportation for a longer as well as for a shorter distance.

(*e*) Since these are available in large variety one can chose conveyor suitable for application.

(*f*) Used for damp out fluctuations and control flow rates.

(*g*) Saves labour cast.

Disadvantages of Conveyors

(*a*) Considerable space.

(*b*) Fixed path.

(*c*) High investment.

(*d*) High level of maintenance is needed for some cases.

(*e*) Marshalling is difficult.

Applications of Conveyors

Conveyors are used in mass production industries where unit loads are uniform, the required movement of the material is continuous, path and rate of movement of material is not likely to charge.

Conveyors are used :

- When material is to be moved frequently between specific points.
- To move materials over a fixed path.
- When there is a sufficient flow volume to justify the fixed conveyor investment.

A good system of conveyors besides bringing apart low cost transportation can also be employed for :

(*a*) Processing activities performed during transportation.

(*b*) Work-handling devices on a moving workstation.

(*c*) A medium for providing storage.

(*d*) Inspection of the product in transit.

(*e*) A device for pacing operations and reducing production and control activities.

Although conveyors can be used in almost any kind of industry here are some important industries in which conveyors are used.

(*i*) Construction and mining.

(*ii*) Power plants.

(*iii*) Cement industries.

(*iv*) Automobile industries.

(*v*) Mass production industries.

8.1 BELT CONVEYORS

A belt conveyor system consists essentially of an endless belt of resilient material connected between two flat pulleys and moved by rotating one of the pulleys by the electric motor. Normally material is fed on the belt near the other end pulley. The moving belt carrying the material towards the driver pulley is likely to sag between the two end pulleys due to its self-weight and pay load. That is why the belt is supported both on the carrying side and tile return side by a number of rollers, called the idlers. As the belt is always under tension, it is prone to elongation resulting in slackness over the pulleys and loss of tension and power. That is why some kind of device for tensioning the belt is incorporated in the system. This tensioning device is known as take-up arrangement. The system is inherently very simple.

8.1.1 Introduction to Belt Conveyor

Belt conveyors have attained a dominant position in transporting bulk materials due to a number of inherent advantages like the economy and safety of operation, reliability, versatility and practically unlimited range of capacities. In addition, they are suitable for performing numerous processing function while supporting a continuous flow of material between operations.

Belt conveyors are most versatile and widely used mechanical conveyors. They are used for transportation, feeding, discharging bulk solid material. This conveyor is suitable to almost dry type of material free flowing solid to met stocky and lumpy powders. Normal belt may be conveyed at angle of inclination of upto 22°. The bulk material on the belt conveyor is discharged over the drive pulley of other conveyors.

The capacity of belt conveyor depends on speed, shape and width of belt, method of feeding and solid properties. Belt conveyor are the most common form of material handling system used in manufacturing and mining industries thermal plants and in construction projects.

Heavy industries like steel, fertilizer, chemical and cement etc. cannot function without the belt conveyors. In construction project the belt conveyors are used for handling. The material in asphalt plant, crushing and screening plants, aggregates mixing plants etc. In coal mining and other mining industries these are used for handling coal or other raw material. The belt conveyors are capable of conveying large quantities of material continuously over considerably long distances at a fast speed.

For transporting the material for short distances conveyors may be a portable or a fixed unit. When a belt conveyor system is to be used to transport for a considerable distance upto several kilometers in such cases, the system consists of a number of flights. Each flight is a complete conveyor unit discharging its load on the tail end of the successive unit. Belt used in the system are costly and hence needed to be carefully maintained. The belts are generally made of rubber cover cotton or rayon laid up in the plies. The strength of the rubber belt depends on number of plies, canvas plies. Wooden wire mesh and steel belts are also used depending upon the strength required for the conveyor to conveying various types of materials, their quantity to be transported etc. Belts are specified by width, number of plies and weight of each layer of ply.

8.1.2 Types of Belt Conveyors

According to position :

(*a*) Stationary.

(*b*) Mobile.

(*c*) Portable.

According to Path of Belt Conveyor :

(*a*) Horizontal.

(*b*) Inclined.

(*c*) Mixed type.

According to their Design and Application :

(*a*) General purpose conveyors.

(*b*) Special purpose conveyors for ores, conveyor for carrying food.

(*c*) Underground conveyors.

(*d*) Special purpose conveyors such as oil resisting, heat resisting, submerged.

(*e*) Above water.

(*f*) Air supported.

(*a*) The conveyors may be troughed and flat, may be inclined upward or downward with or without curvature in vertical plane.

Troughed belt conveyor is that in which the belt forms a trough on the carrying side while running over idler rolls which are either in set of 5-rolls, 3-rolls or 2-rolls.

Flat belt conveyor is that in which the belt runs flat on the carrying side over straight roller type idler or a set of idlers.

(*b*) The fix or portable belt conveyors are used according to the requirements of the production process.

8.1.3 Essential Components of Belt Conveyors

Essential components of belt conveyors are as listed below :

Figure No 8.1 shows the general arrangement of a belt conveyor system. The essential features of a belt conveyor system than be described with reference to. It consists of a driving pulley (1) at the head end and take up pulley (2) at the tail end, both being accommodated in a steel frame (3) an endless belt (4) connects the two pulleys on both runs, and supported by suitable roller-type supports, called the idlers. At the upper side of the conveyor, also called the carrying side, because the pay load is carried on this side, the belt is supported by idler (5). At the other side, also called the return side because usually there is no pay load, the belt is supported by a stationery runway instead of idlers. In some special installations both the upper and lower sides transport loads (usually unit load) simultaneously in opposite directions. The drilling pulley (1) receives its rotation from drive unit (7). The material to be conveyed is loaded on the belt by one or more feed-hoppers (8), mounted over the conveyor belt line, normally places at the tail-end side. The load conveyed by the moving belt is discharged over the driving pulley (1) into discharge spout (9) However, discharge can be arranged at any point along the conveying run by means of special discharge devices. At the tail-end side, a device, called the take-up arrangement (10), is attached with the take up pulley (2) to keep the belt under a minimum tension, so that the bell does not slip over the pulleys due to extension of its length. Belt cleaner removes material adhering to the outer belt face. It is usually located near the driving pulley (1).

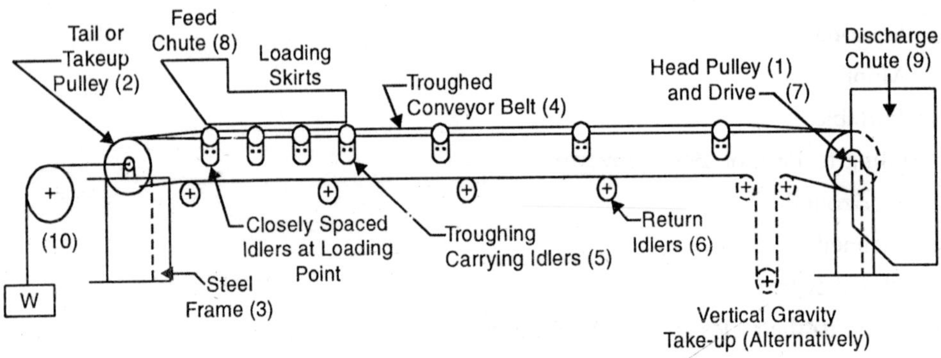

Figure 8.1. *Important parts of belt conveyor.*

8.1.4 Typical Layouts or Travel Path of Belt Conveyors

Figure 8.2 shows typical layouts of conveyors. Belt conveyors can be designed for practically any desired path of travel.

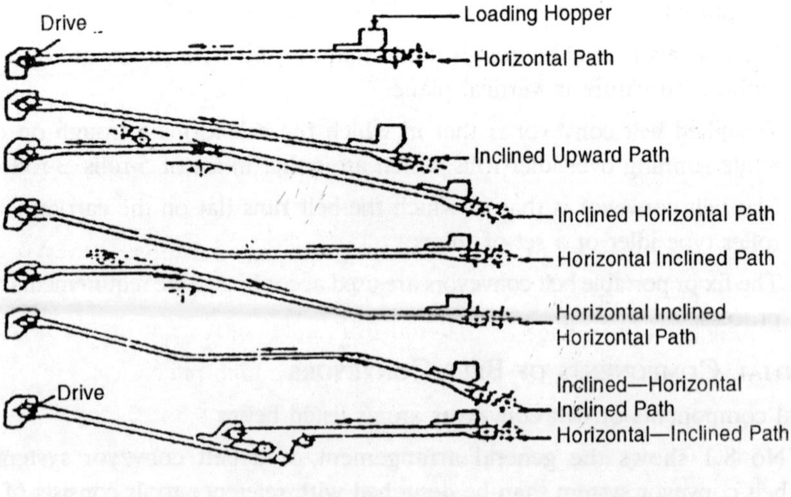

Figure 8.2. *Typical layouts or travel path of conveyors.*

In general, a conveyor belt consists of three elements top cover, carcass and bottom cover.

Table 8.1 : Important Parts and their Function of Belt Conveyor System

Sr. No.	Name	Function/Use	Design/Type/Material
1.	Frame	To support belt and rollers	Designed depending on type of conveyors, generally of grey cast iron.
2.	Belt	To transport material over its surface	– Endless belt specified by width number of plies, length and thickness. – Belt materials vary widely ranging from polyster covered with rubber and plain cr coated canvas, to woven wire or steel ribbon.
3.	Idlers	To support belt over it	– 1000 to 1200 idlers per km. – Upper idlers and lower idlers. – Size depends on material properties to be conveyed.
4.	Drive/Pulleys	For driving belt over rollers or idlers	– Two pulleys at each end. – Driving pulley is at the discharge ends. Run generally by electric motor. – Motors are generally squirrel cage or wound rotor type.
5.	Take ups	To adjust the length of belt.	– Screw or pulleys
6.	Hold backs	To prevent backward running of belt when power failure.	– Mechanical type.
7.	Feeders	To deliver material on belt.	– These should be designed to reduce impact loading. – May be from top of belt or from side of belt.

8.1.5 Flat Belt Conveyor

- These types of conveyors are simple in design are simple uses flat belts supported on a single idlers.
- The width and thickness of belt depends on the material to be conveyed generally used for shorter distance.
- Major problem is a spillage of material from a belt surface at higher speeds.
- Upper rollers have low pitch as lower rollers have high pitch.
- These can be of various types such as shown in figure 8.3 *a, b, c.*

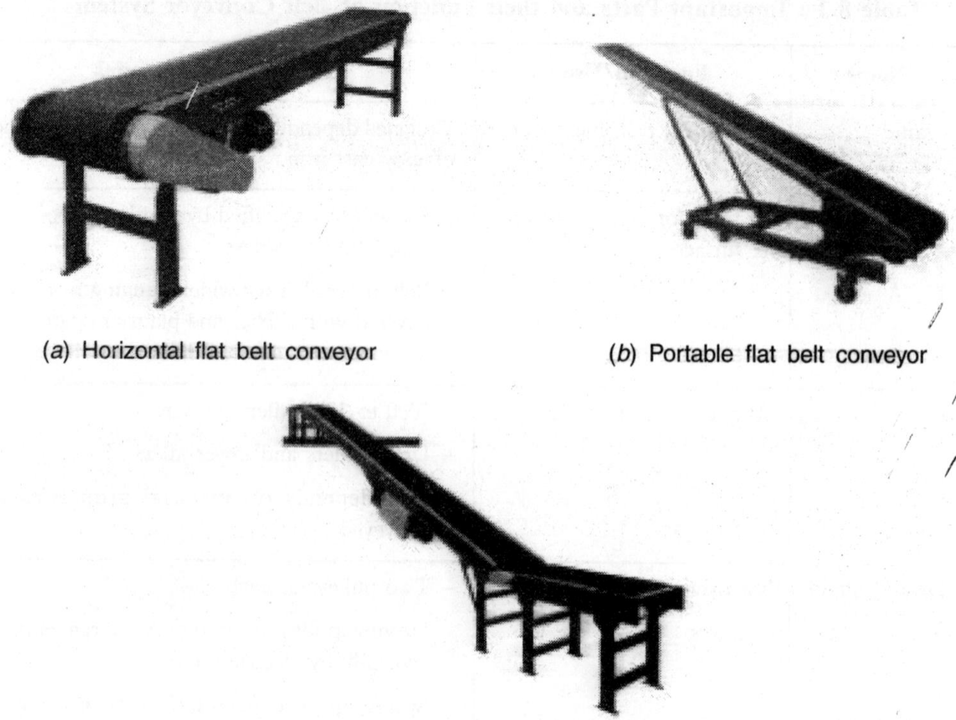

(*a*) Horizontal flat belt conveyor

(*b*) Portable flat belt conveyor

(*c*) Inclined flat belt conveyor

Figure 8.3

8.1.6 Troughed Belt Conveyors

This type of belts are characterized by angled or elevated edges of the belt as shown in the figure 8.4, 8.5 and 8.6. In this type of belt is supported by idlers which may vary from 3 to 5 depending on material to carry. The trough angle varies from 20° to 30°. The capacity of such type of conveyor depends on speed and width of belt, depth of trough. The figure shows banked troughed belt conveyor design details.

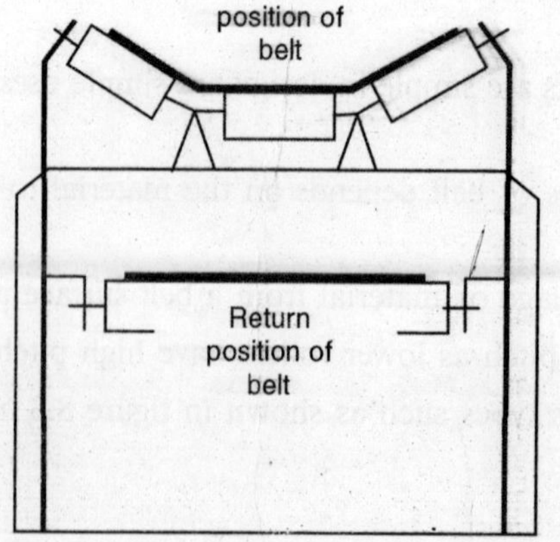

Figure 8.4 *Layout of troughed belt conveyor with 3-idler rollers.*

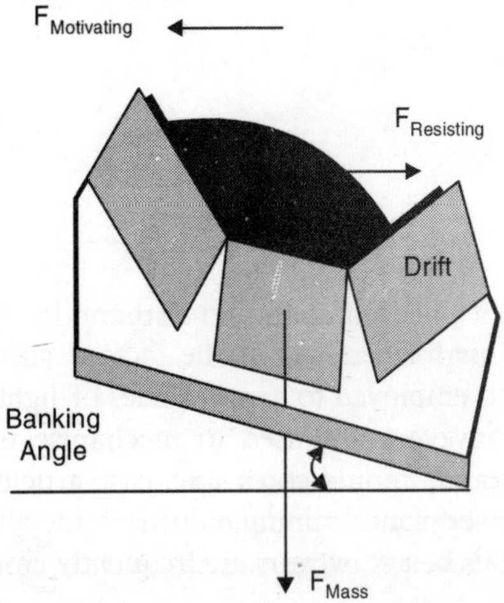

Figure 8.5. *Banked Troughed belt conveyor.*

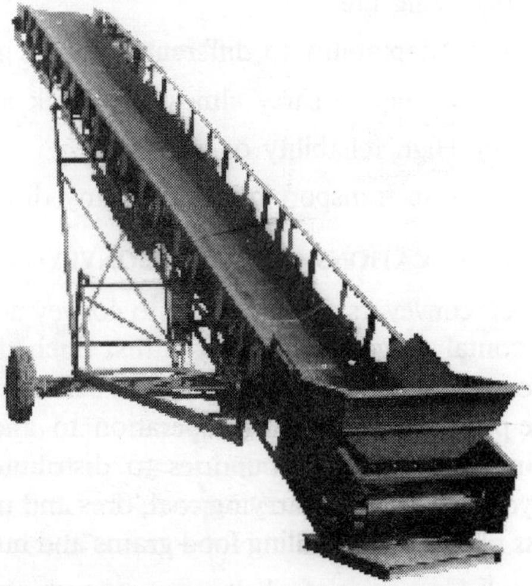

Figure 8.6. *Portable troughed belt conveyors.*

Applications of Troughed Belt Conveyors

Stone, gravel, sand, coal, salt glass, plastic, cans, commingled recyclables, top soil, wood chips, mulch, sawdust, almost any bulk material.

8.1.7 Timing Belt Conveyors

The Timing Belt Conveyor System (figure 8.7) is designed especially for product indexing applications where a cleated, or fixtured belt is required, or for wider timing belt applications. The conveyor is ideal for special conveying requiring positioning or loading of products. Depending on the product and the application, custom fixtures and other belt widths are available.

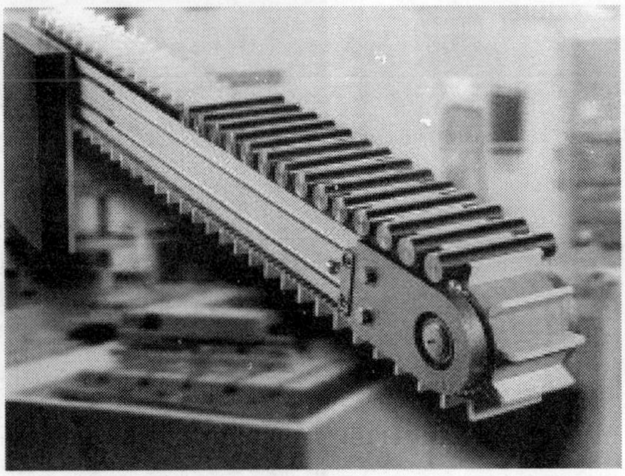

Figure 8.7

8.1.8 Advantages Belt Conveyors

(*a*) Noiseless operation.

(*b*) Large length of conveying path.

(*c*) Lower power consumption.

(*d*) Long life.

(*e*) Adaptability to different types of goods.

(*f*) Ability to carry almost any bulk material.

(*g*) High reliability of operation.

(*h*) Can transport material in any direction.

8.1.9 Applications of Belt Conveyors

Belt conveyors are employed to convey a great variety of bulk materials and also unit loads along a horizontal or gently inclined paths. Such duties can be performed within the factory shed or in outdoor work. Within the factory shed belt conveyors are also employed to carry articles of lightweight in line-production from one operation to another. Belt conveyors are used to mechanise material handling operations in foundries to distribute moulding sand, mould-cores and cast articles. Belt conveyors are used for carrying coal, ores and minerals in power plants, mining industries, metallurgical process plants. For handling food grains and building materials belt conveyors are frequently employed.

(*a*) Capacities of a belt conveyor system may range from 500 to 5000 m^3/hr and more. The distance covered for a belt conveyor system be as small as 50 m (for indoor work) to 5 km. for supplying raw material from **mine** area. Used for conveying various bulk and unit loads along horizontal or slightly inclined paths.

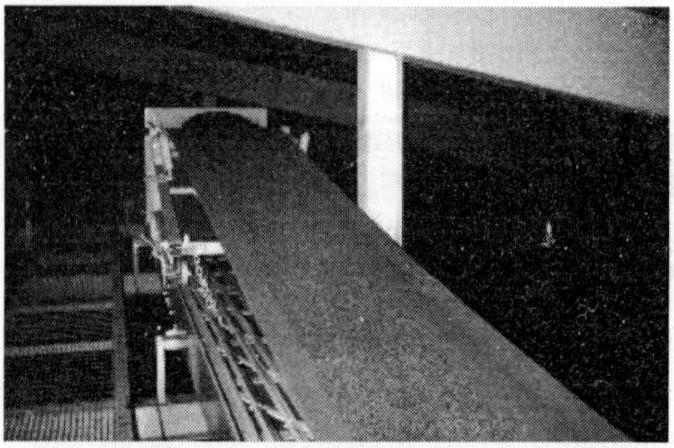

Figure 8.8

(*b*) Used in foundry shop.

(*c*) Used for supplying coal.

(*d*) Transferring building materials, fossil minerals, grains store.

(*e*) Used in wheel excavators.

(*f*) Ice-cream plant.

(*g*) Power plants.

(*h*) Bakeries.

(*i*) Confectionery.

(*j*) Feed and flour mills.

(*k*) Laundries.

(*l*) Textile.

(*m*) Ceramic.

(*n*) Chemical plant.

(*o*) Ship yards.

(*p*) Post office.

(*q*) Supply depots.

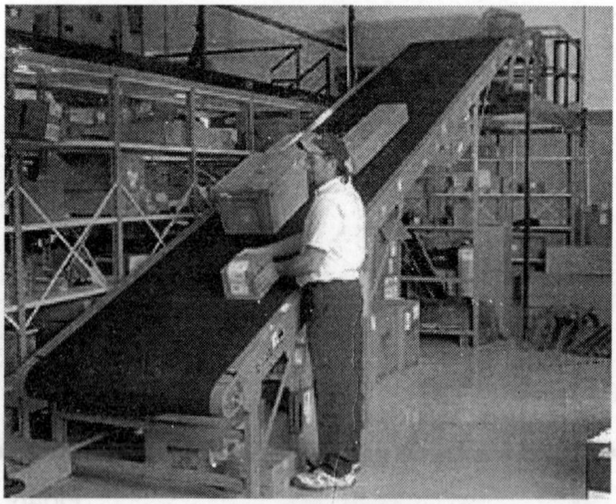

Figure 8.9

8.1.10 Limitations of Belt Conveyors

(*a*) Accumlation difficult.

(*b*) complicated marshalling.

(*c*) The loss of light weight bulk material carried away as dust or spilled from the belt along its path is another objectionable features.

(*d*) Continuous or periodic monitoring of belt is necessary.

(*e*) Heat affects the material of belt.

8.2 SCREW CONVEYORS

Screw conveyors are conveyor systems that transport bulk materials from one location to another. In certain applications, screw conveyors may also prove advantageous in the breakdown and agitation of materials. Screw conveyors consist of a rotating, spiral screw within a trough. The rotation of the screw pushes the materials through the trough to the intended destination. Screw configuration make screw conveyors useful in transferring powders, sludge, fuels, ash, feed, cement, ice and certain minerals. Because the screw will cut through materials during operation, sticky materials, materials that do not breakdown easily, or those materials not intended to be broken down are usually not ideal materials for screw conveyor systems to process.

8.2.1 Introduction

Screw conveyors are found in numerous industrial applications in the agricultural, food and beverage processing, pharmaceutical, chemical, mining, oil and gas, and construction industries. Screw conveyors are cost-effective and easy to install. The selection of a screw conveyor depends upon the industrial application; material and material consistency; and additional operations required of the screw conveyor system.

Consists of a tube or U-shaped stationery trough through which a shaft-mounted helix revolves to push loose material forward in a horizontal or inclined direction and a drive unit for running the shaft.

Screw conveyors are accepted in most industries as the usual means of conveying powders and granules from point 'A' to point 'B'. Designs can be applied to most applications in the food, pharmaceutical, chemical, mineral and sewage treatment industries.

Compact in design, screw conveyors are easily adapted to suit the needs of the location and can be mounted horizontally, vertically or at an angle, and include planes as necessary.

Machines are available in trough or tubular form; trough conveyors are generally used in horizontal applications and at inclines up to 20°, while tubular conveyors are selected for elevating purposes at angles greater than 20°.

Screw conveyors are very versatile in that as well as conveying they can be used as feeders and to control material flow in processing operations, and also as mixers, agitators or stirrers within a variety of processes. Screw conveyors can be sealed to prevent leakage, can be jacketed to provide a drying or cooling medium and can be manufactured in special steels to resist corrosion, abrasion or high temperatures.

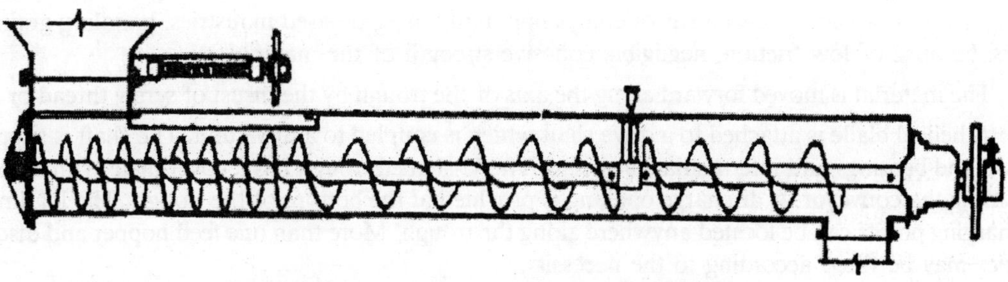

Figure 8.10 *Variable pitch screw conveyor.*

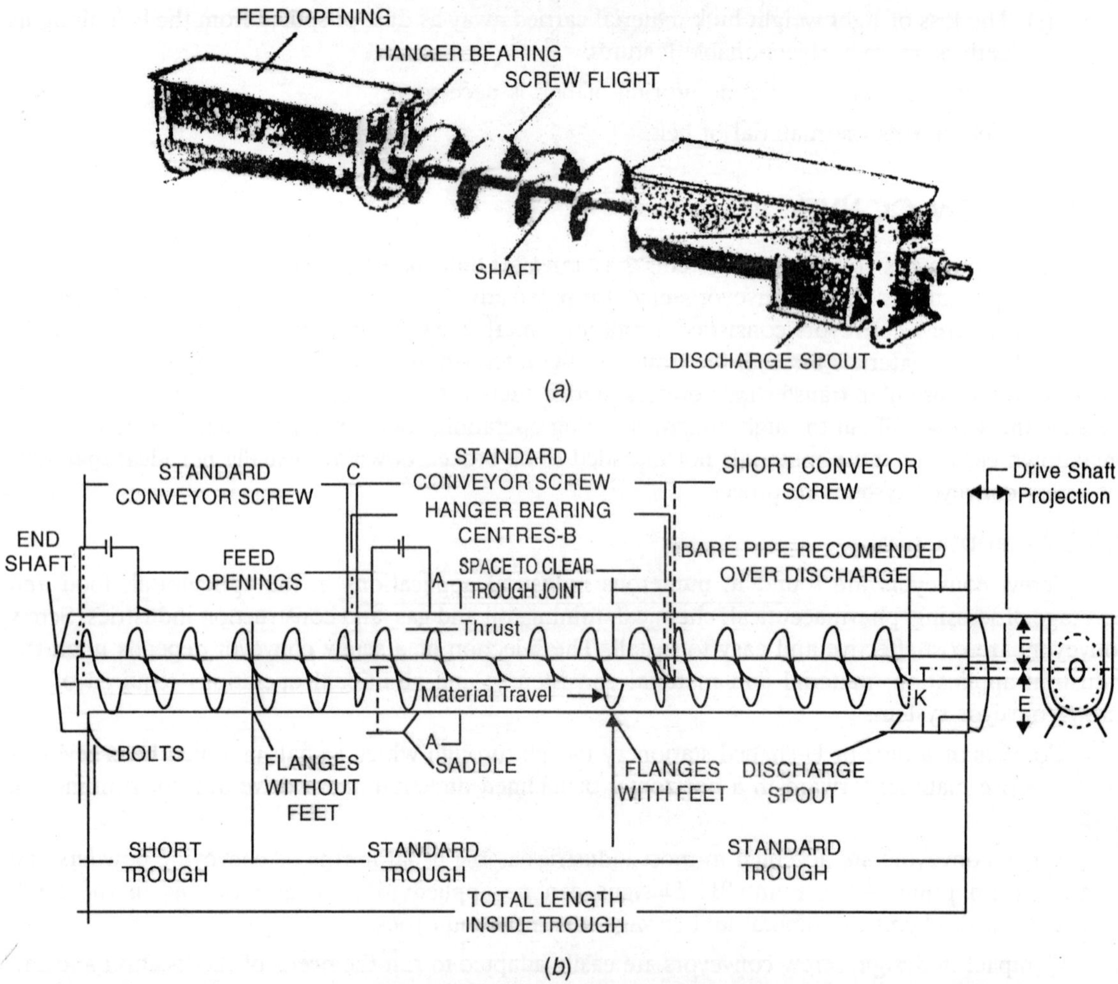

Figure 8.11. *Parts and terminology.*

Screw conveyors serve the manufacturing industry in a wide variety of applications : conveying, distributing, collecting, mixing, heating, cooling, elevating, batching, blending, aerating and providing crystallization or coagulant action.

Screw conveyors are used principally for transmitting material in the form of powder or paste with the application of rotating screw.

Flourmill industry was probable the first to employ horizontal screw conveyors to convey corn and flour. These screws were made by fitting wood blades to a central shaft rotating inside a cylindrical casing. Even now such screw conveyor enjoys popularity in agro-based industries, handling grains and flours, because of low friction, negligible cohesive strength of the material.

The material is moved forward along the axis of the trough by the thrust of screw thread or flight.

A helical blade is attached to a drive shaft which is coupled to a drive unit. The shaft is supported by two end bearings and intermediate bearings. The U-shaped trough has a cover plate with a opening for lading the conveyor. A discharge opening is provided at the bottom of the trough. The loading and discharging points can be located anywhere along the trough. More than one feed hopper and discharge hopper may be fitted according to the necessity.

The principle of material movement along the trough is similar to the sliding motion of a nut along a rotating screw when the nut is not allowed to rotate. The weight of the material and the friction of the material against the wall prevent the load from rotating with the screw.

8.2.2 Essential Components

The Conveyor Screw imparts a smooth positive motion to the material as it rotates within the trough.

Couplings and Shaft connect and transmit motion to subsequent screw conveyors. Held in place by self-locking Through-U-lockbolts.

Hangers provide support, maintain alignment and serve as bearing surfaces.

Troughs and Conveyors completely enclose the material being conveyed and the rotating parts. Covers are available in various types and are secured to the trough by Spring, Screw, Tight-Seal or quick-acting Barron Clamps depending on the trough being used.

Inlet and Discharge Openings may be located wherever needed, discharge spouts may be without slides or fitted with either flat or curved slides. These slides may be operated by hand, rack and pinion gears or by power.

8.2.3 Types of Screw Conveyors

- **Flexible screw conveyors** move solids and powders in all directions. Flexible conveyor systems contain a flexible screw capable of self-centering itself within the conveyor tube, which prevents scraping of the screw against the tube wall. Because the conveyor motor powers the screw directly, the need for extra components is eliminated, reducing upkeep and maintenance and increasing cost-effectiveness.
- **Jacketed screw conveyors** use substances such as water, oil, or steam to provide heating or cooling to materials transported through a screw conveyor. Jacketed screw conveyors consist of an enclosure surrounding the conveyor trough.
- **Portable screw conveyors** can be transported from one location to another, which increases convenience and the efficiency of the industrial application.
- **Trough screw conveyors** are common in the food processing industry. Trough conveyors contain an open channel duct through which the materials are conveyed.

  Single Flight Standard Pitch

  Trough configurations allow for easy cleaning, which reduces the incidence of future product contamination.
- **Tubular screw conveyors** protect against material contamination during conveyance, which remains advantageous in industrial applications involving sensitive materials. Tubular screw conveyors consist of an enclosed tube-shaped duct through which the materials are conveyed configuration may take cleaning difficult.

8.2.4 Types of Screw Flight

The screw of the conveyor may be right-hand or left-hand, the right-hand type being the usual design. The threads of the screw may be single, double or triple.

The flight of the screws may be made in either of the two ways : (*a*) as helicoids, (*b*) as sectional flight. (*a*) Helicoid flight are formed from a flat bar or strip into an edge. (*b*) Sectional flights are formed from a flat disc and the thickness of the thread is uniform throughout. A continuous helix is made by joining a number of sectional flights together on a piece of pipe and butt-welding them. Various styles of screw flights are in use, depending on the service required:

(*a*) **Single Flight Standard Pitch :** Conveyor screws with pitch spacing equal to the diameter are considered standard and are suitable for most conveyors of conventional horizontal design.

Flights with pitch progressively increased are used to draw fine, free flowing materials uniformly from the entire length of the feed opening.

(*b*) **Single Tapered Flight Standard Pitch :** Tapering flight screws are generally recommended to draw friable lumpy materials from bins or hoppers.

(*c*) **Double Flight Standard Pitch :** Double flight produce a smooth gentle flow of certain materials are recommended for augers or post hole digging and similar applications.

(*d*) **Short Pitch or Continuous Flight :** If the conveyor is required to handle dry granular or powered material that do not pack, this style of flight may be selected. It is of regular construction and recommended for inclined conveyors having a slope of 20 degrees or more, including vertical conveyors. This style is extensively used feeder screw. Short pitch screws are recommended for inclines over 20 degrees slope to vertical conveying. They retard flushing of free flowing materials.

(*e*) **Half pitch conveyors** are recommended for inclined or vertical conveying and for materials which are very free flowing.

(*f*) **Ribbon Flight :** If the conveyor is to handle lumpy, clinging, sticky, gummy or viscous substances, this type flight may be selected. It consists of continuous helical flight formed from steel bar and secured to the pipe by supporting lugs. Ribbon flights are excellent for sticky or gummy materials where the gap between the screw and the pipe prevents material build-up.

(*g*) **Cut Flight :** In this type of flight screws have notches cut in the periphery of the flight. These notches supplement the conveying with moderate mixing action. They are recommended for conveyors required to handle light, fine, granular or flaky materials. The screw flight periphery is notched to supplement the conveying action with a moderate mixing action.

(*h*) **Cut and Folded Flight :** This type of flight is characterized by notches as in cut flight, together with folded segments. This type of flight creates agitation and aeration resulting in better mixing. This type of flight is used to handle light or medium weight materials having fine, granular or flaky characteristic.

(*i*) Some screw conveyors have cut flight with paddles mounted at regular intervals. The flow of material past the flight resulting in greater agitation and mixing. The paddle conveyor is made of steel paddles on threaded rod shanks inserted through the pipe. Conveying action can be controlled by adjusting the paddle angle. Excellent for stirring, blending or dry or fluid materials.

(*j*) Sometimes screws are made of stainless steel to suit special requirements, like the sanitation requirements for handling foods, drugs and other hygienic materials.

8.2.5 Feed and Discharge Arrangements

Feeding and discharging hoppers may be placed conveniently along the length of the conveyor the various possible configuration. According to necessity any arrangement may be adopted.

8.2.6 Standard Dimensions of a Screw Conveyor

Screw conveyors are usually designed and manufactured with standard dimensions of nominal diameter of the helical screw, pitch of screw, diameter of screw shaft, height of the conveyor above the centre of screw and thickness of trough. In this regard usually IS : 5563-1985 is followed.

8.2.7 Effect of Lump Size

The size of a screw conveyor depends on two factors : (*i*) The capacity of the conveyor and, (*ii*) The lump size of the material to be conveyed. The size of the lump indicates the maximum dimension of the particle.

Usually there are three ranges of lump sizes which are considered for selection of screw size. These are :

1. A mixture of lumps and fines in which not more than 10% are lumps ranging from maximum size to one half of the maximum; and 90% are lumps smaller than one half of maximum size.
2. A mixture of lump and fines in which not more than 25% are lumps ranging from the maximum size to one half of the maximum; and 75% are lumps smaller than one half of the maximum size.
3. A mixture of lump only in which 95% or more are lumps ranging from maximum size; and 5% or less are lumps less than one tenth of the maximum size.

The allowable size of a lump in a screw conveyor is a function of the radial clearance between the outside diameter of the central pipe and the radius of the inside of the screw trough, as well as the proportion of the lumps in the mixture.The lump size of the material affects the selection of screw diameter D, which should be at least 12 times larger than the lump size of a sized material and four time larger than the largest lumps of an unsized material.

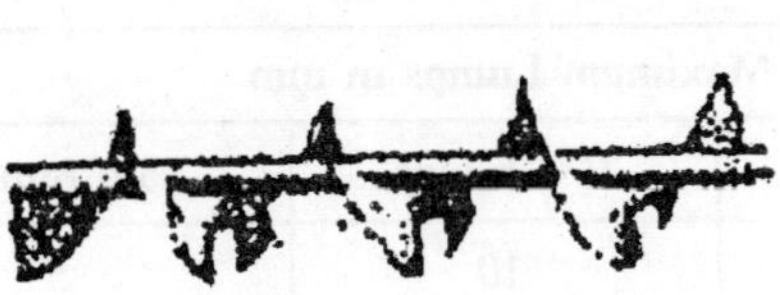

Cut Flight Conveyor Screw

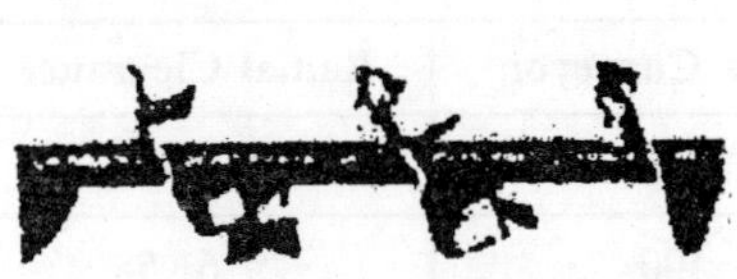

Cut and Folded Flight Conveyor Screw

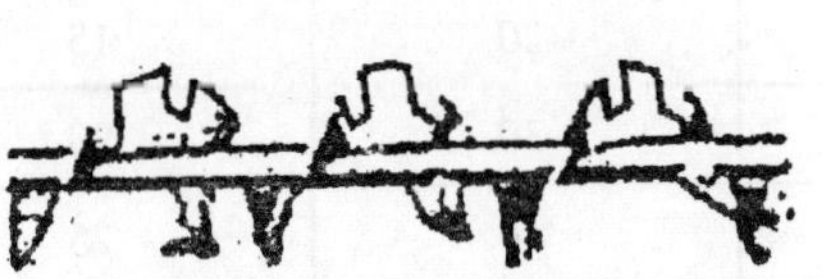

Cut Flight Conveyor Screw with Paddles

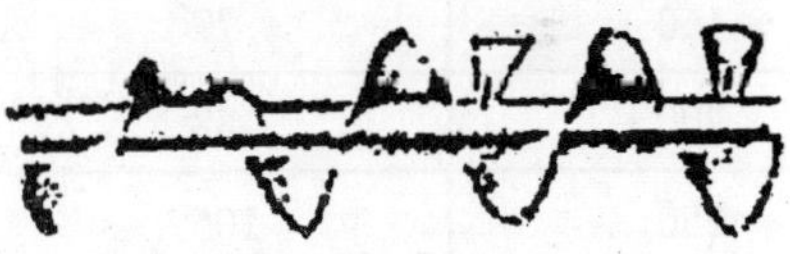

Continuous Screw with Paddles

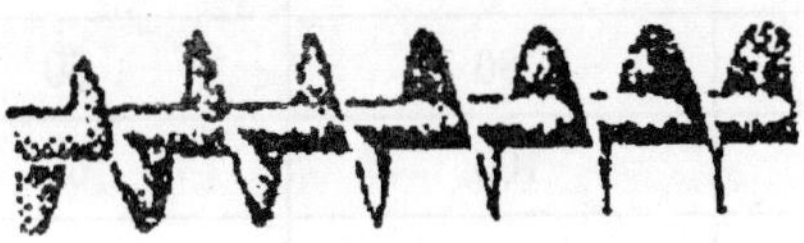

Short Pitch Conveyor Screw

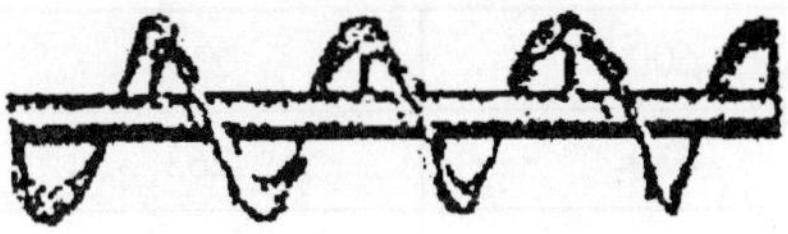

Ribbon Flight Conveyor Screw

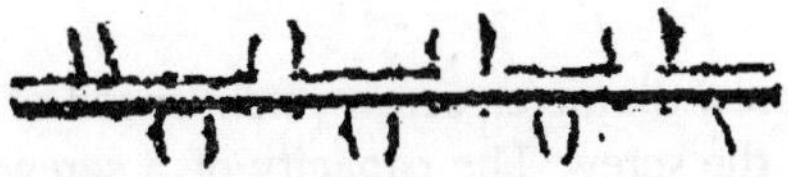

Paddle Conveyor Screw

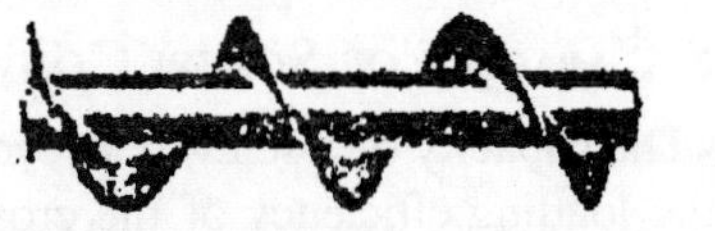

Stainless Steel Conveyor Screw

Figure 8.12

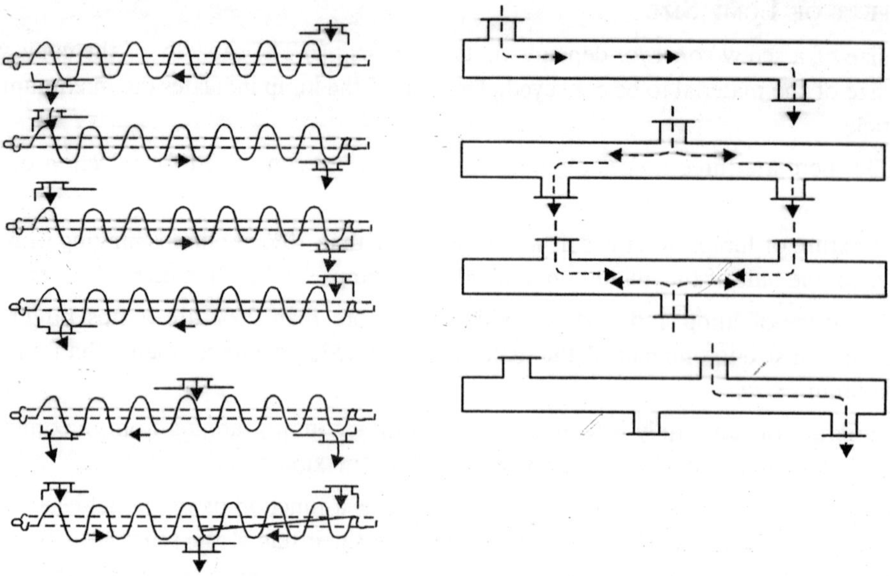

Figure 8.13

Table 8.2 Screw Conveyor Maximum Lump Size

Screw Conveyor	Radial Clearance	Maximum Lumps in mm		
mm	mm	10% lumps	25% lumps	95% lumps
100	61.5	35	10	5
125	70.0	40	15	10
160	79	45	20	15
200	88	50	30	20
250	105	60	40	25
315	131.5	75	50	30
400	175	100	65	40
500	219	125	80	50
630	263	150	100	65
800	306	175	150	85

8.2.8 Capacity of Screw Conveyors

The capacity of a screw conveyor depends on the screw diameter, screw pitch, speed of the screw and the loading efficiency of the cross sectional area of the screw. The capacity of a screw conveyor with a continuous screw may be expressed as :

$$Q = V.p. = 60\frac{\pi}{4}D^2.s.n.\ \psi\ p.c.$$

Where,

V = Volumetric capacity in m^3/hr.

P = Bulk density of the material, kg/m^3.

D = Nominal diameter of screw in m.

S = Screw pitch in m.

n = r.p.m. of screw.

ψ = Loading efficiency of the screw.

C = Factor to take into account the inclination of the conveyor.

8.2.9 Power of Screw Conveyors

The power requirement for a screw conveyor may be estimated from the sum of the power required to turn the empty screw and additional power required to move the solids.

$$\text{Empty power} = 2.45 \times 10^{-6}\, L\, N\, F_D\, F_B \text{ kW}$$

Where,

L is the screw length in m.

N is the rotating speed in rpm.

F_D is a function of the screw size, usually about 5.

F_B is a bearing factor, about 4.

$$\textit{Material motion power} = 5.39 \times 10^{-6} LCWF_M F_F F_P$$

Where,

C is the capacity in m^3/h.

W is the material density in kg/m^3.

F_M is a material factor (usually 1.0).

F_F and F_P are a flight factor and a paddle factor respectively, both usually 1 for standard screws.

C the capacity can be estimated from

$$\frac{\pi}{8}(D_S^2 - D_A^2)\, N \times (P-t)$$

(Assuming the screw is half full) D_S is the screw diameter.

D_A is the axle diameter.

P is the pitch of the screw (normally equal to the diameter of the screw) and t is the thickness of the blade. In addition if the screw is inclined there will be a term for work done against gravity. The length of the screw is limited by the torque available at the drive. The pitch sometimes decreases from the feed end to the discharge end to condition the powder. Some control of flow rate can be achieved by setting the speed of the screw.

8.2.10 Screw Pitch

Commonly the screw pitch is taken equal to the diameter of the screw D. However, it may range from 0.75–1.0 times the diameter of the screw. Screw diameter : The nominal screw diameter may be taken from table 8.1 r.p.m. of screw : The usual range of r.p.m. of screw is 10 to 165. It depends on the diameter of screw and the type of material. For further details see Table 8.4 on capacities of screw conveyor.

However, a guideline may be had from Table 8.2.

Table 8.2

Screw Diameter in mm		150	200	250	300	400	500	600
Speed .r.p.m	**Max**	150	150	118	118	95	95	75
	Min	24	24	24	20	20	20	15

8.2.11 Loading Efficiency

The value of loading efficiency should be taken large for materials which are free flowing and non-abrasive, while for materials which are not free flowing and, or, are abrasive in nature, the value should be taken low :

ψ = 0.12 to 0.15 abrasive material.

= 0.25 to 0.30 for mildly abrasive material.

= 0.4 to 0.45 for non-abrasive, free flowing materials.

The inclination factor c is determined by the angle of the screw conveyor with the horizontal. Approximate values may obtained from the following Table 8.3.

Table 8.3 : Values of Inclination Factor C

Angle of Screw with the horizontal	0°	5°	10°	15°	20°
Value of factor C	1.0	0.9	0.8	0.7	0.65

The capacity of various size screw conveyors for four different loading efficiencies and for various classes of materials are shown in Table 8.5.

Table 8.4 : Capacities of Screw Conveyor

Material Class Code IS:	**Nature of Material**	**Loading Efficiency**	**Screw Diameter mm**	**Maxm. r.p.m.**	**Capacity Cubic meter/hour at one r.p.m.**
A15	Very fine		100	165	0.017
A25	Fine, Very Fine		125	165	0.033
B15	Flowing free		160	165	0.07
B25	Free	45%	200	150	0.136
C15	Flowing.		250	118	0.265
C25	Non Abrasive		315	118	0.530
			400	95	1.085
			500	95	2.120

8.2.12 Advantages of Screw Conveyor

The screw conveyor is one of the most cost-effective methods for transporting bulk materials. Screw conveyors are used in thousands of applications in almost every industry. The screw conveyor has many advantages over other types of bulk material handling equipment. Some of the advantages are :

(*a*) Screw conveyors are capable of handling a great variety of bulk materials from sluggish to free flowing.

(*b*) Screw conveyors can have multiple inlet and discharge points. Bulk materials can be conveyed and distributed to various locations as required. Slide gates or valves can be added to control the flow into and out of a screw conveyor.

(*c*) When a screw conveyor is used as a metering device, it is considered a screw feeder. Screw feeders are used to initiate a material process by metering product from a bin or hopper.

(*d*) Screw conveyors are very compact and adaptable to congested locations. Screw conveyors do not have a return similar to a belt or drag conveyor.

(*e*) Screw conveyors are totally enclosed to contain the product and prevent spillage.

(*f*) Screw conveyors can be utilized in the horizontal, vertical or any inclined position depending upon the characteristics of the product being conveyed.

(*g*) Screw conveyors can be used for mixing various products together and for breaking up large lumps.

(*h*) Screw conveyors can be designed without a centre pipe. This type of conveyor is called a shaft less screw conveyor and is designed for conveying wet, sticky, sluggish products such as industrial sludge and bio-solids.

(*i*) Screw conveyors can be used to cool, heat or dry products in transit. Depending on the heat transfer requirements, a screw conveyor can be jacketed, or a hollow-flight design utilized to provide the necessary heat transfer for the application.

(*j*) Screw conveyors can be designed to be vapour-tight or hold an internal pressure. This is very important when conveying toxic or hazardous products such as those in the chemical industry.

(*k*) Screw conveyors can provide an air lock between upstream and downstream equipment.

8.2.13 Screw Conveyor Choice

Determining the most efficient method to move material from point A to point B is a matter of simple deduction. Just match the material flow characteristics, process parameters and space limitations to the performance windows and dimensional restrictions of different conveyors. Then, factor in purchase, maintenance, labour and cleaning costs and the conveyor choice is clear.

Because the performance, size and cost of conveying methods vary, only a general knowledge of the different types is necessary to decide which is best. Armed only with the fundamentals, anyone can determine whether or not a flexible screw conveyor can handle an application most efficiently.

In its basic form, a flexible screw conveyor consists of a spring-steel or stainless-steel, flexible screw that is enclosed in a flexible or rigid steel tube. A small-diameter, plastic inner tube may also run the length of the screw. The conveyor is driven by an electric motor normally located at the discharge end.

The intake end of the conveyor tube is typically equipped with a charging adapter. This U-shaped trough connects to the outlet of a hopper or directly to the outlet of process equipment such as a bulk bag discharger, bag dump station, grinder, crusher, screener, blender, reactor or storage vessel of any kind.

The flexible screw passes through the charging adapter trough, exposing a section of the screw to material flowing into the trough from above. When rotating, the exposed section of screw feeds material into the outer tube. It is then propelled through the tube by the enclosed portion of the screw.

Because of its loose fit, the screw automatically self-centers within a material-filled tube, providing clearance between the screw and tube wall. This clearance between the screw and tube wall can eliminate or minimize grinding, crushing and impact damage.

8.2.14 Limitations of Screw Conveyor

Because every conveying possesses has inherent limitations, each must be identified and ruled unimportant or less important than corresponding attributes, before deeming a flexible screw conveyor to be the best alternative for an application.

(*a*) Screw conveyors, like other mechanical conveyors, do not generally evacuate material completely when operating in a forward direction. Residual material can be fully evacuated, however, by removing the cleanout cap and reversing the screw rotation.

(*b*) Although these conveyors tend not to grind or crush, not all materials can be conveyed damage-free. Some large, brittle flakes or large, soft particles may break, particularly through long runs or steep inclines in a flexible screw conveyor. But, proper system engineering can eliminate or minimize these effects.

(*c*) Flexible screw conveyors are often used where capacity requirements are less than 50 tph (per conveyor). Thus they are infrequently employed for off-loading of ships or railcars.

(*d*) Screw conveyors do not always get favorable consideration due to high friction of the material against screw and trough, probability of material degradation due to crushing of material and due to higher power consumption resulting from higher friction. Therefore, screw conveyors have limitation in capacity and distance of conveying. They are suitable for low and medium capacities for short distance. Capacity may be as high as 100 m^3/hr and distance of conveying is of the order of 30 to 40 m typically.

8.2.15 Applications of Screw Conveyors

(*a*) Screw conveyors are employed for handling a great variety of materials, which have relatively good flow ability. Sticky and stingy materials are unsuitable for screw conveying.

(*b*) Dusty, hot materials, or materials having ordour are often conveyed by screw conveyors, because, it can be easily made dust-proof by jacketing the trough.

(*c*) Typical applications include grain storage plants, feed mills, cereal processing plants, chemical plants, sand preparation shop of a foundary for delivery of clay to the mixing machines.

(*d*) Screw conveyors are very suitable for being adopted as a device to control the volume of materials flow from the bottom of bins, hoppers, storage silos etc. When screw conveyors are used for this purpose they are termed as screw feeders. Screw feeders have wide applications for the operation of processing units, like driers, hammer mills, oil expellers and other innumerable mills.

(*e*) Screw conveyors may be used to convey materials which can be subjected to thermal operation during transit. The conveyor trough is provided with jacket through which the heating or cooling medium may be circulated to effect the thermal operation.

(*f*) Screw conveyors are sometimes used for blending of several ingredients to prepare a finished products. The screw profile of the conveyor are specially designed for this purpose.

(*g*) Conveying between storage vessels.

(*h*) Bulk bag dischargers.

(*i*) Manual bag dump stations.

(*j*) Drum dump stations.

(*k*) Crushers, screeners and blenders.

(*l*) Reactors and packaging equipment.

(*m*) Injection molding and extrusion machines.

(*n*) All industries in particular mining, food, chemical, paper and pulp, waste, building, wine, power generation etc.

(*o*) Controlling material from the bottom of bins, hoppers of the like.

8.3 BUCKET CONVEYOR

Used to move bulk materials in a vertical or inclined path. Buckets are attached to a cable, chain, or belt. Buckets are automatically unloaded at the end of the conveyor run. Bucket conveyors are used to move the granular, powered or liquid materials. The bucket is mounted on chain or a belt. The movement may be vertical or flat. The vertical movement may be continuous wherein bucket are hooked in a sequent circular manner, or describe where bucket are hooked or lifting.

In a bucket elevator, an endless belt or chain has a series of buckets attached to it filled with material. These buckets are lifted vertically to a head pulley or sprocket where the materials are dumped. The buckets are then returned back down to a tail pulley or sprocket at the bottom. Bucket elevators are not self-feeding. To avoid overfilling the buckets and damaging the machinery, the bucket elevators are fed at a controlled rate. They are usually arranged where the chain or belt path is vertical or steeply inclined in a single plane. Special chain supported bucket systems that can travel in two and three planes have been developed.

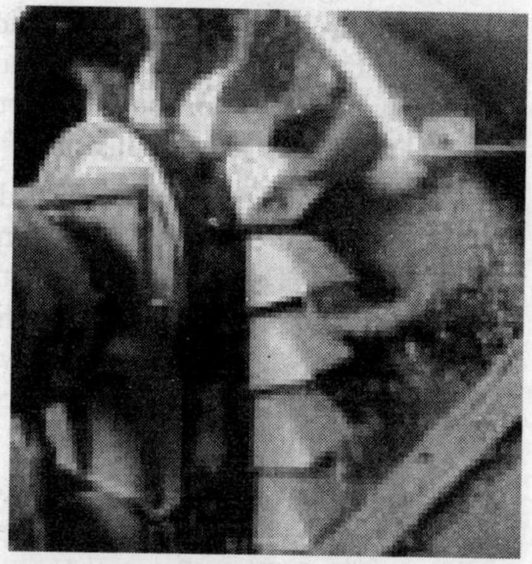

Figure 8.14

There are four broad classifications of bucket elevators : centrifugal, continuous, positive, and internal discharge. The most commonly used are the centrifugal and continuous discharge elevators. The positive discharge and the internal discharge elevators are used for special applications.

Bucket elevators are used for lifting (maximum incline of 10 degrees), and are composed of a conveyor belt equipped with buckets and stretched vertically between two pulleys.

Bucket elevators have the advantage of easy lifting to a great height (70 m). They use little energy, take up little space, and are moderately priced. This is a stationary piece of equipment, however, with a relatively high installation cost (for digging the ditch).

Other elevators that work faster or slower can be used, according to the specific weight and characteristics of the grain. Indicatively, their average speed is 2.5–3 m/s, allowing for continuous work, and their maximum speed is 6–8 m/s.

8.3.1 Construction

Bucket elevator consists of a vertical trough built up of steel or cast iron. Driving element is completely enclosed in this trough. Buckets are rigidly attached to a driving element. The driving element is continuous and passes over the upper and lower pulley. Upper pulley is driver and lower one is known as take up. The casing has three portions upper portion is known as head, middle portion and bottom portion is known as boot. The drive is essentially a motor drive with gear box. The driving motor and gear box is connected to the driving element and installed on head. The bulk material is fed to the buckets at the bottom, buckets carries the material towards head and discharges it at the top. The tension is adjusted using the take ups. Stops at appropriate places were provided to avoid backward movement of driving element.

8.3.2 Classification of Bucket Conveyor

These are classified using various ways as below.

(*a*) Vertical or inclined.

(*b*) Stationary and mobile.

(*c*) Transport or process elevators.

(*d*) By driving element belt and chain driven.

(*e*) By type of loading and unloading they are classified as high speed and gravity.

8.3.3 Important Parts of Bucket Conveyor

Following are the important parts of the bucket elevator.

(*a*) **Casing.** It contains all the elements and constructed in three portions head, middle and boot. Boot contains feeding spout, manholes for inspection and take-up. Mid portion is constructed using steel plates and angles. Head consists of drive motor.

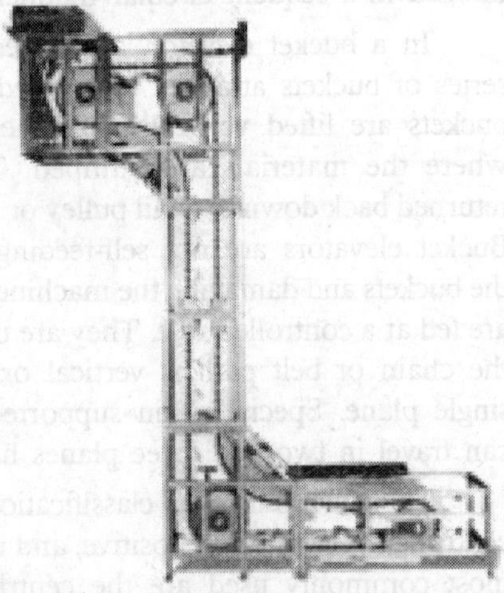

Figure 8.15

(*b*) **Driving Element :** It is belt driven or chain driven. The width of belt 25-150 mm greater than that of buckets. The belt must have atleast four plies to have sufficient strength to hold the buckets. Chain are generally laminated, chains with a pitch of 100-630 mm or welded round link chains from steel rods 16–28 mm in diameter with heat treated links.

(*c*) **Buckets :** The speed of buckets ranges from 0.4 to 2.6 m/s. There are several types of buckets deep (D type), shallow (S type), V shaped with guiding side walls (V type) and round bottom (R type). Other than these there are trapezoidal.

Buckets were bolted to the driving element. The type, size, weight, capacity, spacing, and projection of the buckets are critical factors in determining the correct bolt for carrying the load and holding the buckets. The number, size and location of the bolts are also important. Three different bucket configura-tions are available, depending on various throughput require-ments and type of product. Buckets are constructed of polypropylene.

The buckets are made of a food grade reinforced polycarbonate, molded in one piece and a can handle temperatures from –5 to 215 degrees Fahrenheit. Anti-static and stainless steel buckets are optional.

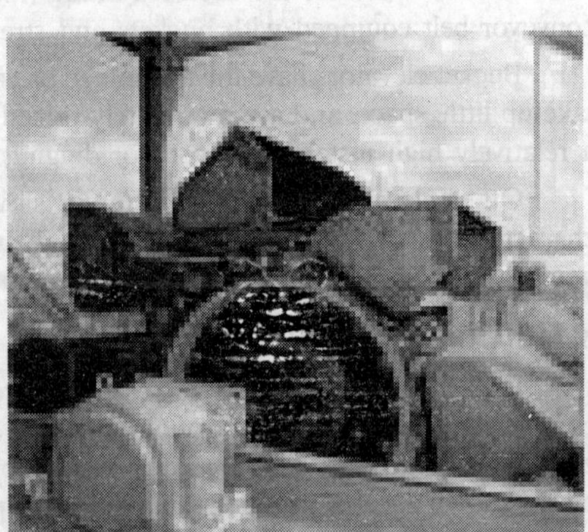

Figure 8.16

The buckets have a quick release mechanism to facilitate fast installation or removal of buckets. All elevators are equipped with an automatic chain tensioning device and built-in overload protection.

(*d*) **Drive :** It contains motor drive with reducer gear box.

(*e*) **Take-ups :** It is used for tensioning of driving element.

(*f*) Safety guards and stops.

8.3.4 Loading and Unloading Methods of Bucket Conveyor

Loading is done by scooping the material on the bottom of elevator or pouring the material into the buckets.

Buckets are unloaded using (*a*) centrifugal discharge or (*b*) gravity discharge.

(*a*) Centrifugal discharge is mainly due to speed of revolving pulley or sprocket at head. The material is thrown out of buckets due to centrifugal force.

(*b*) In gravity type, buckets were made inclined and material flows out of bucket through gravity after it passes over the pulley or sprocket.

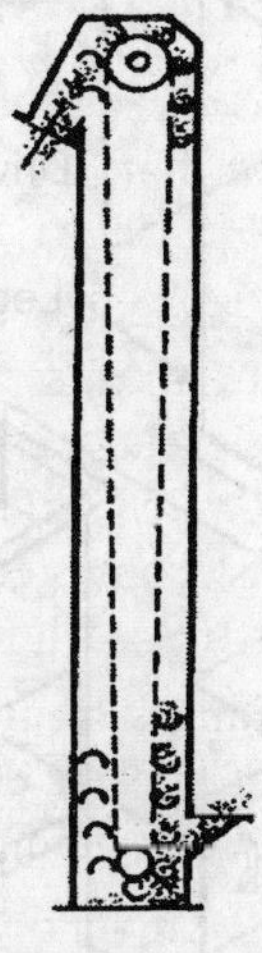

Figure 8.17

8.3.5 Applications of Bucket Conveyor

(*a*) They are used for conveying only bulk materials.

(*b*) Bucket conveyors are used to move the granular, powered or liquid materials.

(*c*) The bucket elevators allow for controlled feeds at one or more locations and can easily interface with various types of feeding devices.

(*d*) Bucket elevators are very well suited for a broad range of bulk products in the food, agriculture, pharmaceutical, cosmetic, chemical, ceramic, glass, recycling, building and metallurgical industries and others requiring gentle handling.

(*e*) Standard models are available.

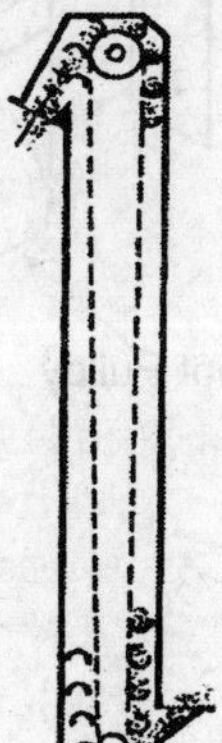

Figure 8.18

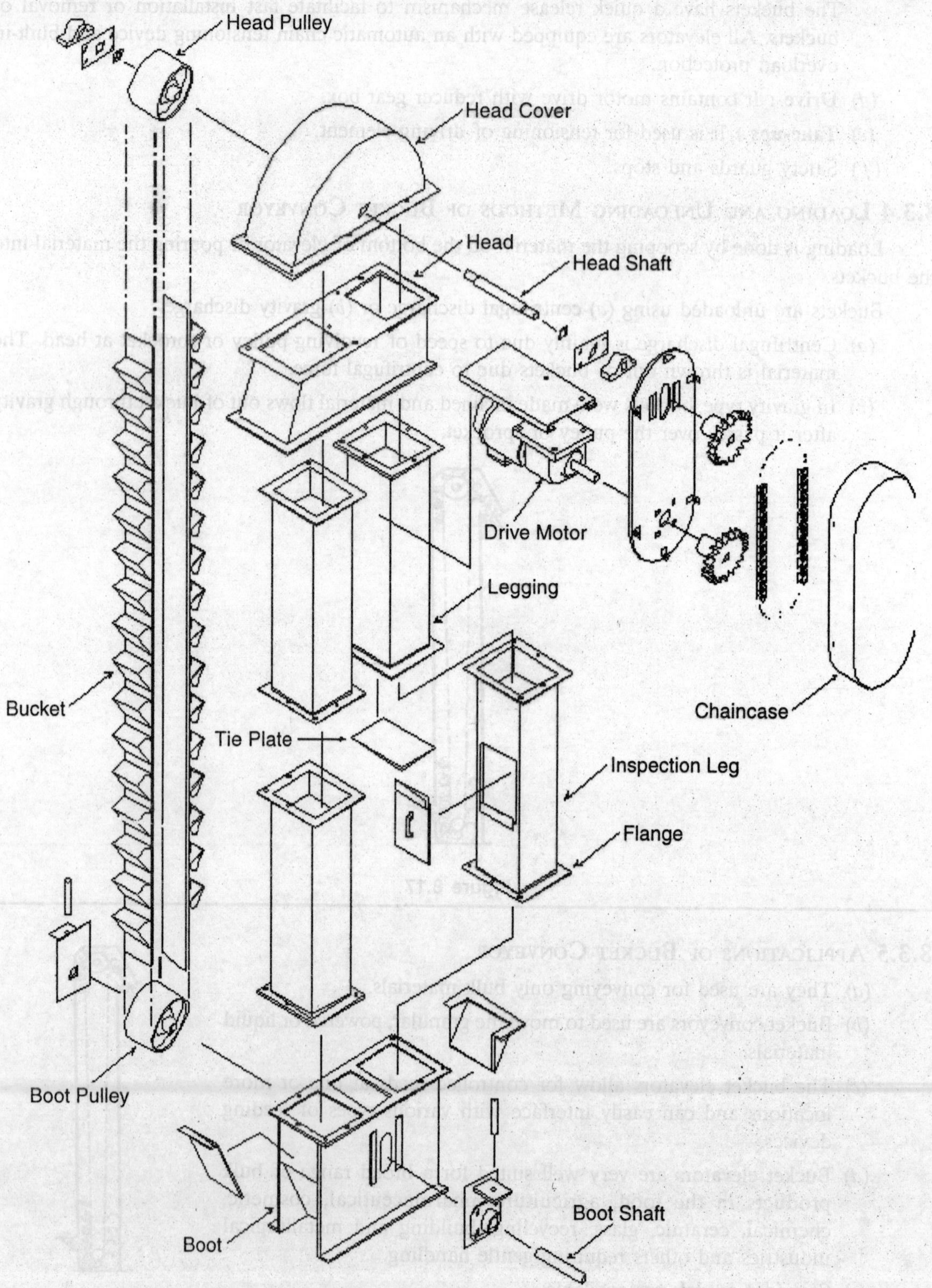

Figure 8.19. *Dis-assembly of bucket elevator.*

Coal Bucket Elevator

The bucket elevator was designed to feed coal from a tipping hopper to an existing overhead storage hopper.

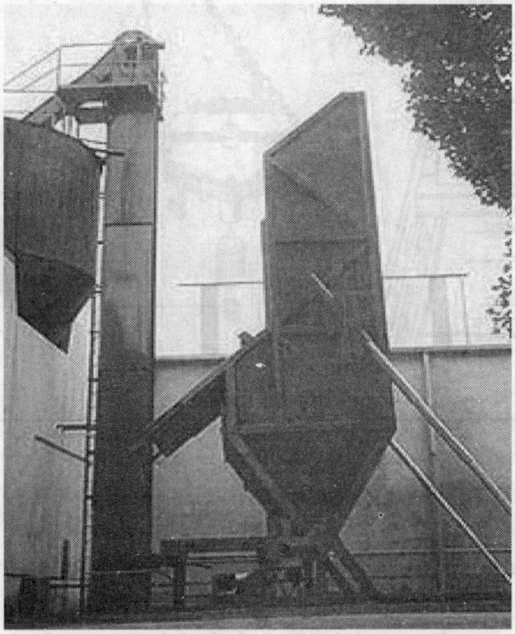

Figure 8.20

8.4 CHAIN CONVEYOR

Chain conveyors employ continuous chains wrapped around head and tail and sprockets. Load is carried on the chain running into the frames driven by motor. It essentially consists of following parts:

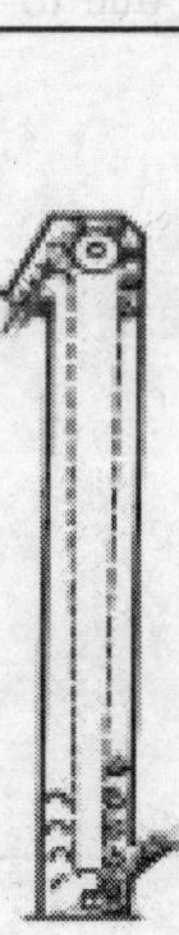

Figure 8.21

8.4.1 Parts of Chain Conveyor

Driving Sprocket

It is directly connected to the motor. Motor drives the sprocket. Motor power depends on the load to be conveyed and distance.

Endless Chain

Chain envelopes sprockets is driven continuously by motor. It actually carries load over it. Chain may be single strand or multi strand with constant pitch or variable pitch depending on the load and type of the unit. Chain is generally made up of steel, usable iron, chilled iron, manganese steel, cast steel, welded steel and some times of plastics. The capacity of a chain conveyor is the product of the unit's available cross-sectional area multiplied by the chain speed. The speed of the chain conveyor is based on the material that should be conveyed. Abrasive materials are conveyed at speeds of less than 3–7 m/min.

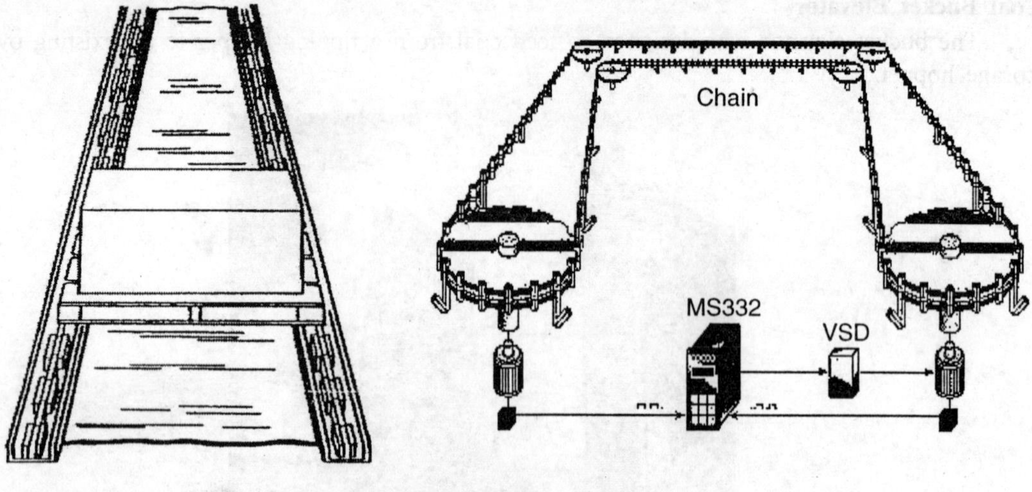

Figure 8.22

Chain Types

There are several types of chains which are used in the chain conveyor according to the load and path of conveying.

Chain with and without roller—They cannot follow the turn due to this usually used for straight and longer distances.

Bowling chain—These are the special type of the chains used to carry the material in any direction due to the fact that they can negotiate and follow the curves exactly.

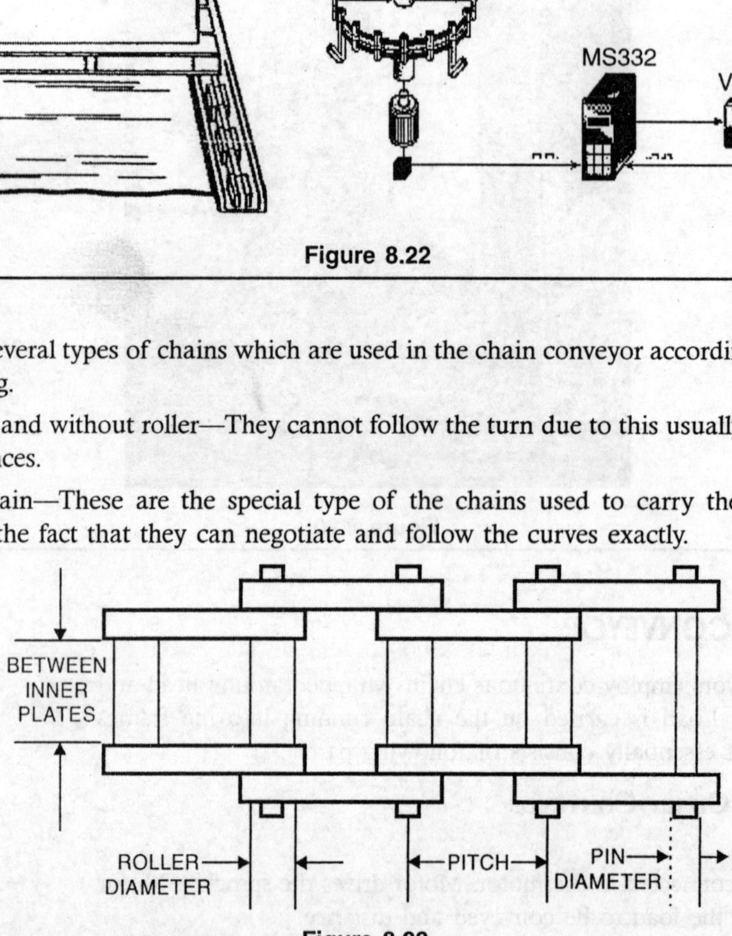

Figure 8.23

Frame

Chain slides or rolls inside the frame with predetermined speed. It also supports the drive motor and sprockets.

8.4.2 Multi-Strand Chain Conveyor

Multi-strand chain conveyor is a transportation conveyor. The strands of chain serve as the carrying surface for the load. The end drive unit connects the multiple strands of chain by means of a drive shaft and sprocket configuration. The conveyor offers significant application versatility and can also be used to convey loads that, due to bottom surface configuration, cannot be conveyed on a roller bed conveyor. This is a 3-strand chain deck of a typical bale feed system used by Pulp Mills.

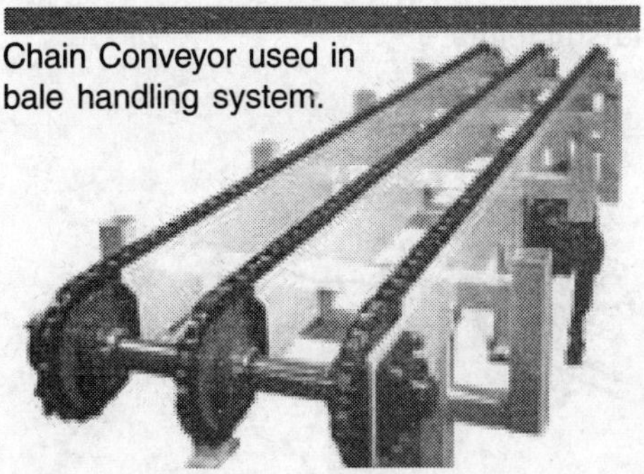

Chain Conveyor used in bale handling system.

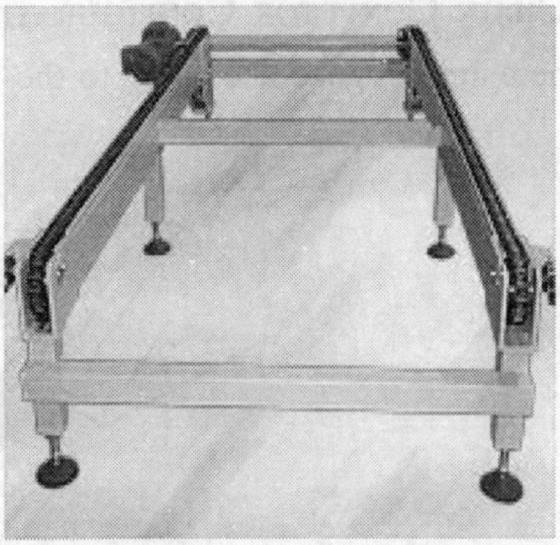

Figure 8.24

8.4.3 Advantages of Chain Conveyor

(*a*) Positive conveyance of a wide variety of loads.

(*b*) Minimal maintenance requirements.

(*c*) Quiet operation.

(*d*) Material can be transported to any direction.

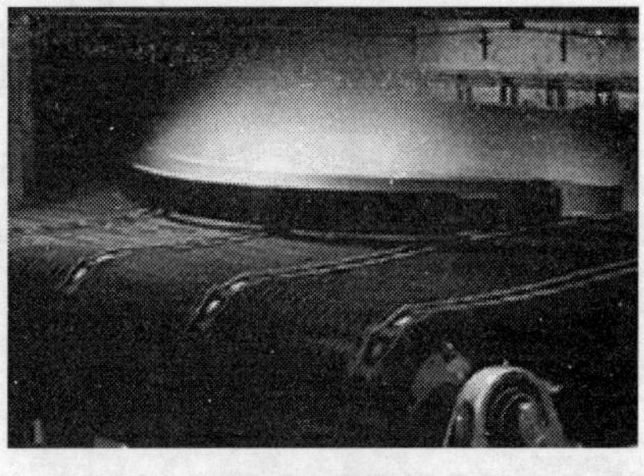

Figure 8.25

8.4.4 Disadvantages of Chain Conveyor

(*a*) Accumulation is difficult.

(*b*) Need of pallets for multi strand operation.

(*c*) Due to sliding of chain wear of frames may occur.

8.4.5 Applications of Chain Conveyor

(*a*) These are widely used over both short or long distances carrying goods which are unsuitable for roller conveyors.

(*b*) Because of their ideal transport surface characteristics, chain conveyors are used in the pulp and paper and forest products industries to move medium to heavy loads of wood chips and hogged waste wood. Also in the furnaces.

(*c*) They are also used in the Reclaim Industry to move tire chips, etc.

(*d*) This is used for moving barrels and heavy boxes.

(*e*) Unpackaged products.

- Foodstuff.
- Semifinished products.

(*f*) Packed products.

- Bottles.
- Boxes.
- Crates.
- Bags.

(*g*) Product carriers.

Systems for, *e.g.,* assembling or processing.

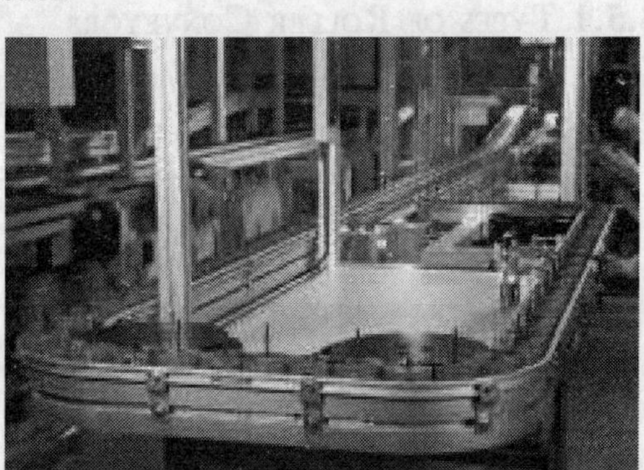

Figure 8.26

8.4.6 Chain Conveyor Configurations

These are designed according to the individual and suitability.

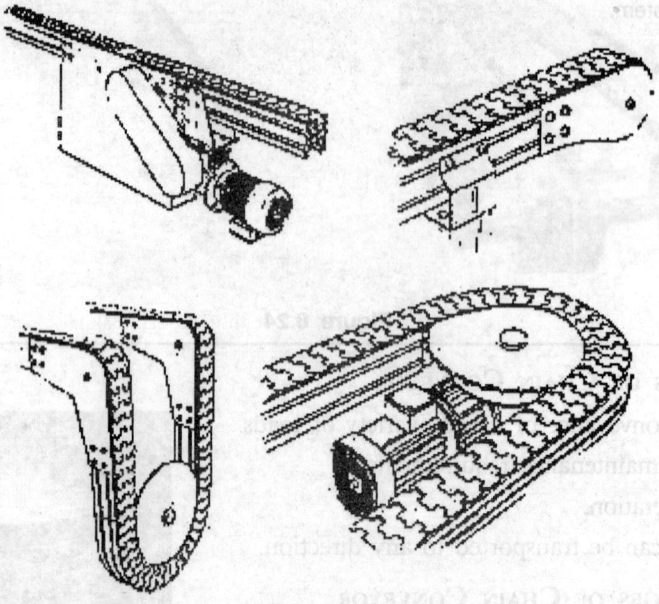

Figure 8.27

8.5 ROLLER CONVEYOR

This type of conveyor is used to transport various shapes of products such as boxes or materials which extend over several rollers. This type consists of cylindrical rollers supported in frames over which materials are allowed to move. They are driven by power or gravity. These are of different varieties and can move material in horizontal direction as well as from the upper floor to lower floor.

These have a long life readily assembled and installed, easy to adjust and used for handling large variety of loads. Load with rigid and smooth base are moved on the roller conveyors. Other items can be moved after placing them on flat boards. Pitch of the rollers should be such that minimum of 3 rollers must remain under load at all times, spacing is generally kept at 8 cm, 10 cm or 15 cm.

8.5.1 Types of Roller Conveyors

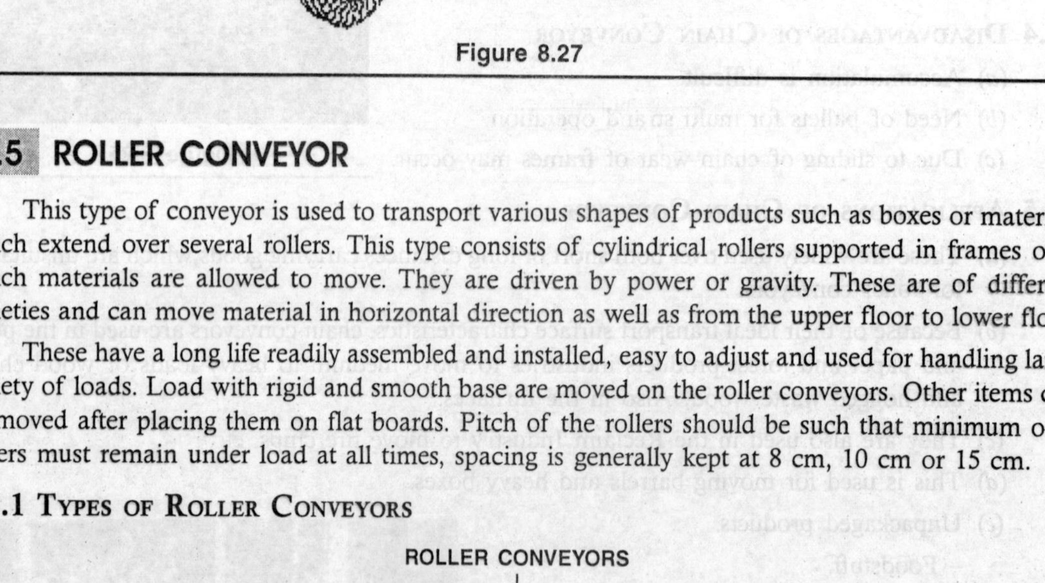

Figure 8.28. *Classification of roller conveyors.*

Rollers generally range from 2 cm to 8 cm in diameter with length governed by load. The rollers are made of cylindrical tubing with a bearing on each end. Bearings are installed in the supporting frames. Generally tapered rollers are provided at the turns. Roller conveyors may be flat, circular or spiral geometry. Roller conveyors are used for relatively short distances as compared to belt conveyors.

8.5.2 Gravity Type Roller Conveyor

In this type of the conveyors rollers were not given power, it is set at an appropriate incline, goods move down by gravity. The angle of inclination depends on the material to be moved. following table shows the relation between goods type and inclination.

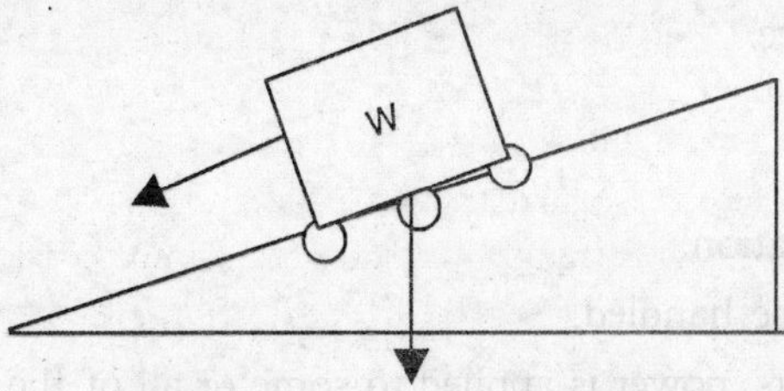

Figure 8.29. *Gravity conveyor.*

Table 8.6 Relation between Goods Type and Inclination

Sr. no.	Goods type	Angle of inclination
1	Heavy pallets and wooden bins	2.5–3.5 %
2	Cardboard boxes 10–20 kg	3–5%
3	Cardboard boxes 0.5–1 kg	10–13%

In this type of conveyors handling speed varies with the weight and the nature of the goods.

The inclination is always expressed with respect to the horizontal in three different ways : angle of incline, percent of grade, rise or fall in cm per m of conveyor length.

This type of conveyors is again classified in two ways stationary and mobile. In stationary type whole conveyor unit works in a fixed line where in the mobile type, conveyor can move and is mounted on a wheeled undercarriage. A conveyor is assembled from individual sections. A section comprises a frame supported by stands and set of rollers mounted on it. Stands may be made adjustable to vary angle of inclination. See following figures.

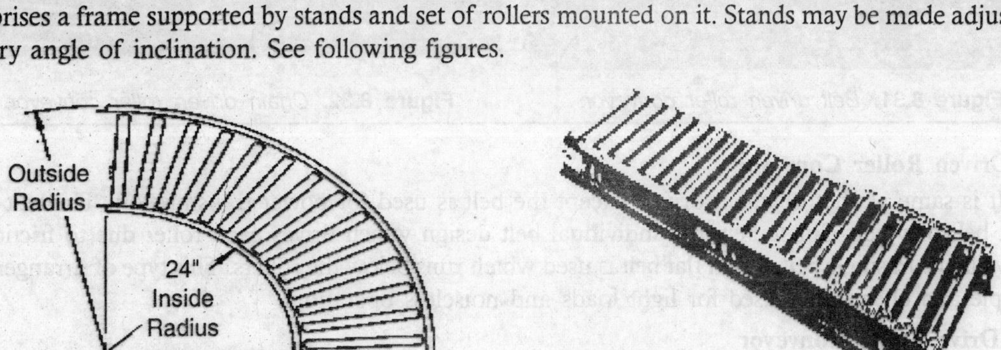

Figure 8.30. *Straight roller on straight and curved path.*

Unpowered roller conveyors are generally used for short distances and where there are no speed control is required. They allows accumulation and cheap compared to powered roller conveyors.

They are used in industrial warehouses, in depots and stores. Applications of this roller conveyors is limited, difference in input and output levels restricts the application of this roller conveyors.

8.5.3 Powered Roller Conveyors

In this type of the conveyors power is supplied to the rollers that's why it is known as powered roller conveyor. These are classified according to nature of power transmission system.

A drive becomes necessary when following requirements should be met :

(*a*) Long transport distance.

(*b*) Horizontal transport.

(*c*) Controlled speed.

(*d*) High degree of automation.

(*e*) variety of material to be handled.

In powered roller conveyors, power is applied to some or all of the roller to propel the loads. In powered roller conveyors, the power is generally applied by chain on sprockets, belts, gears etc. power is supplied by electrical motor in all cases. These are used for moving objects on level runs, up slight grades (upto about 10°) or descent on down grades (upto 17°). These are more expensive than gravity roller conveyors or belt conveyors but are more rugged than belt conveyors.

Chain Driven Roller Conveyors

In this type of conveyors power is supplied by the chain. It consists of the motor which is connected to a rollers through a sprocket. Chain drive may be given to the individual rollers or to the all rollers. Sometimes double chain operation can be used to increase the driving capacity.

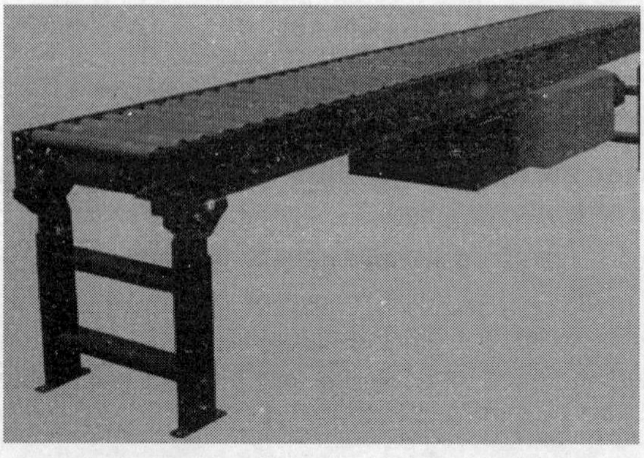

Figure 8.31. *Belt driven roller conveyor.*

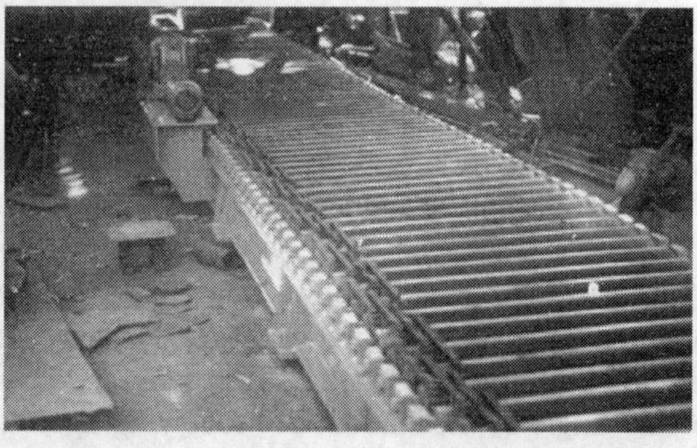

Figure 8.32. *Chain driven roller conveyor.*

Belt Driven Roller Conveyor

It is same as that of chain driven except the belt is used for power transmission. The belt may be flat belt or v-belt. Single belt and individual belt design which drives each roller due to friction is also available. In single belt design flat belt is used which runs below the rollers. This type of arrangement is simple and cheap. It is used for light loads and noiseless operation.

Gear Drive Roller Conveyor

In this longitudinal shaft is provided over the conveyor length on either sides of the frame. Power is supplied to motor which drives longitudinal shaft. This shaft carries pair of bevel gears depending on the number of rollers. Generally one pair is used for one roller. This type of design is costly and complicated, speed ratio can be varied using variable speed motor.

8.5.4 Roller Design

Any roller conveyor consists of number of rollers, supporting frame and stands and drive in case of powered roller conveyor.

The size and number of rollers depends on the length of the transportation, width and load of material to be transported. The pitch of the roller is selected according to the application. It generally depends on smallest item to be conveyed. Rollers may have cylindrical shape or disc shape. Single roller, double and triple roller design is also available. Tapered rollers are used in some cases where turning is required on curved path. Roller bodies are generally made by tube sections. They should be light in weight, wear resistant, should posses certain elasticity and high compressive strength. Rollers are made from steel, plastic steel and synthetic material.

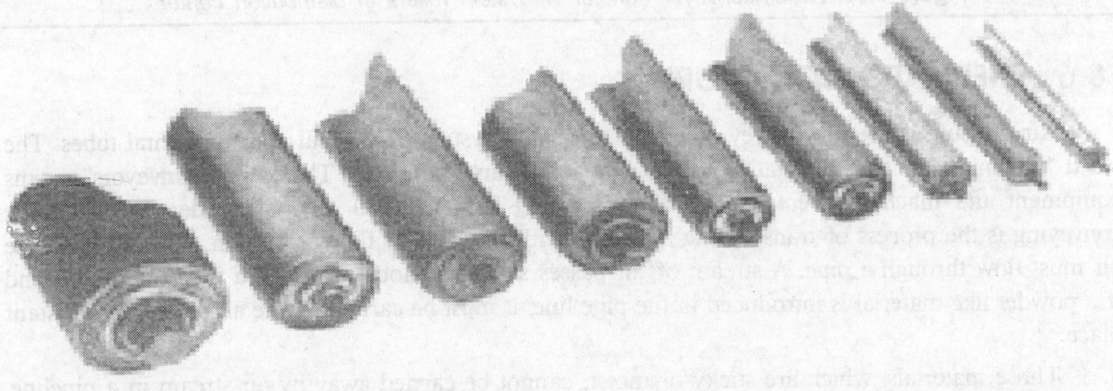

Figure 8.33. *Variety of rollers.*

8.5.5 Application Examples of Roller Conveyors

Roller conveyors combined with belt conveyors, stops and pusher stations.

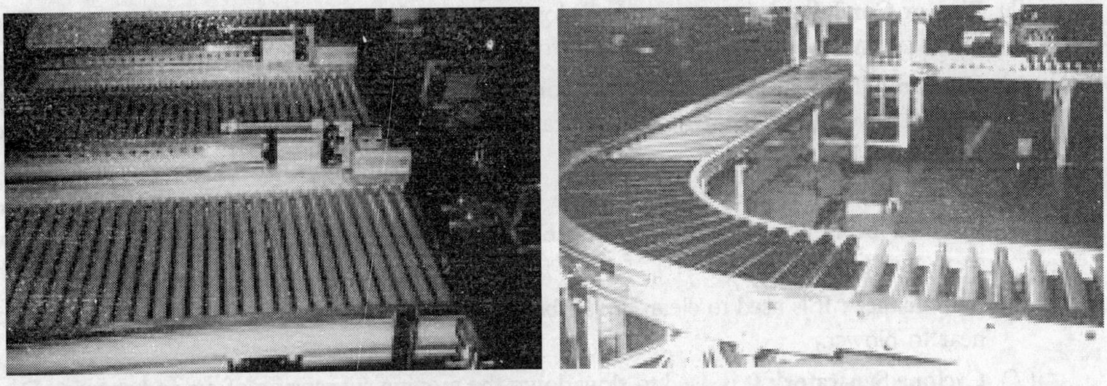

Figure 8.34. *Roller conveyors combined with belt conveyors, stops and pusher stations.*

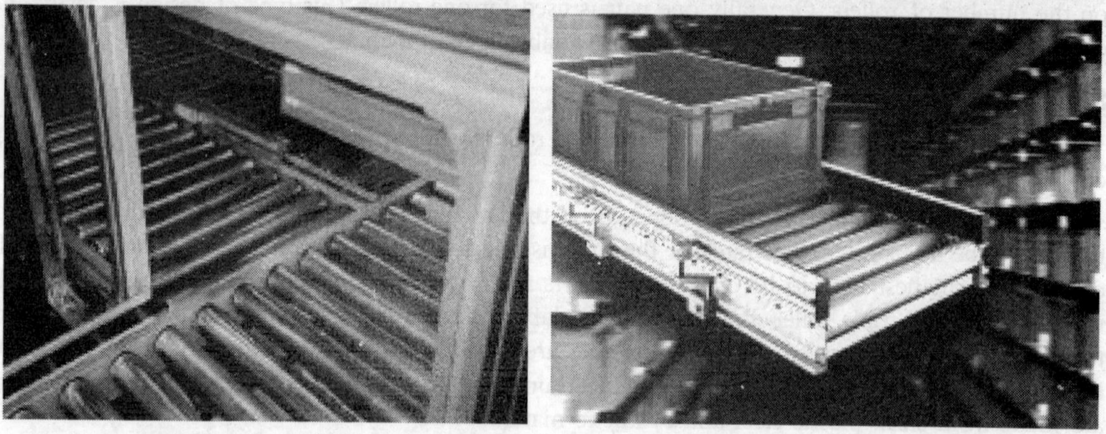

Figure 8.35. *Roller conveyor transfer with steel rollers in distribution center.*

8.6 PNEUMATIC CONVEYORS

Air pressure is used to convey materials through a system of vertical and horizontal tubes. The word "Pneumatic" means of relating to or using air or some other gas. The word "Conveyors" means equipment and machinery, employed for transporting bulk material on mass scale. So pneumatic conveying is the process of transporting materials with the help of flowing stream of air or gas. The air must flow through a pipe. A stream of air passes at some velocity through a pipe or conduit and if a powder like material is introduced in the pipe line, it must be carried by the air stream to a distant place.

Those materials, which are sticky or moist, cannot be carried away by air stream in a pipeline. Those materials, which are powder-like, free flowing, are ideally suitable for pneumatic conveying.

8.6.1 Important Parts of Pneumatic Conveyors

(*a*) **Blower :** It is used to blow the air in the conveyor line. The blower has specified capacity in m^3/min or m^3/hr for the particular application.

(*b*) **Pipe or Conduit :** It is connected to blower and carries all the bulk material inside it. The pipe size varies according to material to be conveyed.

(*c*) **Hoppers :** These are used to feed the conveyor line. The feed rate depends on the capacity of the blower. Position of the blower can be changed according to the type of pneumatic system used.

(*d*) **Valves :** These are used to control the flow of air and particles. They are placed at the specified locations in the conveyor line. Sometimes it is used to change the direction of flow or in the pipe bends.

(*e*) **Air Cleaner :** It is used to clean the air before entering into the system. Generally it is fitted near to blower.

(*f*) **Cyclone Separator :** It is used to slow down the moving suspension of air and powder. Due to reduction in velocity of air powder gets deposited and air is moved to the dust collector.

(*g*) **Storage Silos :** From the cyclone separator powder is transferred to the storage silos from where it can be used.

(*h*) **Dust collector :** Before discharging into atmosphere air gets cleaned in to the dust collector.

8.6.2 Classification of Pneumatic Conveying Systems

Pneumatic conveying systems can be classified on the following basis :

(*a*) On average particle concentration (modes).

(*b*) On air pressure (types of systems).

(*c*) On air supply arrangement.

(*d*) On solid feeder type.

8.6.3 Classification based on Particle Concentration

Depending on the mass flow ratio, defined as the ratio of mass of particles conveyed to the mass of fluid used to convey. Pneumatic conveying system may be classified into two modes:

(*a*) Dilute Phase (*b*) Dense Phase.

If the mass flow ratio is low, the system is said to operate in dilute phase. Whereas if the mass flow ratio is high, the system is said to operate in dense phase. The range of mass flow ratio is 0–15 for dilute phase and over 30 for dense phase.

(*a*) In a dilute phase system the material is carried through the pipeline by a large volume of air having high velocity but relatively low pressure. The stream of air or gas carries the material in suspension in the pipeline as discrete particles owing to lift and drag forces acting on each particle. The distribution of particles over the cross-section of the pipe is fairly uniform. In order to keep

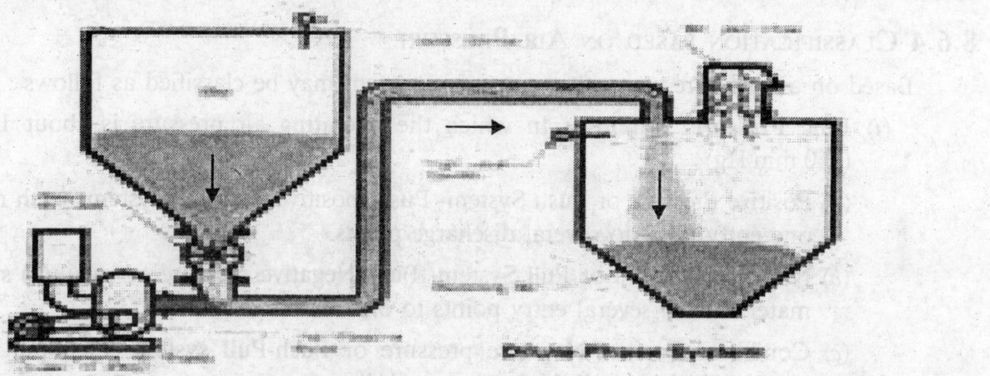

Figure 8.36. *Dilute phase system.*

the particle in suspension, the air must pass a minimum velocity, called the *Pick-up velocity*. The Pick-up velocity for a particular material depends on many factors like the shape, size, and specific weight of material and inclination of the pipeline.

If the velocity of air is gradually lowered down what is required to keep the particles in suspension, the particles gradually settle down and form the dune at the bottom of the pipeline all along the length of the pipe. In this condition the material is about to choke the pipeline and if the pressure is increased these dunes and plugs of material may move along the pipeline causing a dense phase flow.

(*b*) In a dense phase flow usually the velocity of fluid is much lower than the minimum velocity required for a dilute phase flow. The distribution of particles over the cross-section of the pipe is non-uniform. At the lower part of the pipe there are a slow moving dunes or plugs of material and the upper part of the cross-section of the pipe is filled with certain proportion of finer particles in suspension in a state of dilute phase. The maximum mass flow-ratio achievable for a dense phase flow ratio depends on many factors like the nature of materials and air velocity. It is usually greater than 30.

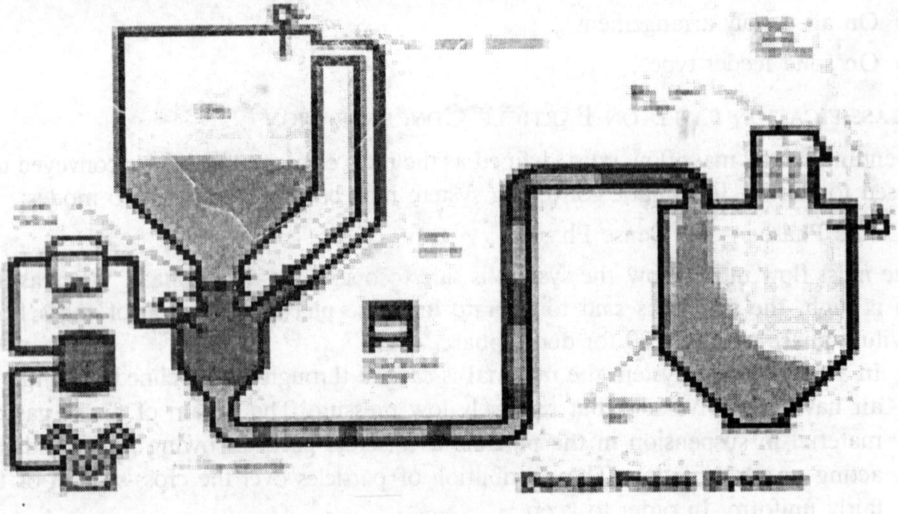

Figure 8.37.

8.6.4 Classification based on Air Pressure

Based on air pressure pneumatic conveying system may be classified as follows :

(*i*) **Low Pressure Systems :** In which the operating air pressure is about 1 atmosphere (760 mm Hg)

(*a*) Positive Pressure or Push System- Push (positive pressure) systems push material from one entry point to several discharge points.

(*b*) Negative Pressure or Pull System- Pull (Negative pressure or vacuum) systems move material from several entry points to one discharge point.

(*c*) Combined Positive-Negative pressure or Push-Pull system- Push-pull systems are combinations with multiple entry and discharge points.

(*ii*) **Medium Pressure System**

(*iii*) **High Pressure System**

(*i*) Low Pressure Systems

(*a*) Positive Pressure or Push System

Positive pressure system is normally employed to pick-up material from one source and delivers the same to so many points. The essential component of a positive pressure system consists of a blower which sucks air through an air inlet-filter and delivers the air at higher pressure to the conveying line. The material which is stored in a storage silo or hopper is fed into the conveying line through rotary air-lock feeder.

In general, a positive pressure system is not suitable for multiple pick-up points because the air-leakage through several rotary valve may be considerable high as compared to total air requirements. The material is then carried in suspension along with air through the conveying pipe line and delivered to one or more than one receiving hoppers. Receiving hoppers are nothing but a kind of silo where the material is stored before use. If more than one receiving hoppers are to be served by the same conveying line, diversion valves are used in the pipe line to deliver the supply of material to the desired receiving hoppers.

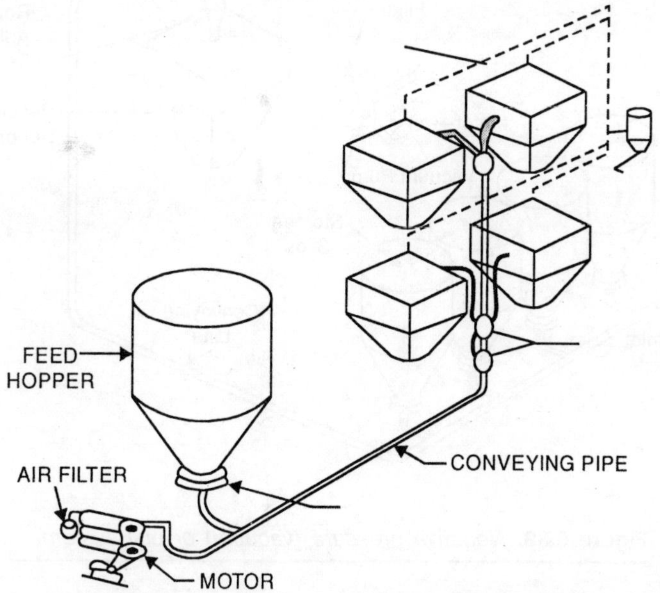

Figure 8.38. *Positive pressure or push system.*

Separation of solid and the air is usually achieved with the help of a cyclone separator and a bag fitter unit, which are fitted on the top of the receiving hoppers. A cyclone separator is essential a cylindrical vessel with its axis vertical and the conveying pipe line is connected with it in such a way that the mixture of solid and air enters the cyclones tangentially and receives a circular motion within the vessel.

This produces a vertex flow of the mixture of solid and air so that the heavier solid particles are collected at the centre of the cyclone and drop down by gravity. Although the major work of separation of solid from air is achieved in cyclone separator, still the air contains certain amount of dust particles to be filtered in bag-filter unit. The relatively cleaner air is taken out from the separator from the top and led into another cylindrical vessel where a bank of vertical bags, of cotton or other artificial fibers, are arranged in such a way that the air is made to pass through these bags before going out into the atmosphere. The dust particles carried by air from the cyclone separator are now filtered before being discharged into atmosphere.

An important merit of the positive pressure system is that the blower or fan, which is the heart of the conveying system, is always free from the dust of the material to be conveyed.

(*b*) Negative Pressure (Vacuum) or Pull System

A negative pressure system is employed to pick-up material from multiple sources from multiple sources and delivers the same to one located point. The system consists of a vacuum pump, which is in communication with the conveying pipeline via the separation unit (cyclone separator) and the bag filter unit. Because of the vacuum produced in the system, air is drawn in the conveying pipeline through the air inlet nozzle. The material from the storage hopper is introduced in the conveying line in which a negative pressure is maintained, through suitable feeders.

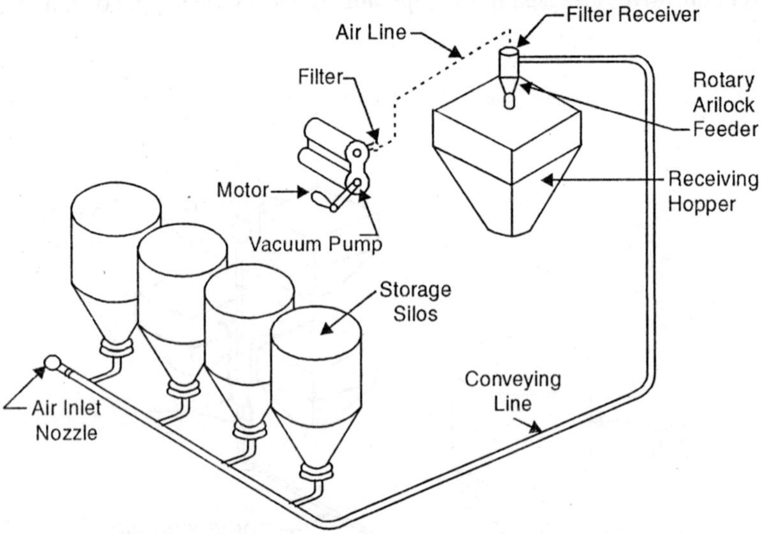

Figure 8.39. *Negative pressure (vacuum) or pull system.*

As the material is introduced in the pipeline at negative pressure, the feeders used in this type of system need not be complex in design. A number of storage silos may be connected with the same conveying line, although when one feeder delivers material to the conveying line, usually other feeders are cutout. However the material is conveyed along the pipeline to the cyclone separator where the major portion of solid is separated from the air. The relatively clean air from the separator is allowed to pass through the bag filter unit, which further cleans the air. The cyclone separator and the bag both deliver materials to the receiving hopper where the material is stored for use.

The most notable point in favour of this system is that all air leakage is from the surrounding into the system, so that pollution of the surrounding by the material being conveyed is practically eliminated. Hence this type of system is particularly suitable for toxic, explosive and powder like materials. But a serious limitation of the system is (*i*) that, as the vacuum pump is located at the discharged end, Unless the solid-air separation unit and bag filter unit is specially efficient, there is a possibility of air dust entering the vacuum pump and thereby damaging the life of the same. It is due to this reason that in a vacuum system multiple receiving hoppers cannot be used. (*ii*) That in practice only about 0.7 bar pressure drop can be allowed in the whole system and hence long distance conveying and high tonnage of conveying rate is not possible with a negative pressure system.

(*c*) Positive-Negative Pressure or Push-Pull System

This is a combination of the above two methods. In many industries pneumatic conveying system has to be used in a fairly general condition. There are number of storage hoppers or pick-up points and the material has to be conveyed to a number of deliver points for use. For such requirements the positive pressure system and negative pressure system must by synthesized into a positive-negative pressure system.

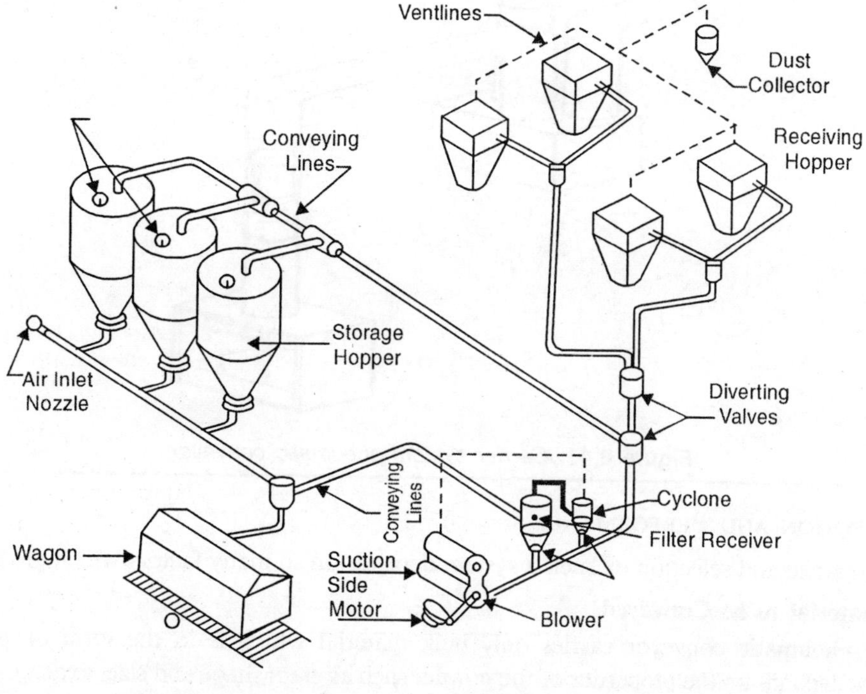

Figure 8.40. *Positive-negative pressure or push-pull system.*

(*ii*) Medium Pressure System

If the pressure of the air in the conveying line is required to be in the range of 1 to 3 atmosphere (760 to 2280 mm of Hg) the system is said to be operate in the medium pressure. The phase shift from dilute phase to dense one is possible just by increasing the air pressure. Essential component in this system is screw pump with variable screw pitch.

(*iii*) High Pressure System

High pressure system is used when material is to be transferred at a distance as high as 50 m. Again this system uses variable pitch screw pump but it is also having the blow tanks. The efficiency of this system largely depends on type and performance of blow tanks.

8.6.5 Some Other Pneumatic Conveying Apparatus

(*a*) Air Slides

When powder is placed on the porous membrane and air is passed through the holes of the membrane powder gets lifted. Air and powder forms a mixture (gets fluidised). Now this mixture behaves as solid body and when the slight slope is given to the membrane it slides down and discharges material at some distance.

(*b*) Carrier System Pneumatic Conveyor

This system is designed for specific applications. It contains conduit in which carrier is placed. This carrier carries material to be conveyed. As the air velocity in the conduit increases carrier moves and discharges at a specified point. Carriers are used to transport items or paperwork (*e.g.,* money from drive-installs at banks).

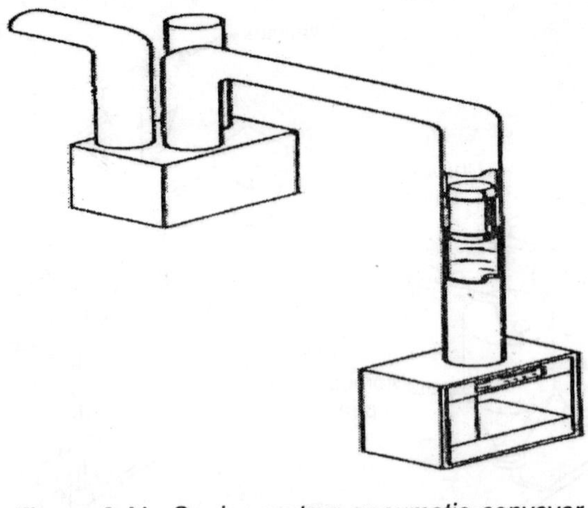

Figure 8.41. *Carrier system pneumatic conveyor.*

8.6.6 Selection and Performance

Performance and selection of such a system depends on so many factors which are listed below:

Type of Material to be Conveyed

Since pneumatic conveyor carries only bulk material which is in the form of powder their performance depends on the properties of the powder such as grain shape and size, toxicity, abrasiveness, density, moisture content and flowability.

Air Velocity

This is a most critical factor in the pneumatic conveying because small varation in this will affect the performance of the pneumatic conveyor. Generally these variations comes into picture due to leakage into or out of system, blockage of the passages of the blower, deposition of the powder at the bottom of the conduit, power variation and quality of maintenance. Air velocity is important in the different sense as well because particle movement depends on it. Particle movement is due to lift and drag forces coming on to the individual particles. This force depends on the location of the particle in the suspension and the distance from the edge or surface of the conduit. Generally particles at the centre of the pipe moves faster.

Size of the Conveying Pipe

Generally pipe size depends on the material to be conveyed. Optimum pipe size should be selected so as to maintain air stream velocity. It should have minimum bends and gradual inclinations as far as possible.

Environmental Aspects

Whatever be the type of equipment we use environment always plays the important role. For pneumatic conveyors dry environment is best with constant air pressure or velocity.

Selection of pneumatic conveyors depends on space availability and application for which it is used.

8.6.7 Advantages of Pneumatic Conveying System

(*a*) Major advantages are that material is completely enclosed.

(*b*) It is easy to implement turns and vertical moves that is material can be moved vertically, horizontally or along inclined path.

(*c*) Materials can be picked up from different points in a plant and can be delivered to a single point or many points in the same plant or in a different plant.

(*d*) Low maintenance cost and lesser manpower for operation.

8.6.8 Disadvantages and Limitations of Pneumatic Conveying System

(*a*) System gets affected due to air leakage from and into system.

(*b*) Pneumatic conveyors usually involve high-energy consumption per tonne of material conveyed. The specific energy consumption for a horizontal pipe line is of the order of 1–10 KW–h/km.

(*c*) Cannot transport Sticky, mist powders.

(*d*) Problem of wear and abrasion are very high in pipe line used for pneumatic conveyors due to toxicity and abrasive action of powder.

(*e*) Conveying distance for pneumatic conveyor is limited. Without booster stations, vacuum or low pressure conveyors are workable up to 500 m in length and for high pressure systems up to 2 to 3 km.

(*f*) Pneumatic conveyor design requires a high degree of skill.

8.6.9 Applications of Pneumatic Conveying System

Cement, flour, powdered milk, powdered ingredients for medicines, explosive powder for ordnance factories, to mention only a few, are basically powder like, free flowing, dry and non-stick substances which are to be transported in a concealed condition in order to avoid contamination and hazards. So, for mass scale handling of such materials pneumatic conveyors are essential, irrespective of cost factor.

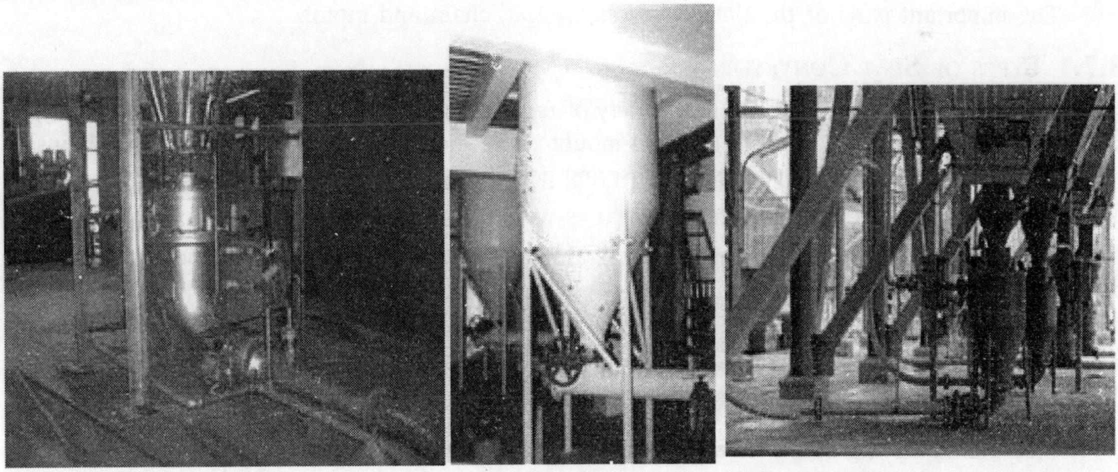

Fly ash handling *Hoppers* *Cu concentrates handling*

Figure 8.42

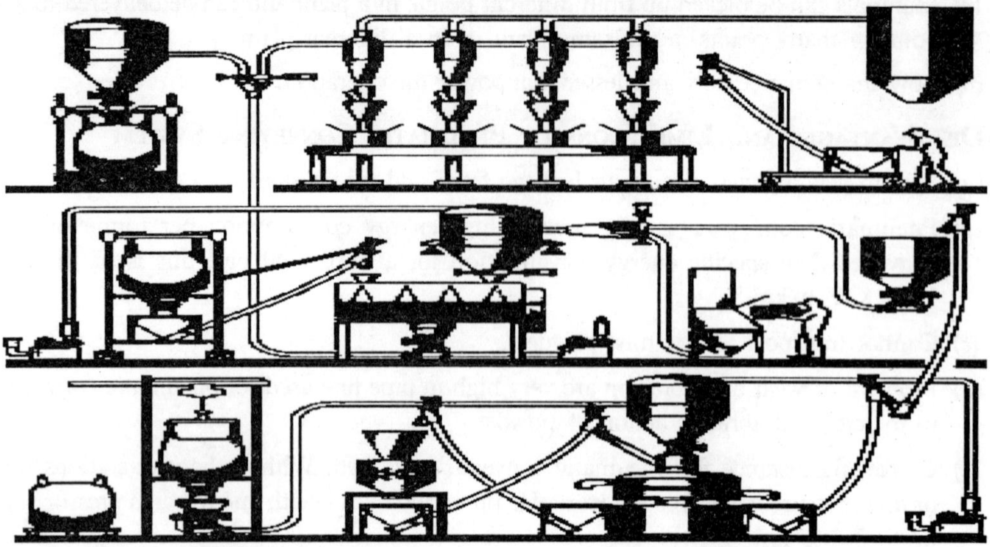

Figure 8.43. *Pneumatic System Layout at Work.*

8.7 SLAT CONVEYORS

Slat conveyors are suitable primarily for moving special materials and goods between fixed locations or stations which are likely to be permanent for some considerable time. Their function is basically similar to that of belt conveyors and they can negotiate inclines where continuous transport without accumulation is acceptable. It can handle heavy materials.

The important parts of the slat conveyors are slat, chain and motor.

8.7.1 Types of Slat Conveyors

Although Slat Conveyor come in a variety of types all consist of flat or shaped lengths (part) of steel, wood or some other material (rubber) mounted on and following the movement of chains to form a continuous hard, rigid surface for carrying goods.

Slat conveyors can be used for horizontal as well as on inclined paths.

Conveyors of this group known as by several names depending on the application they are used for :

(*a*) MS Slat Conveyor—uses steel slats.

(*b*) Multi line Slat Conveyors—used for variety of products.

(*c*) Crate Conveyor—used to carry crates may be open or close.

(*d*) Can Carrying Slat Conveyor—used for cans such as milk cans.

(*e*) S Type Slat Conveyors—it is In the form of "s" alphabet.

(*f*) Bottle Slat Conveyors—used to carry bottles of different shapes and sizes.

Slat conveyors normally have two chains connected to slats. The chains run on tracks with sliding or rolling friction. The chain is driven by the motor which is connected at discharge end. Commonly wooden or steel slats were used.

Wood is the most commonly used slat material, at least for moving general mixed cargo. It has good friction properties is cheap, corrosion resistant and is used to convey pallets and heavy mixed goods. Material and goods items which are very abrasive or very heavy require steel slats. Steel pallets were preferred when material to be carried.

Both wooden and steel slats are utilized in these conveyors. Steel slat are preferable to wooden ores for conveying articles with abrasive contacting surfaces. Steel slats present a smoother surfaces than those made of wood is an advantage in side-loading and diverting.

8.7.2 Specification of Slat Conveyor

These are specified according to their length, width, capacity, speed etc.

Speed	–	up to 100 m/min
Capacity	–	20 kg/m to 3000 kg/m
Standard length	–	6 m to 100 m
Width	–	300 mm to 1250 mm.

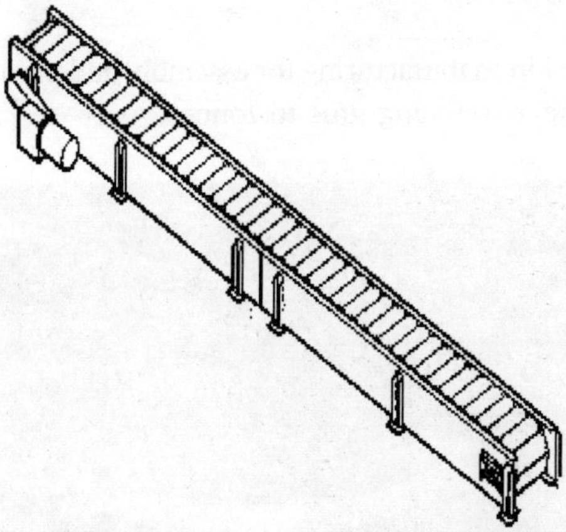

Figure 8.44. *Slat conveyor lay out.*

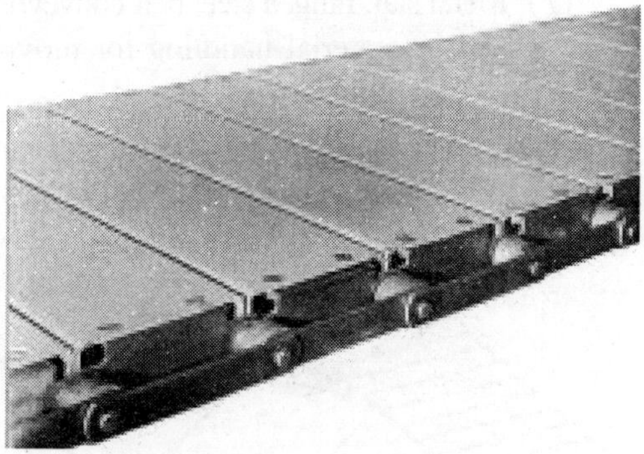

Figure 8.45. *Steel slat.*

8.7.3 Advantages of Slat Conveyor

(*a*) If the slats are set sufficiently high above the frame, the article can overhang without dragging.

(*b*) They can stand up under rough usage.

(*c*) Handle wide variety of product.

(*d*) No friction between goods and conveyor.

(*e*) Horizontal and inclined routes.

(*f*) Tolerance to heavy loads.

(*g*) Replaceable slats.

(*h*) Reliability.

(*i*) Goods flexibility.

8.7.4 Limitations of Slat Conveyor

(*a*) Slat conveyors are intended for handling small item.

(*b*) Not for bulk material.

(*c*) Goods accumulation is difficult.

(d) High noise level.

(e) Low speeds.

(f) High power consumption.

8.7.5 Applications of Slat Conveyor

Slat conveyors are suitable primarily for moving special material and goods between fixed location or stations which are likely to be permanent for some considerable time. Used to shape and store.

(a) Medium to large sized product foundry, chairing and bottling industry.

(b) Cotton, rolls of paper industry, handling newsprint in rolls.

(c) Conveying bag, parcel.

(d) Tilt slat conveyor used for sorting operation.

(e) Bottling and canning plants use flat chain or slat conveyors because of wet conditions, temperature and cleanliness requirements.

(f) Metal slat, hinged steel belt conveyors are used in manufacturing for assembly and production and in material handling for moving packages weighing gms to tonnes.

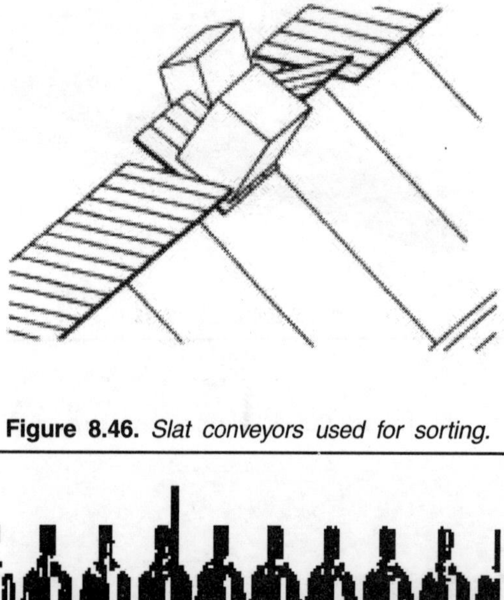

Figure 8.46. *Slat conveyors used for sorting.*

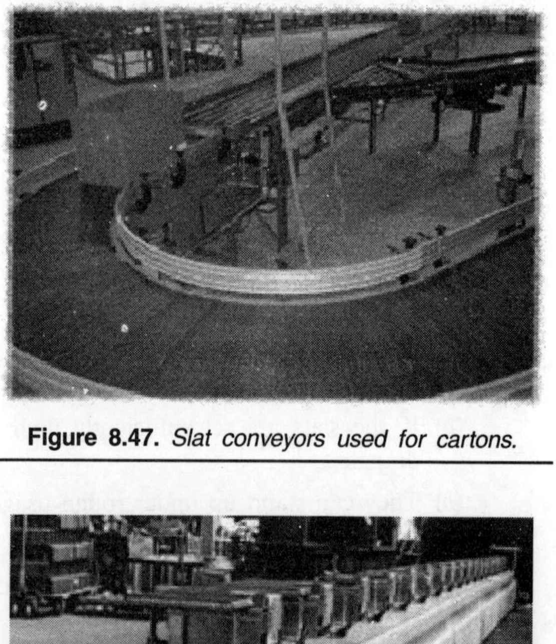

Figure 8.47. *Slat conveyors used for cartons.*

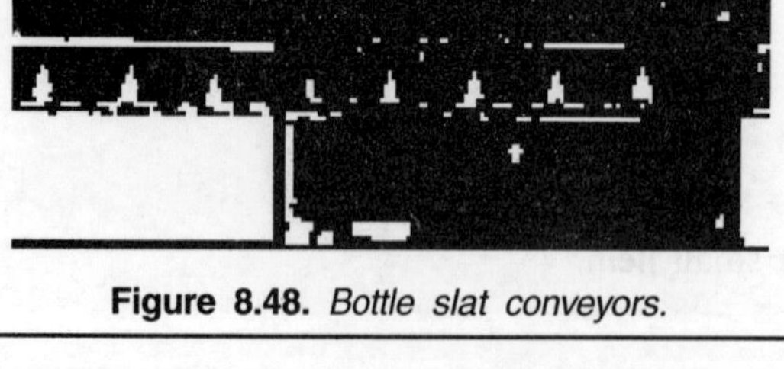

Figure 8.48. *Bottle slat conveyors.*

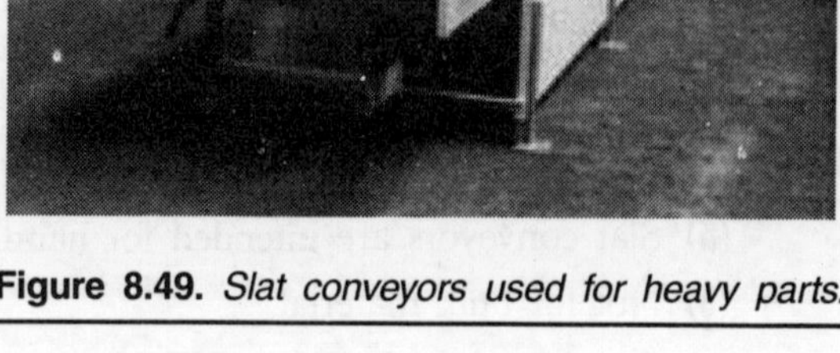

Figure 8.49. *Slat conveyors used for heavy parts.*

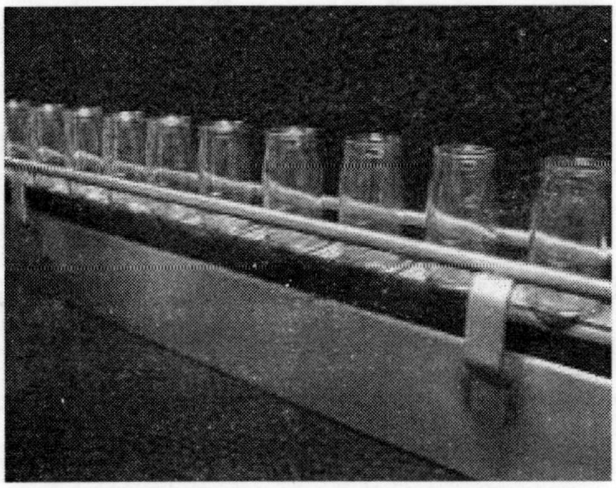

Figure 8.50. *Glass slat conveyors.*

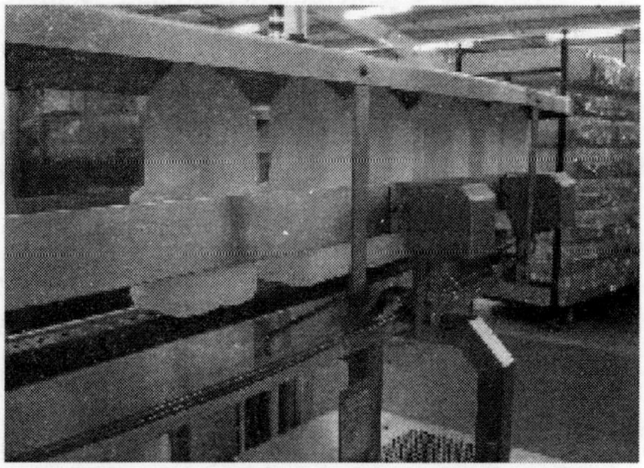

Figure 8.51. *Slat conveyors used for jar.*

8.8 VIBRATORY OSCILLATING CONVEYOR

A vibrating conveyor consists of a trough supported by tuned springs and/or hinged links having a drive system. The drive system is arranged to oscillate the trough, causing solid particles to be moved along the trough. Thus these conveyors are sometimes called oscillating conveyors. There are two types of oscillating conveyors : reciprocating and vibrating. An **oscillating conveyor** is similar in construction, but vibrates at a lower frequency and larger amplitude (not as gentle) in order to convey larger objects such as hot castings.

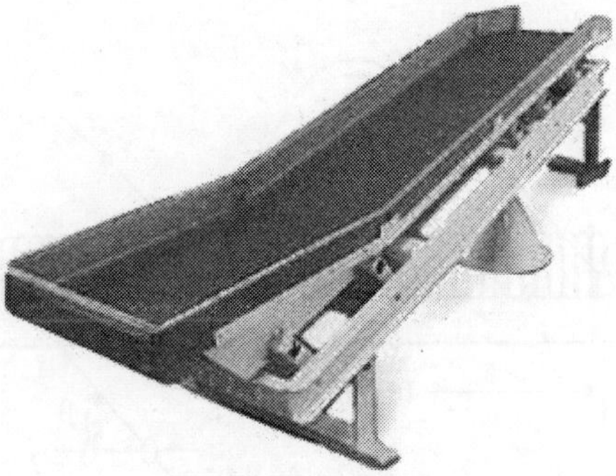

Figure 8.52

The vibrating conveyor consists of a two mass oscillating system, driven at its natural frequency of oscillation to minimize power consumption and vibrational stress.

Vibrating conveyors can be supplied in all shapes and sizes for conveying a range of different materials.

The vibrating conveyor consists of the following elements :

(*a*) Watertight stainless steel tray riveted to a rigid mild steel frame.

(*b*) Swinging arms.

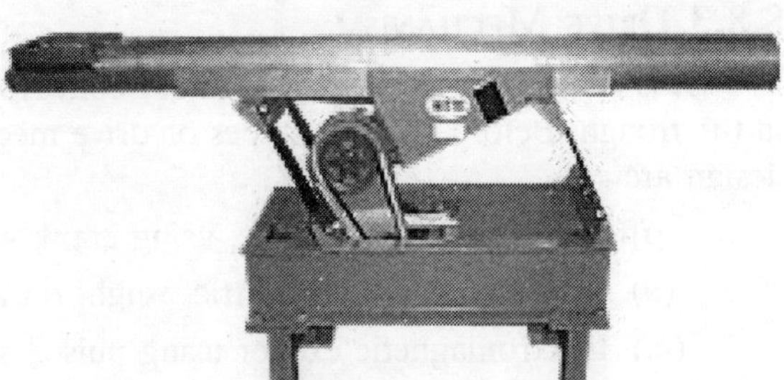

Figure 8.53

(*c*) The counterbalance weight is driven by the eccentric shaft through a connecting and suspension drive head.

(*d*) Manual or pneumatically operated trap door (optional).

The fundamental action of the vibrating troughs on the bulk material loading on it is to through the particle upward in the forward direction, so that the material performs series of shots hopping movement and propagated at a certain speed.

8.8.1 Classification of Vibratory Conveyor

Vibratory conveyors may be classified on the basis of three considerations :

(*i*) The pattern of motion of the trough or load.

(*ii*) Drive mechanism.

(*iii*) Frequency of excitation.

8.8.2 Pattern of Motion of the Trough or Load

By the pattern of the trough or load, vibratory conveyors may be classified into :

(*i*) Reciprocating or inertia-type conveyors.

(*ii*) Vibrating or oscillating conveyors.

(*i*) In reciprocating or shaker conveyors as the trough moves forward during its forward motion, it carries the material with it.

(*ii*) In a vibrating or oscillating conveyors the load is periodically thrown above the trough (jumps) with the vertical component of the trough acceleration greater than the acceleration due to gravity.

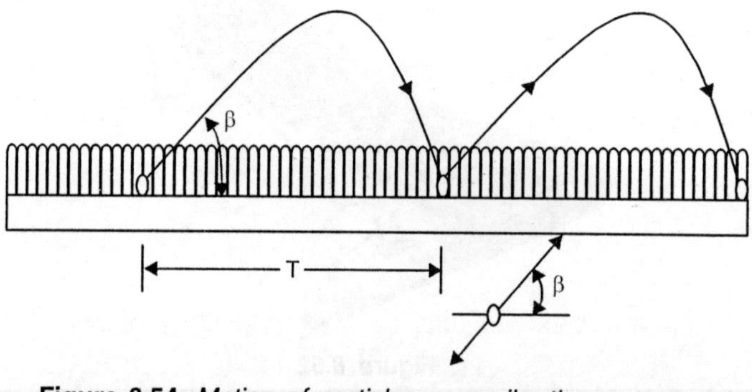

Figure 8.54. *Motion of particles on a vibrating conveyor.*

8.8.3 Drive Mechanism

It is convenient to classify vibratory conveyors according to the method used to excite vibrations in the trough. Before principal types of drive mechanism commonly employed in vibratory conveyor design are :

(*i*) Direct mechanical type, using crank and connecting rod.

(*ii*) Single or double eccentric weight on a rotating shaft.

(*iii*) Electromagnetic exciter using pulsed single-phase ac supply.

8.8.4 Frequency of Excitation

According to the range of frequency of excitation of the trough, the vibratory conveyors are classified as (*i*) sub-resonant, (*ii*) resonant and (*iii*) super-resonant system.

(*i*) In resonant system, the elastic elements of the trough may be so adjusted and the drive system runs at such a speed that the frequency of the excitation force of the vibration exciter, ω, becomes almost equal to ω_0, the natural frequency of oscillators of the elastic system of the conveyor. In practice during steady operation the ratio of frequencies ω/ω_0 may range from 0.85 to 1.1.

(*ii*) In the sub-resonant system, the frequency of the excitation is much less than the natural frequency of oscillations of the conveyor. That is, $\omega << \omega_0$. Sub-resonant systems have limited application due to the presence of damping source which consumes large energy.

In a super-resonant system, the frequency of excitation of the trough is much higher than the natural frequency of the oscillations. In other words, $\omega >> \omega_0$.

(*iii*) In a super-resonant system, the elastic elements (springs or other re-silent materials) have low stiffness. That is why the conveyor requires low starting torque, but consumes more power during steady operation at high frequency. But the real difficulty in using such a system is that during passing through the resonant condition, at the time of starting and stopping, the soft elastic elements are subjected to high stresses for which frequent failures of these elements may result. Therefore, the super-resonant systems are used in light duty conveyors.

8.8.5 Analysis of the Motion of Material along the Trough

Mathematical analysis of the motion of the particles on a vibratory conveyor is extremely complex as it involves many factors related through motion and material properties. It is therefore that most designers depend on empirical relation and experimental data. However, a basic mathematical analysis provides an insight of the interrelationship between the most important criteria is an average flow velocity of material for which an expression is to be established.

The theoretical approach to calculate the transport velocity of the material is based on the following assumptions :

(*i*) Bulk solids behave as a solid body and may be presented as a single mass.

(*ii*) Rotation of bulk solids may be ignored.

(*iii*) Air resistance neglected.

(*iv*) Impact between bulk solids and trough bottom and side wall neglected.

(*v*) Oscillatory path of trough is straight.

(*vi*) Coefficients of static and dynamic friction are numerically same.

8.8.6 Material Transport Velocity

The actual velocity of the particles in the trough varies with time; hence it is the average conveying velocity that is significant. This is the most important single factor that governs the design and selection of vibratory conveyors.

The average velocity of the bulk solid along the trough is often expressed as a function of this theoretical maximum value.

$$V_8 = \eta\, X_{max} = \eta$$

8.8.7 Other Design Parameters

- Frequency and Amplitude of Vibration
- Throw Angle β
- Load Carrying Elements

➤ Throughput Capacity

➤ Power of the Motor for Electrical Drive.

Technical Data

Width range : 300, 400, 500, 650, 800, 1000, 1200, 1400mm

Conveying speed : 0.1–0.3 m/s

8.8.8 Advantage of Vibratory Conveyor

The following points may be considered as advantageous in selecting a vibratory conveyor :

(*a*) Hot and abrasive materials can be handled.

(*b*) Cooling, drying and dewatering operation can be done during transport.

(*c*) Scalping, screening or picking can be done.

(*d*) Units can be covered and made dust tight.

(*e*) Simple construction and low head room.

(*f*) Can be made leak proof.

(*g*) Low maintenance.

(*h*) Low spillage of fines.

(*i*) Controlled feeding and material alignment.

(*j*) Heavy duty drive assemblies are fitted to all conveyors.

(*k*) Optional screening section can be built in for separation of fines and sawdust.

(*l*) Optional fiberglass section for metal detection can be built into conveyor.

8.8.9 Disadvantage of Vibratory Conveyor

Such conveyors suffer from the following disadvantageous :

(*i*) Relatively short length of conveying (about 50 m maximum).

(*ii*) Limited capacity, about 350 tonne per hour for length of conveying of 30 m.

(*iii*) Some degradation of material takes place.

8.8.10 Applications of Vibratory Conveyor

(*a*) Vibrating conveyors gently and safely feed, move and screen dry, granular, free-flowing, and non-free flowing products efficiently and cost-effectively while protecting product integrity.

(*b*) Vibratory conveyors find wide spread applications in the transportation of dusty, hot, toxic and chemical aggressive bulk material through a closed trough or piping chemical, metallurgical, mining industries and manufacturing of building materials.

(*c*) Vibratory conveyors are also employed for the transportation of steel chips in machine shop, hot knocked-out stand, wastes and small castings in foundry shops. Vibratory feeders are also in use for delivery or small machine parts like screws, rivets, etc.

(*d*) Vibratory conveyors are hardly employed for handling common bulk loads, such as sand, gravel, coal etc., as the same can be done efficiently by belt conveyors.

(*e*) Can be used to convey almost all granular, free-flowing materials.

8.8.11 Vibratory Conveyors are Ideal For

- Food/Dairy
- Plastics
- Chemicals

- Pharmaceuticals
- Minerals
- Pet Products.

8.8.12 Application Examples

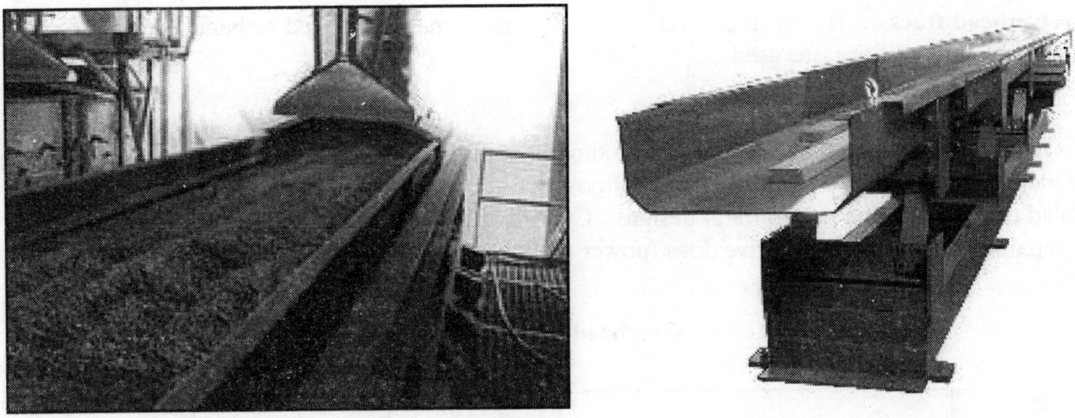

Figure 8.55. *Vibrating conveyor used for soil and timber transport.*

8.9 OVERHEAD CONVEYOR

An overhead conveyor describes all conveyors with roof-mounted drives. They are known as overhead conveyors because load is carried on the moving carriages which move on overhead tracks. In simple manner loads can be transported by carriers running on rails suspended from the roof or carried on the floor by vehicles with roof level drives. They can be used for any complicated configuration and are restricted to in house handling.

8.9.1 Design Details

The main parts of this type of conveyors are

Rail : It is mounted on the roof and is used to support the rollers or wheels which run over it.

Carrier : One end of this contains rollers and at the other end it holds attachment.

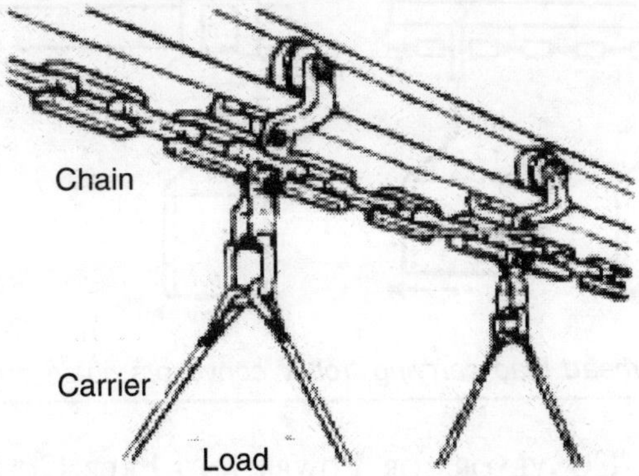

Figure 8.56. *Main parts overhead conveyor.*

Drive : Depending on the load and configuration. It is selected chain drive is popular. Power is supplied by electric motors.

Suspension attachment : This type of attachment depends on geometry of load and its properties.

Speed controller : Gear drives are used to vary the speed or variable speed motors are used. and **Chain Tensioner** (only in case of chain drives).

Overhead tracks : The track of overhead conveyors is made of rolled or bent sections of steel. Generally I beam sections are used.

8.9.2 Types of Overhead Conveyor

Overhead conveyors may be divided into three main types, depending on the drive method. The usual method is by chain. Worm drives and self-powered carriers are two other options. Chain operated overhead conveyors come in two main designs : Continuous direct drive or with a disengaging feature via a separate chain fitted with drive dogs (power and free type) to facilitate accumulation - switching etc.

Overhead Conveyor

- *Chain Operated*
- *Worm Operated*
- *Self Powered*

Depending on the method of the connection of the hauling or driving chain with load carrying trolley attachment and the method by which the load is moved they can be divided as

(*a*) Overhead load carrying trolley conveyors.

(*b*) Overhead pusher conveyors or power and free conveyors.

(*c*) Overhead push- carrying conveyors.

(*d*) Overhead load- towing (floor type) truck conveyors.

(*e*) Overhead tow- carrying truck conveyors

8.9.3 Overhead Load Carrying Trolley Conveyors

In this trolleys with load carrying attachment are fastened or attached to the drive chain and move on the overhead track, while the chain moves along that track. In other words wheels and attachment is connected to a chain. It carries endless driving or hauling chain.

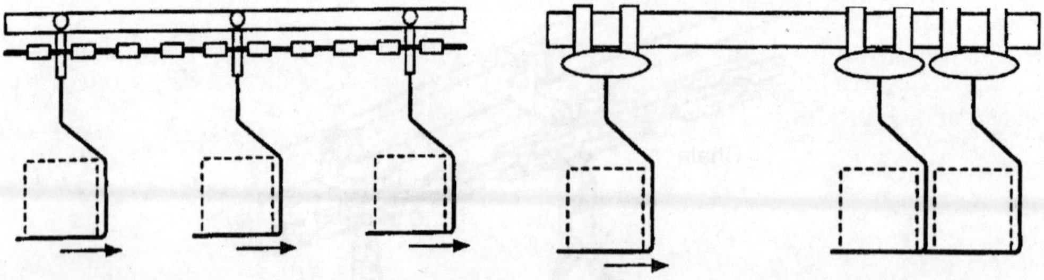

Figure 8.56. *Overhead load carrying trolley conveyors chain and worm operated.*

8.9.4 Overhead Pusher Conveyors or Power and Free Conveyors

In this type chain and trolley runs on separate track. Chain is on one track and trolley is on another track. Compared to the above design it is complicated but load carrying capacity is more.

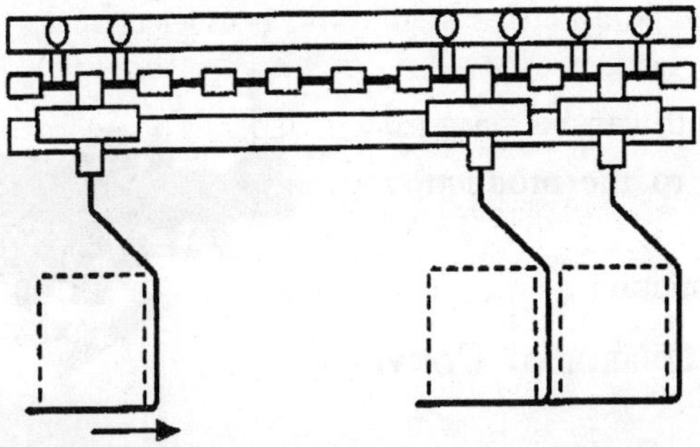

Figure 8.58. *Power and free conveyor.*

8.9.5 Overhead Push Carrying Conveyors

This is essentially a combination of above two types. The load is moved by pushing the chain manually or mechanically. It is provided with the hooks which are attached to the chain, these are used to connect with the attachment which carries the load.

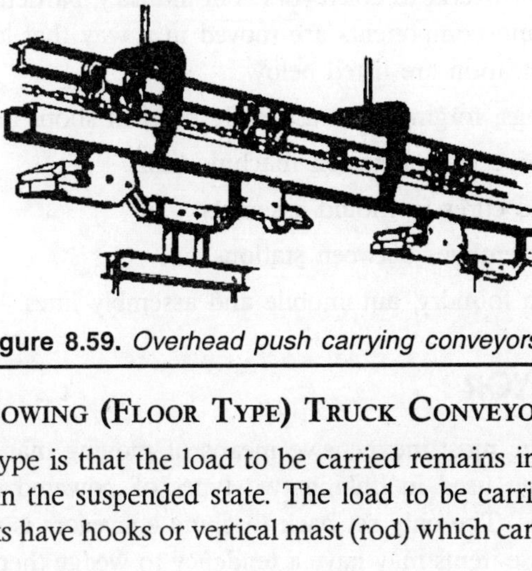

Figure 8.59. *Overhead push carrying conveyors.*

8.9.6 Overhead Load Towing (Floor Type) Truck Conveyors

The difference in this type is that the load to be carried remains in touch with the ground but in the other conveyors it is in the suspended state. The load to be carried is placed in to the small trolleys or trucks. These trucks have hooks or vertical mast (rod) which can be attached to the overhead drive chain.

8.9.7 Overhead Tow-Carrying Truck Conveyors

The only change in this is that the floor trucks can be attached to the trolleys or directly to the chain. In this truck moves in the combined motion means in some portion of conveyor line it is to the ground and in some portion it is moved in the suspended state.

8.9.8 Specifications of Overhead Conveyor

These conveyors are specified as load per carrier, speed and curve radius.

Load per carrier = 25 to 1500 kg.

Speed = 10 to 100 m/min.

Curve radius = It is given in horizontal and vertical terms usually in mm.

8.9.9 Advantages of Overhead Conveyor

(*a*) Free floor area.

(*b*) Layout simplicity.

(*c*) Simple design.

(*d*) Low investment.

(*e*) Any complicated path can be designed.

(*f*) Better adaptability to the production process.

(*g*) Low power consumption.

8.9.10 Disadvantages of Overhead Conveyo

(*a*) Limited headroom.

(*b*) Roof must have load bearing capacity.

(*c*) Speed of operation is slow.

(*d*) Risk of accidents due to moving overhead loads.

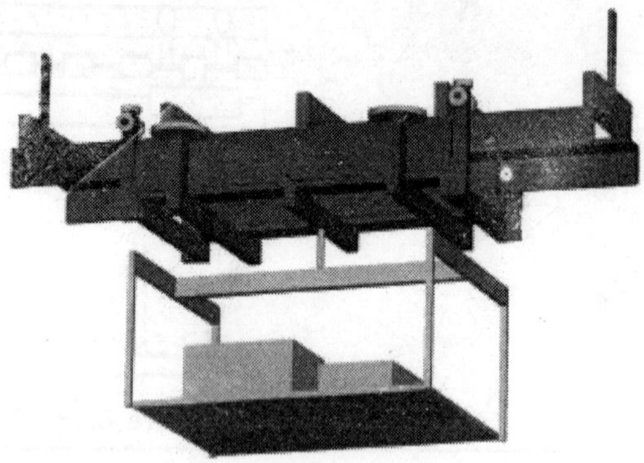

Figure 8.60.

8.9.11 Application of Overhead Conveyor

The most extensive use of overhead conveyors is in industry, particularly in assembly operations where both basic assemblies and components are moved in a way that leaves the floor area free for traffic, etc. some specific application are listed below :

(*a*) Transporting castings, forgings and assembly units in shops and between them.

(*b*) Transporting machine parts between machine tools.

(*c*) Transporting mould cores to mould assembly.

(*d*) Transporting subassemblies between stations.

(*e*) Extensively used in foundry, automobile and assembly lines.

8.10 CHUTE CONVEYOR

The chute conveyor is the most inexpensive means of moving material using a conveyor. This type of conveyor is most often used in linking two types of powered conveyors together and for accumulation in shipping areas. The only set back to chute conveyors is that it is difficult to control the items being conveyed. These items may have a tendency to wedge themselves in the conveyor and cause blockage.

This is a type of conveyor in which material is transported form one place to other by using gravity. They are a powerless conveyor that is they didn't require source of external energy. It can carry unit or bulk material; can be used for any combination. Here are some important points related to chute conveyors.

(*a*) This is one of the simplest devices which have both vertical and horizontal motion.

(*b*) Straight or spiral. A spiral chute could be used to convey items from floor to floor without using up too much space.

(*c*) Large or small, curved or straight, are used to move the material from higher level to lower level.

(*d*) Design depends on particular type of product, distance to be traveled, space availability.

(*e*) Number of different sizes and weight of packages should be restricted otherwise heavier packages will travel very fast while lighter will last to come to rest.

(*f*) Accumulation is possible using chute conveyors.

(*g*) Very inexpensive only construction cost to be incurred.

(*h*) The problem with such type of conveyors is controlling of position of product.

Chute conveyors which broadly used are straight and spiral chute conveyors. Straight chutes are used when horizontal distance is available and distance to be transported is comparatively more. Straight chutes were given slope according to the material and distance over which it is to be carried.

Spiral chutes are used when height is more in a given length. In this, load is lowered following a helical path around a vertical axis. These are designed so that the outer edges is higher than the inner, so as to protect the load being thrown off the chute by the centrifuge force set up during discord.

Spiral chutes have advantages of low, no power requirement, low maintenance cost, economical use of space and controlled speed of descent.

8.10.1 Applications of Chute Conveyor

(*a*) Used to link two handling devices.

(*b*) Used to provide accumulation in shipping areas.

(*c*) Used to convey items between floors.

(*d*) Used in thermal plants, unit handling in warehouses, cement bags handling and at airports.

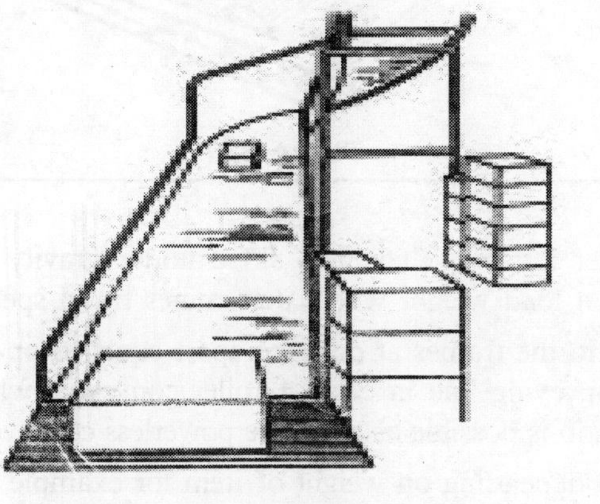

Figure 8.61. *Straight chute.*

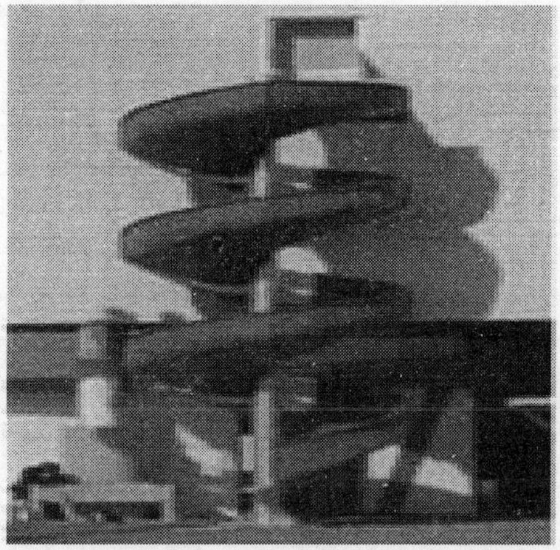

Figure 8.62. *Spiral chute at airport.*

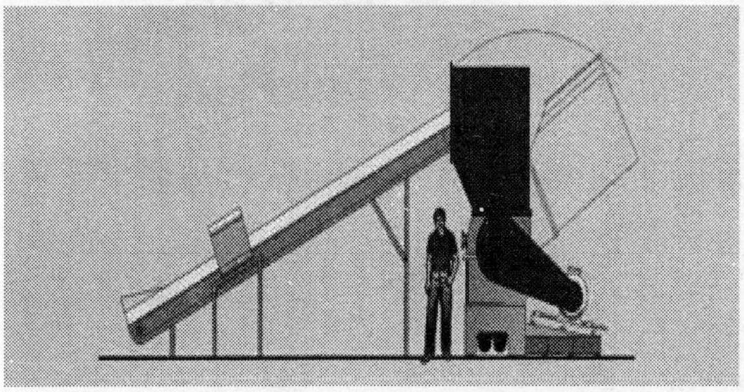

Figure 8.63. *Straight chute.*

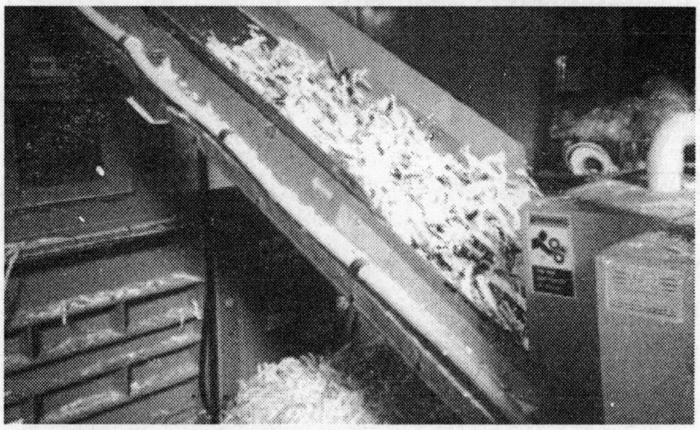

Figure 8.64. *Straight chute in paper industry.*

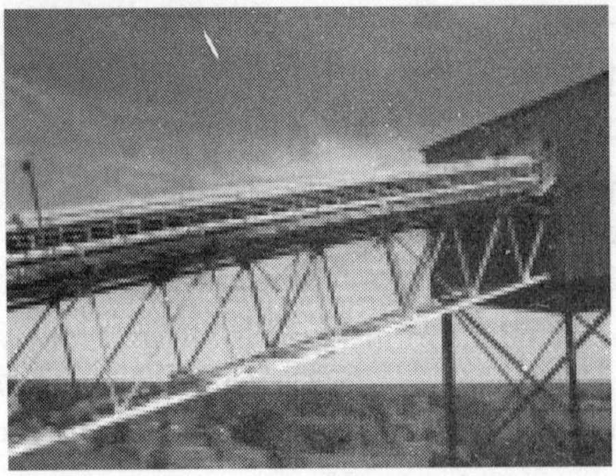

Figure 8.65. *Straight chute in coal industry.*

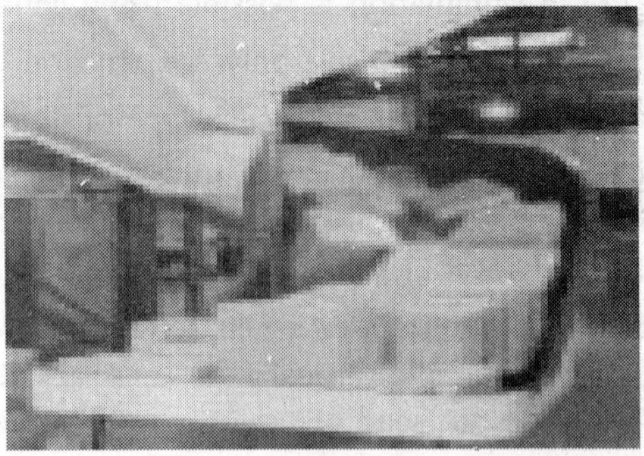

Figure 8.66. *Spiral chute for carton handling.*

8.11 WHEEL CONVEYOR

These are same as that of roller conveyors except in roller conveyors rollers are used but in case of wheel conveyor wheels are used. Wheel pattern depends on the material to be conveyed. Uses a series of skate wheels mounted on a shaft (or axle), where spacing of the wheels is dependent on the load being transported.

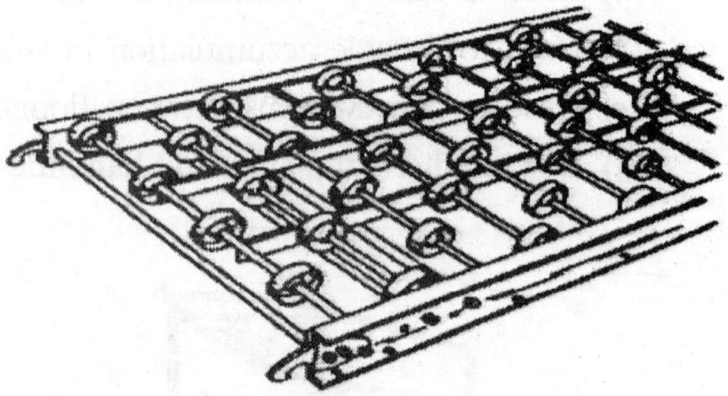

Figure 8.67

These are used on floor unit handling. Wheel conveyors can be set in the staggered pattern and can form the closer configuration. Due to this reason it can be used for smaller and more irregular items. Almost all kind of wheel conveyors are power less *i.e.,* they uses slope under gravity for transportation. Slope for gravity movement depends on load weight which determines travel speed.

For transporting the pallets wheels are mounted in the frames at distance apart equal to or less than width of pallets. When there are bend in the conveying line in case of roller conveyors wheel conveyors often used as a switching device. Accumulation is possible as these are powerless conveyors.

Wheels are made up of plastic, steel or cast iron depending on weight of item for example, for moving light cardboard boxes plastic wheels are used and for moving pallet loads of 400 kg cast iron wheels can be used. Capacity of wheel conveyor range from 600 to 3000 kg/m. wheel conveyors of capacity up to 5000 kg/m is also available.

Recently portable, flexible, expandable versions of wheel conveyors are available in the market as shown in the figure above. These types can adjust their lengths by expanding or contracting its sides and are mounted on wheels. Such type of extendable mobile conveyors is used for loading and unloading purpose.

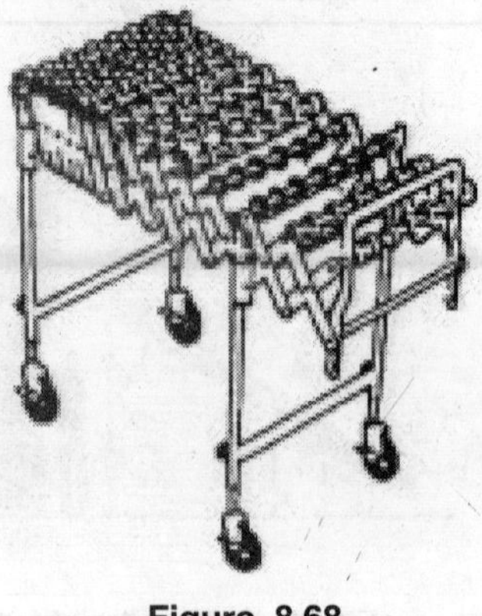

Figure 8.68

Application basically coincides with those described for unpowered roller conveyors except that wheel conveyor are applicable for light weight goods only. These are more economical than the roller conveyor.

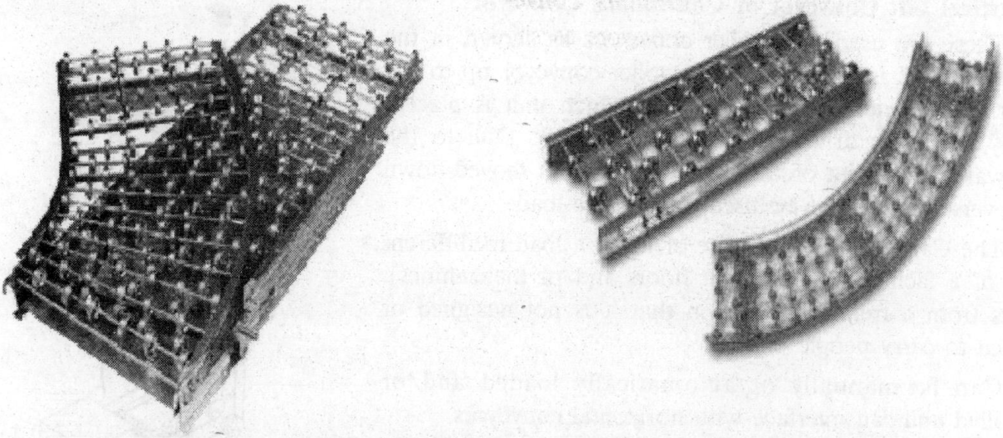

Figure 8.69. *Different types of wheel conveyor design and staggering of wheels.*

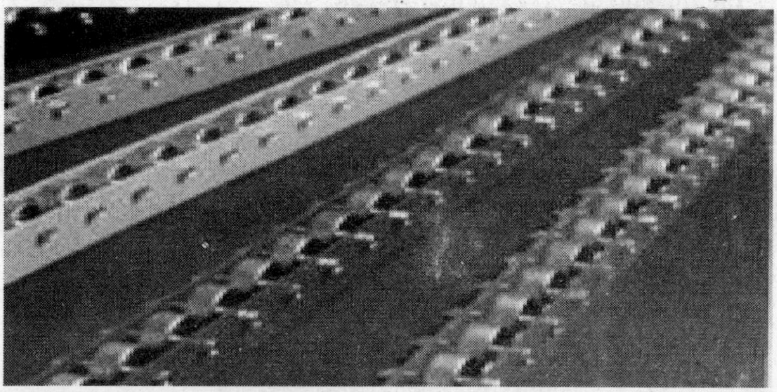

Figure 8.70. *Wheel conveyor mounted in the frames used to transport load on pallets.*

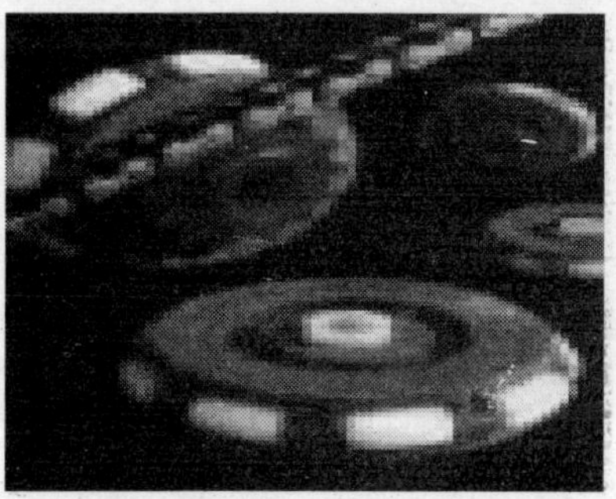

Figure 8.71. *Plastic wheels.*

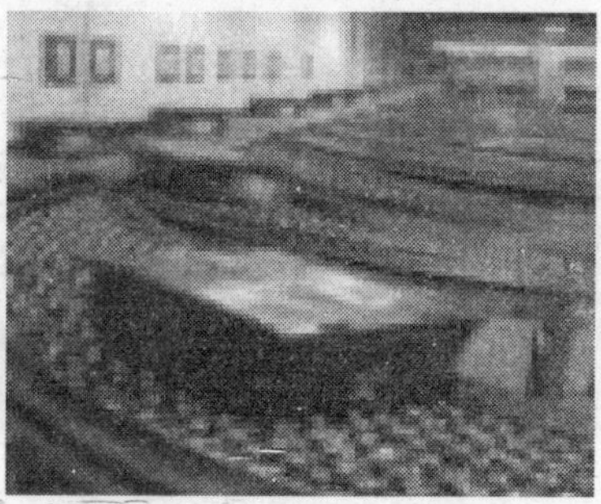

Figure 8.72. *Wheel conveyor used for bin transport.*

8.12 VERTICAL CONVEYOR

Vertical conveyors are used to lift or lower loads between various levels in intermittent flow operations and where horizontal space is limited. There are two types of vertical conveyors, reciprocating and continuous. Accumulation is not possible in this type of conveyors.

(a) Vertical Lift Conveyor or Continuous Conveyor

These are used with roller conveyors as shown in the figure. Material is carried over the roller conveyor up to the carrier. The carrier has platform over which unit is placed, usually carrier is driven by chain and motor. Due to the downward movement of the carrier the load is moved down and reverse process can be used to raise the load.

The Carrier used is to raise or lower a load to different levels of a facility (*e.g.,* different floors and/or mezzanines). Differs from a freight elevator in that it is not designed or certified to carry people.

Can be manually or automatically loaded and/or controlled and can interface with horizontal conveyors.

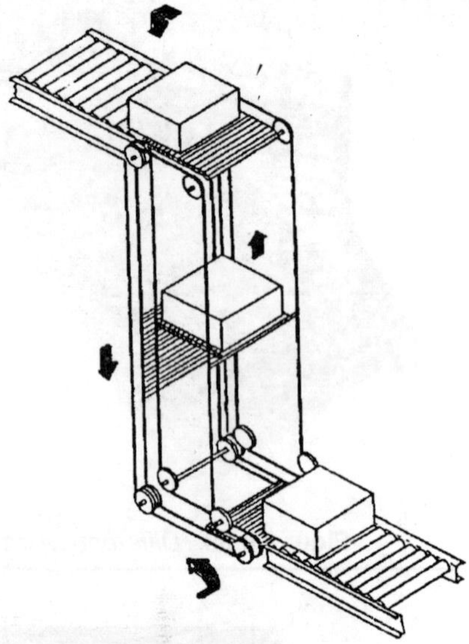

Figure 8.73

(b) Reciprocating Vertical Conveyor

Utilizes gravity-actuated carrier to lowering loads, where the load overcomes the magnitude of a counter weight. The difference between above and this conveyor is that former uses continuous carrier while this uses reciprocating carrier which reciprocates in the frame as shown in the figure.

It is used only to lower a load. Some times this is use as alternative to a chute conveyor for vertical 'drops" when load is fragile and/or space is limited. Can be manually or automatically loaded and/or controlled and can interface with horizontal conveyors.

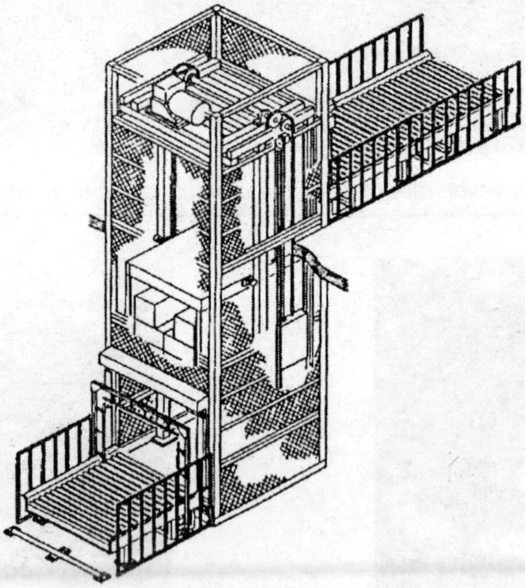

Figure 8.74

8.13 INDEXING CONVEYOR

Indexing conveyors are used for the positioning, orienting the material for the specific purpose or in the specific direction. The purpose may be robotic handling, automated assembly, vision inspection, laser imprinting, and packaging systems. These types of conveyors were specially designed for specific purpose or for specific products. These are generally pneumatically or numerically controlled.

Indexing conveyor provides positioning accuracy of ±0.025 mm. tolerance in four axes etc.

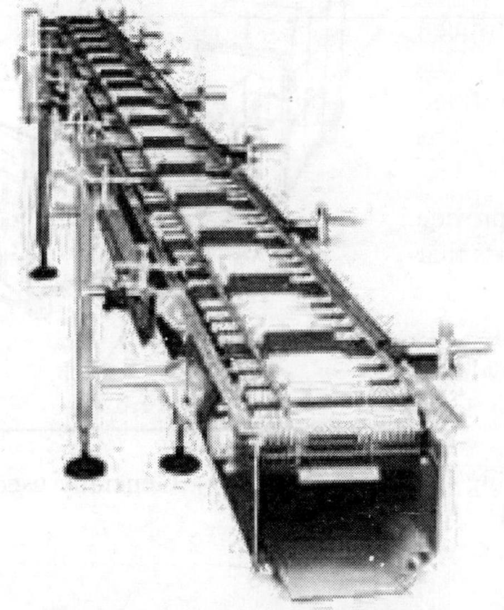

Figure 8.75. *For unit handling.*

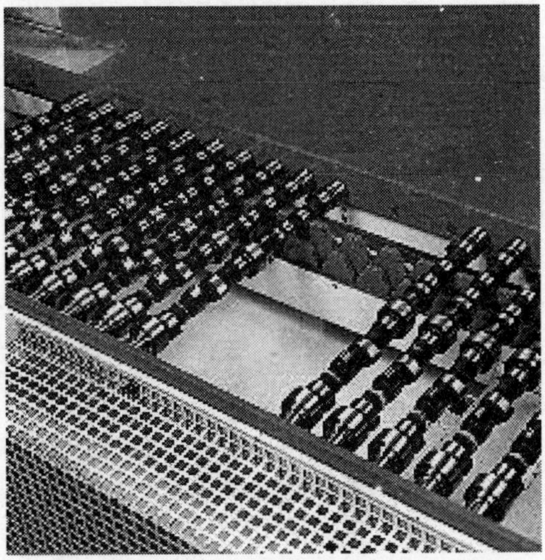

Figure 8.76. *For camshaft handling.*

8.14 CARTS-ON-TRACK CONVEYOR

This conveyor is used to pull a cart along a track. The basic components include a rotating tube, track, and cart. A rotating tube, similar to a screw, transports a cart. A drive wheel is connected to the cart that rests on the tube. Varying the angle of contact between the drive wheel

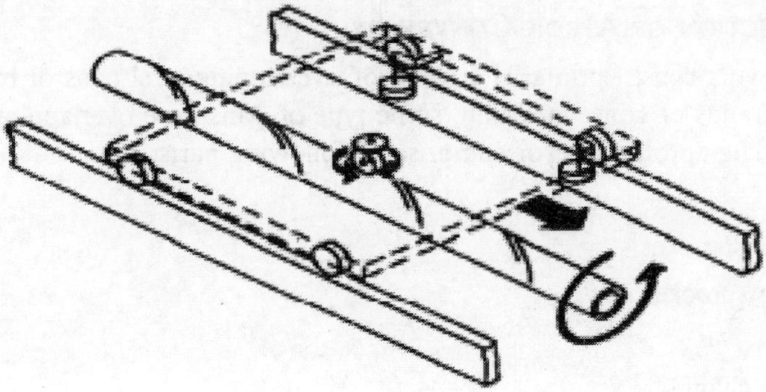

Figure 8.77

and the tube can control the speed. Multiple carts can be located on the tube because they are independently controlled. These carts can accumulate on the tube because they will be stationary when the drive wheel is parallel to the tube.

Used to transport carts along a track. Carts are transported by a rotating tube connected to each cart is a drive wheel that rests on the tube and that is used to vary the speed of the cart (by varying the angle of contact between the drive wheel and the tube). Carts are independently controlled. Accumulation can be achieved by maintaining the drive wheel parallel to the tube.

8.15 TOW-LINE CONVEYOR

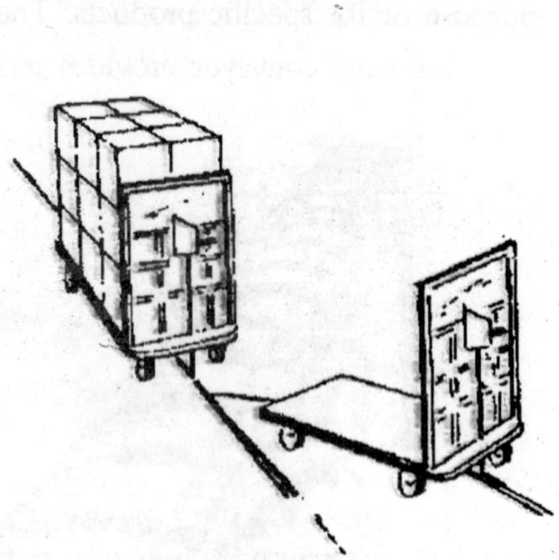

Figure 8.78

Tow-line-conveyors are used to move carts, truck, etc. as conveyors. The carts are attached to the conveyor using a tow-line and they are pulled along the floor by a motor. This tow-line can be located above, in the floor, or on the floor. This conveyor is good if there is a fixed path that your carts move on.

They are used for unit handling. Uses tow-line to provide power to wheeled carriers such as trucks, dollies or carts that move along the floor.

Used for fixed-path travel of carriers (each of which has variable path capabilities when disengaged from the tow-line). Tow-line can be located either overhead, flush with the floor, or in the floor. Selector-pin or pusher-dog arrangements can be used to allow automatic switching (power or spur lines). Generally used when long distance and high frequency moves are required.

8.16 APRON PAN CONVEYORS

8.16.1 Applications of Apron Conveyors

Apron conveyors are used extensively in modern mining, manufacturing and processing industries. Heavy duty apron feeders are used directly under dump hoppers to feed run of the mine over to crusher and have reinforced pans to take the shock of falling lumps.

Apron conveyors can be either horizontal or inclined. Apron conveyors usually operate at slow speeds without any appreciable vibration and as such cause minimum lump breakage.

8.16.2 Construction of Apron Conveyors

Apron conveyors consist primarily a series of overlapping steel pans or buckets carried on two or more endless strands of conveyor chain, some type of pans have overlapping side plates added to confine the load. The apron conveyor comprise of following parts:

(*a*) Frame

(*b*) Driving sprocket

(*c*) Takeup sprocket

(*d*) Apron

(*e*) Pulling chain

(*f*) Drive unit

(*g*) Feed hoppers

(*h*) Discharge output.

Frame

Frame is usually welded of angle iron or channel bar. The terminal sections are separate frames accommodating the drive and take up units. The intermediate sections, supporting the moving parts of the conveyor consists of four to six meter long metal structures. The guide ways for the supporting rollers are made of angle iron and of channel or narrow gauge rail.

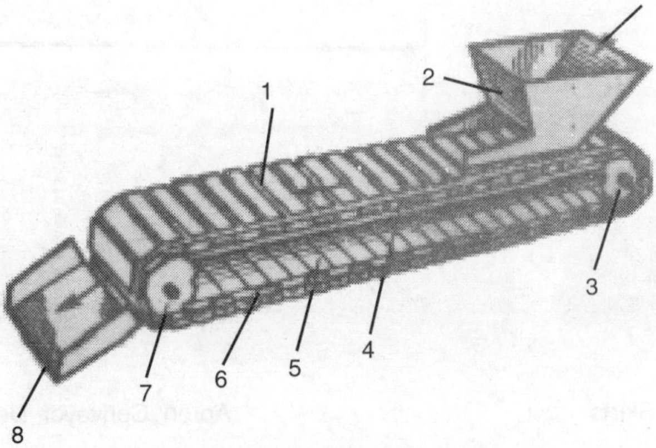

Figure 8.79. *Apron conveyor.*

1—metal plate; 2—hopper; 3—straining sprocket; 4—idle rollers; 5—fixed guides; 6—chain; 7—drive sprocket; 8—chute.

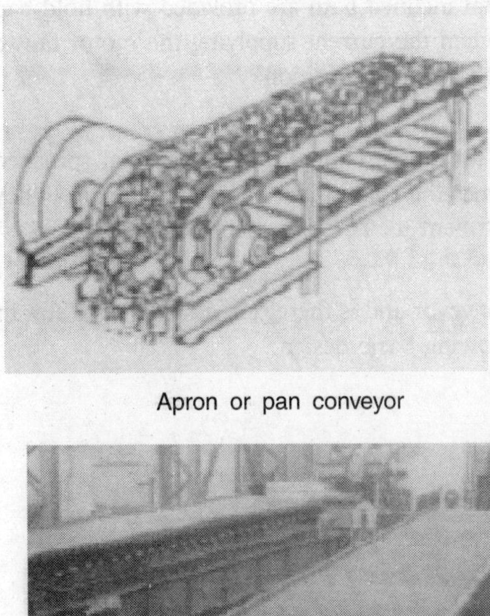

Apron or pan conveyor

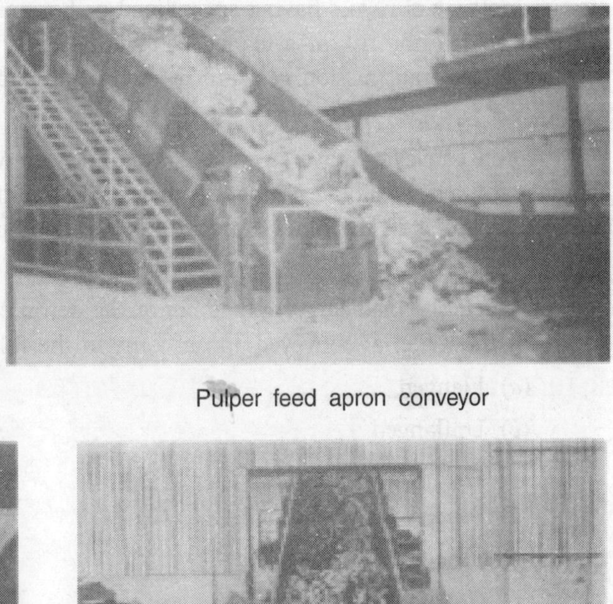

Pulper feed apron conveyor

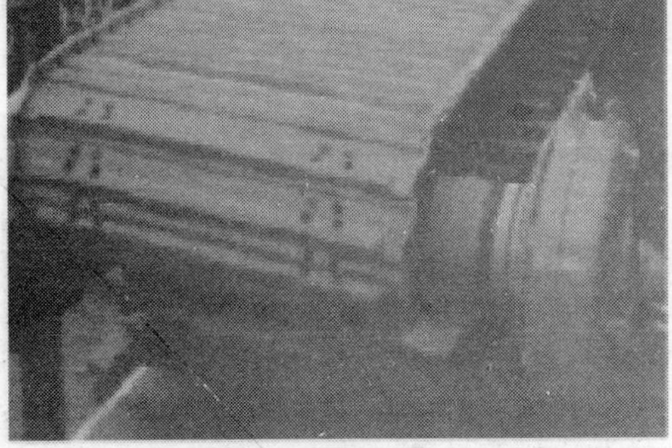

Steel Apron Conveyor

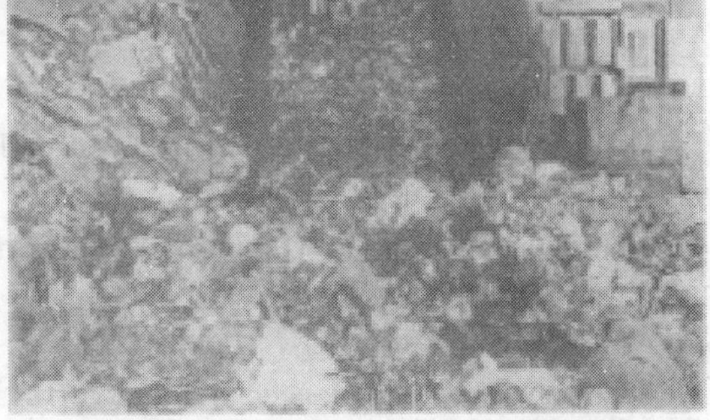

Apron Conveyor for Garbage Handling

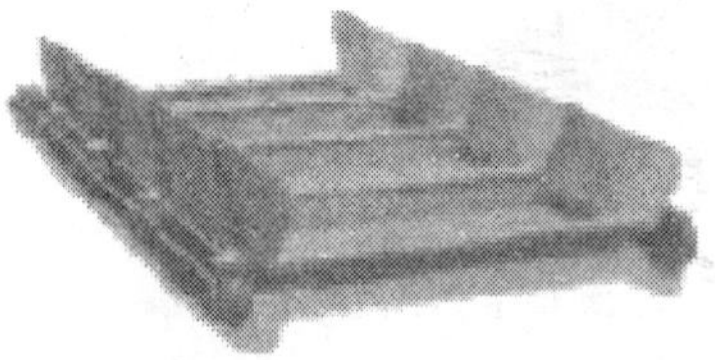

Apron Conveyor Skirts

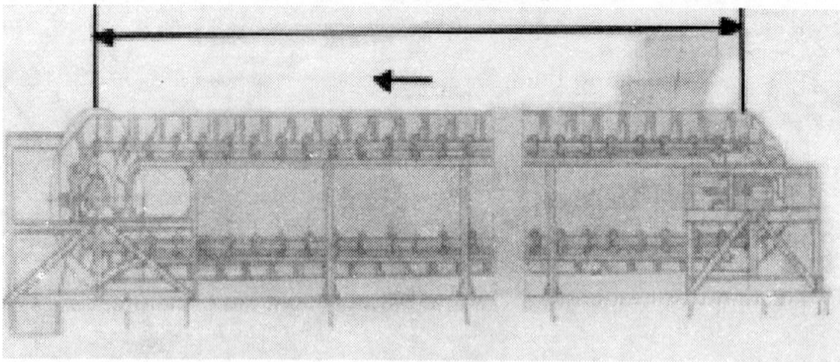

Apron Conveyor Details

Figure 8.80.

Drive Unit

The driving unit of an apron conveyor consists of the driving sprockets, transmission gear and electric motor. Conveyors having an inclined or horizontal inclined path are provided with hold back brake to obviate the risk of a loaded belt sliding back when the current supplying the motor fails or a defect in the transmission gear fail to retain the loaded apron in position.

Takeup

Apron conveyors have screw or spring and screw takeups mounted on the terminal sprockets. The takeup travel is should not be less than 1.6 to 2.0 times the chain pitch one of the takeup pulleys is keyed to the shaft the other is not to permit its adjustment to the chain joint position.

Apron

Apron is the load carrying member of the apron conveyor and is therefore, designated to suit the nature of the material conveyed apron come in the following basic design:

(*a*) Flanged
(*b*) Unflanged
(*c*) Flat
(*d*) Spaced with rollers
(*e*) Spaced without rollers
(*f*) Corrugated
(*g*) Deep
(*h*) Box type
(*i*) Shallow.

Aprons of the corrugated (flanged and unflanged) flat flanged, deep and box types are stamped of four to ten millimetre sheet steel.

Unflanged aprons are commonly specified for unit loads. Toprevent the load from falling off an unflanged apron. Stationary wood or steel strip skirt boards often fringe the apron especially in the loading zone.

Conveyors with unflanged flat and corrugated aprons are often provided with stationary skirt boards if they are intended for bulk materials those with flanged flat and corrugated aprons are intended for bulk loads and piece goods. While flanged deep and box type aprons are specified solely for bulk handling at steep incline 45" to 60" and high capacities.

The plates forming the apron are bolted, riveted or welded to special angles, secured to the link-plates of the pulling chains.

Pulling Member

Apron conveyors are usually equipped with two strands to link plate bush roller chains. The travelling rollers of the extensively used bush roller chain support the apron and load along the guide ways of the conveyor. The roller may be flanged or unflanged may run in sliding bearings or roller bearings the latter are mainly used in heavy duty conveyors. Stationary rollers (idlers) secured on the conveyor frame support the apron of the conveyors employing bush, bush roller and detachable chains, on the loaded run the chain link edges slide on the stationary rollers while on the return the apron is supported like the belt conveyor.

8.16.3 Uses of Apron Conveyor

Apron conveyors are used in conveying hot, sharped lump and piece of materials.

They are widely used in chemical industry enterprises, metallurgy, coal mining, machine building and other branches of industry. They also serve to convey article pastwork stations in line production.

8.16.4 Speed of Conveyor

The speed of the conveyor varies with the class of service and the type of pans. For feeders drawing material from bins or hoppers the speed is usually kept under 10 mts/minutes. Longer conveyors receiving feed loads may be operated at higher speeds of 18 to 30 mt/min. or more depending on the chain, pitch and number of teeth on the sprocket. Conveyors for picking tables and sorting services are usually operated at 10 to 20 mts/min.

8.16.5 Apron Pan Thickness

As per Indian standard specification the recommended metal thickness of apron pans for various conditions are as follows:

(*a*) For Handling light weight metals. That are only mildy abrassive and where corrosion is not a factor to be considered. — 5 mm

(*b*) For handling medium weight materials with probable corrosion and abrasion considerations also where moderate impact is to be encountered. — 6 mm or 8 mm

(*c*) For Handling heavy weight materials which are highly abrasive and corrosive. Also where impact is severe. — 10 mm or 12 mm

8.16.6 Capacities of Apron Conveyor

Apron conveyors are in operation handling upto 3000 tonnes/hr. and there is no practical difficulty in the way of even larger capacities maximum capacities are obtained by using skirt boards, 300 mm to 1200 mm in height to increase the depth of the material in the pan. However these are not suitable for long conveyors.

The volumetric capacity of an apron conveyor depends on width, depth and speed of the moving stream of material and in some cases on the slope and the type of pan the width and depth of material will be governed by the maximum size of lumps to be conveyed.

Lump size (mm)		Width between skirt boards (mn)
Average	Maximum	
75	150	450
100	200	600
150	300	750
200	400	900
250	500	1000
300	600	1200
350	700	1400
400	800	1600
450	900	1800

8.16.7 Advantages of the Apron Conveyors

*1. They are able to handle heavy, large, lumped and hot materials.
2. They have capacity of 3000 tonnes/hr. and over.
3. They are furnished with strong pulling chains, allowing for long distances between the centres.
4. They run smoothly and noiselessly.
5. They may be feed from overhead hoppers.
6. They follow a wide range of complex paths. Transition from one direction to other and smaller curve radii than belt conveyors.

8.16.8 Disadvantages of the Apron Conveyors

1. The considerable own weight of the apron and chains is the disadvantage of apron conveyor.
2. The continuous maintained due to a great number of hinged joints is also a disadvantage for apron conveyor.

8.17 SCRAPER OR FLIGHTS CONVEYORS

Flight conveyors are used to handle various powered, fine, granular and free flowing materials, moving them through a trough by a chain with flights. Elight conveyors differ from apron ones in that the chains are placed in an open trough and instead of plates they have flights.

The length of the conveyor is the total length from the axis of the head and sprocket to the axis of the tail and sprocket with all adjustable sections adjusted to their shortest length measured along the path of the conveyor.

8.17.1 Types of Flight Conveyor

1. Single Strand Conveyors with Scraper Flights

These conveyors consists of malleable iron flights attacked to a single strand of chain and are designed for sliding directly on a steel trough. They are suitable for narrow flights carrying free flowing materials with lumps not larger than 100 mm in one direction only as per figure 1.

2. Single Strand Conveyors with Shoe Suspended Flights

These types of conveyors are similar to above except that mass of flights and chain is carried on malleable iron sliding shoes to reduce wear between flights and trough sliding shoe are attacked to the sides of the flights near the upper edge and slide on flat renewable steel bars. As in figure 2.

3. Double Strand Conveyors with Plain or Roller Chain Suspended Flights

These conveyors consist of steel flights attacked to double strands of rivetless or roller type chain. The chain operate on front renewable steel bars attached to the trough. These conveyors are used for larger capacities and longer paths than the single strand flight conveyors and are suitable for wider flights carrying material containing larger lumps upto 400 mm in size as per figure 3.

4. Double Strand Conveyor in Plain or Roller Chain Suspended Double Depth Style Flights

These conveyors are more or less similar type except that they are designed to carry material containing large lumps on both upper and lower runs as in figure 4.

Scraper conveyors are simple conveying equipment for carrying non-abrasive materials through smooth steel or malleable cast iron troughs. The flights will carry up steeper slopes than belt or apron conveyors. The load can be fed into and discharge from troughs at any point comparatively long conveyor units are practicable. The length is limited only by the working strength of the chain selected to full the flights.

Conveyors will operate satisfactorly at inclination upto 40"–45" the actual capacity decreases as the inclination increases above 20".

8.17.2 Parts of Flight Conveyor

Ordinary solid flight conveyors consist essentially of:

(*a*) Open trough

(*b*) Frame work

(*c*) Putting member

(*d*) Flight

(*e*) Drive unit

(*f*) Take up service

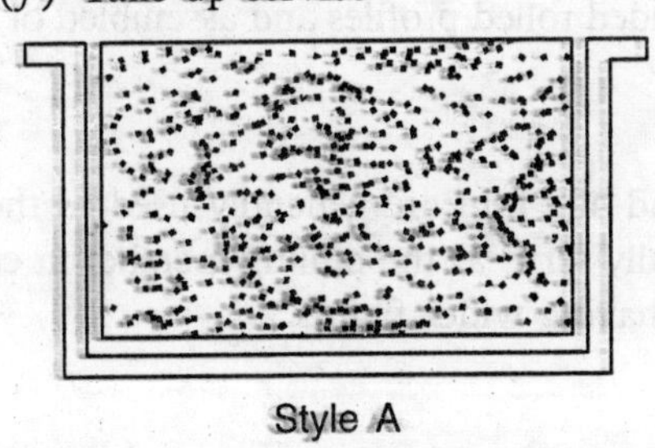

Style A

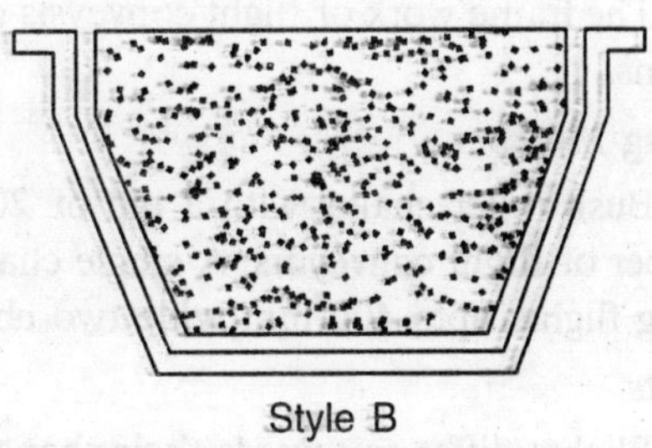

Style B

Steel Flight

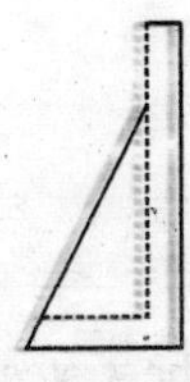

Style B2 Malleable

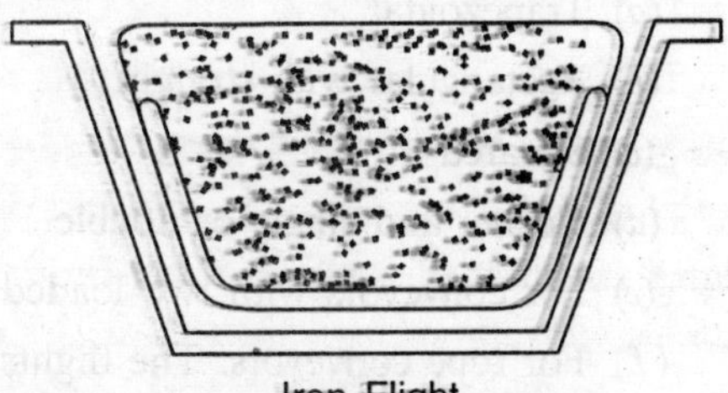

Iron Flight

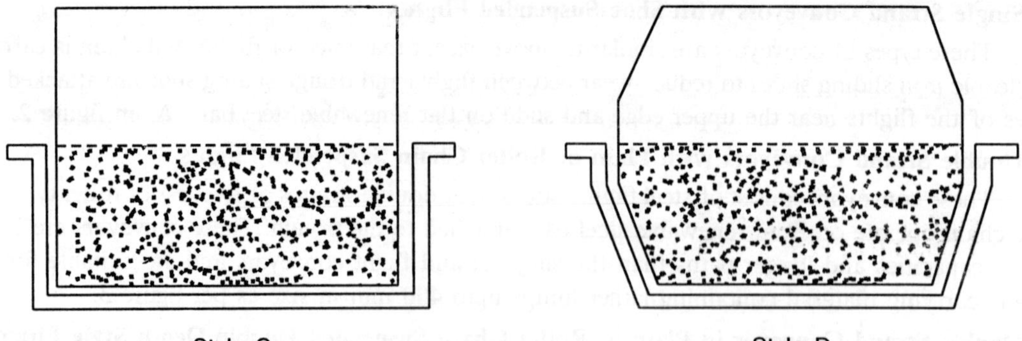

Figure 8.81. *Styles of Flights.*

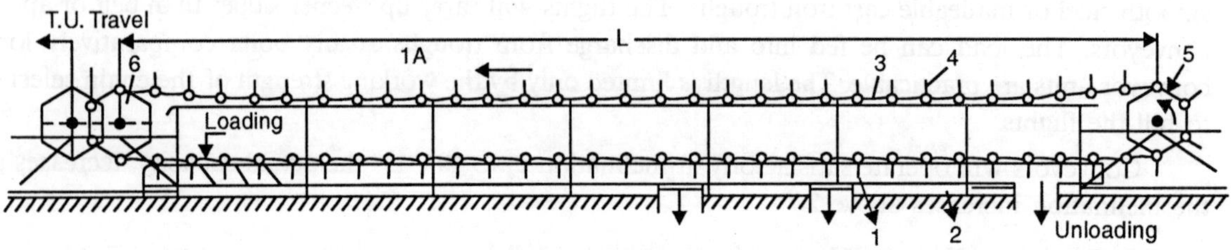

Figure 8.82.

The Trough

The trough is a welded structure of 4 to 5 mm sheet steel of rectangular or trapezoidal section or of rolled profiles trough and may be stamped of steel sheet. In conveyors designed to handle light weight materials such as sawdust, grain etc. The trough may be of wood. The clearance between the flight and trough should be 3 to 6 mm. The trough is assembled of 4 to 6 meter long sections.

The Frame Work

The frame work of flight conveysis generally of welded rolled profiles and assembled of separate sections.

Pulling Member

Bush roller chains with a pin of 200, 250, 320 and 400 mm are generally used as the pulling member of flight conveyors. A single chain or rope usually serve as the pulling member in conveyors having flights upto 400 mm wide two chains in, those having wider flights.

Flights

Flights differ as regards their shape and can be:

(*a*) Trapezoidal

(*b*) Rectangular with straight

(*c*) Sheared

(*d*) Corner and single or double

(*e*) For conveyors with two loaded strands and disk type.

(*f*) For rope conveyors. The flights are made of 3 to 8 mm sheet steel. Flights of A, B_1 and B_2 are suitable for carrying material only in one direction. Flights of style C and D are double depth and are used to carry material on width upper and lower runs an opposite direction. The flight of styles A, B_1 and B_2 can also be mounted on sliding shoes.

Recommended Sizes of Flights

Styles A and B_1 steel plain	*Styles A and B_1 steel with sliding shoes*	*Style B_2 malleable iron*	*Style C steel*	*Style D steel*
300 × 100	300 × 100	300 × 100	300 × 200	300 × 200
300 × 100	300 × 150	300 × 150	300 × 300	300 × 300
375 × 150	375 × 150	375 × 150	300 × 300	375 × 300
450 × 150	450 × 150	—	450 × 300	450 × 300
375 × 200	375 × 200	—	375 × 400	375 × 400
450 × 200	450 × 200	—	450 × 400	450 × 400
600 × 200	—	—	600 × 400	600 × 400
600 × 250	—	—	600 × 500	600 × 500
750 × 250	—	—	750 × 500	750 × 500
—	—	—	—	1000 × 300
—	—	—	—	1300 × 350

Drive Unit

The drive unit is of the usual type commonly supplied with a reducing gear. It has been found expedient to introduce a slip in the transmission gear of the drive of medium and heavy duty conveyors to protect the gear against breakage in case of accidental overwards.

Take up Device

A take up device is provided for all scraper conveyors. When ever practical take ups are located at the foot or loading end of the conveyors. Show take ups are ordinarily used but where tension in the chain is required spring or grarity take up are recommended. In flight or scraper conveyors the take up in addition to its usual purpose serves to ensure stability of the flights by increasing initial tension of the chain.

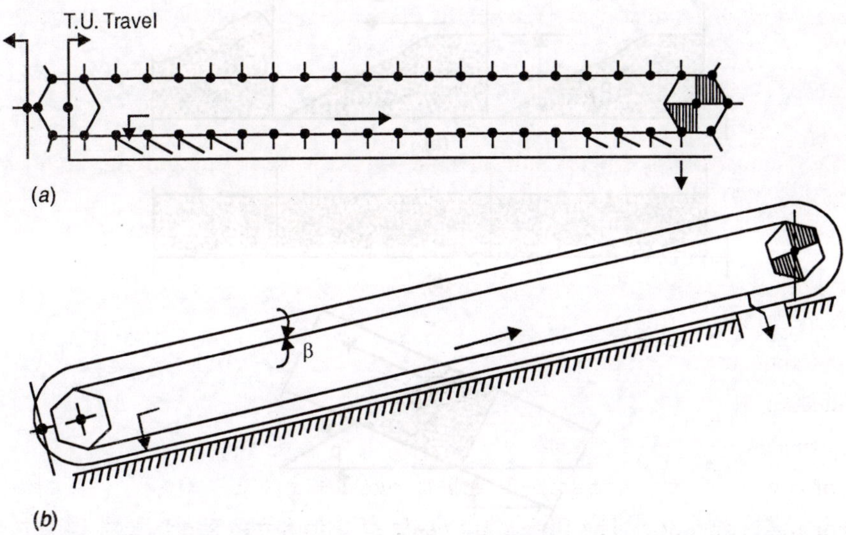

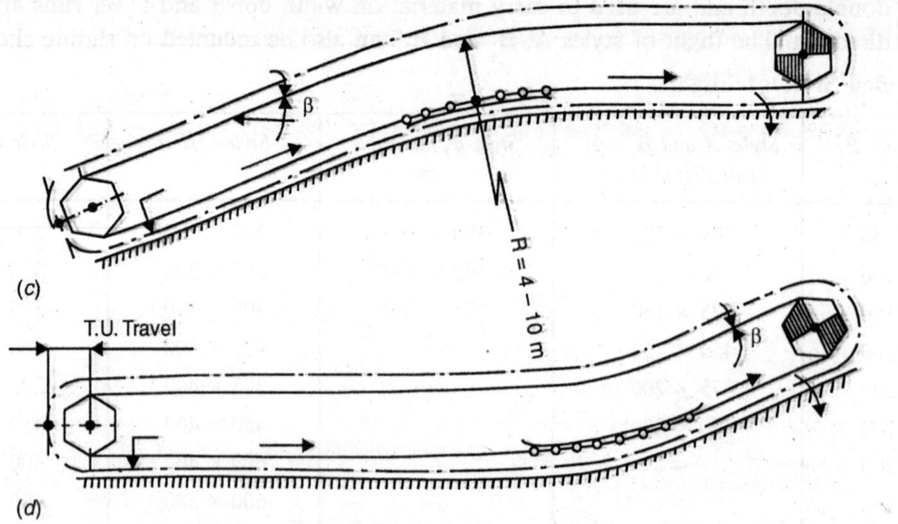

Figure 8.83. *Typical layouts of ordinary flight conveyors.*

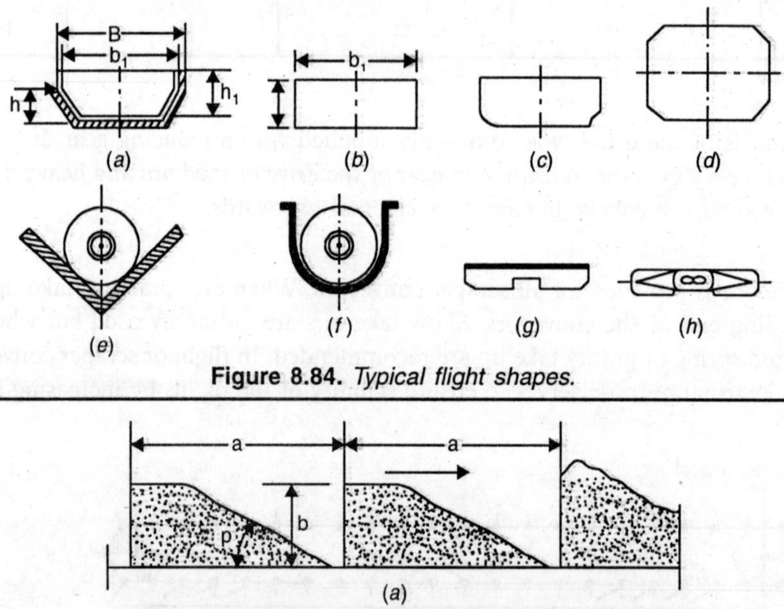

Figure 8.84. *Typical flight shapes.*

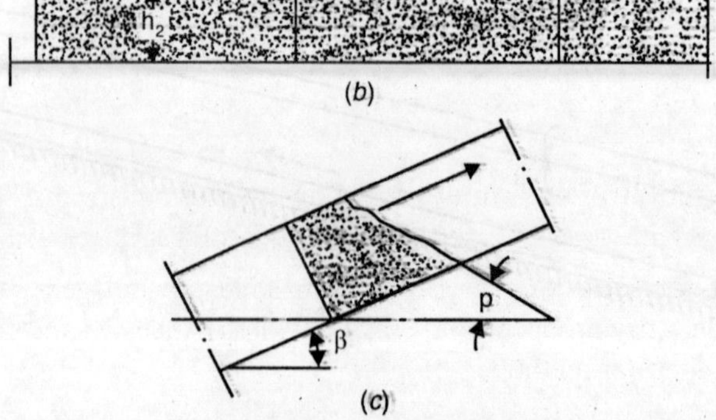

Figure 8.85. *Layout of free flowing material in front of the flights.*

Inclination

Conveyor will operate satisfactorily up inclined upto 40"–50", the actual capacity decreases as the inclination increases above 20".

8.17.3 Capacity of Flight Conveyor

The main dimension required for the calculation of flight conveyors is the cross-sectional area of the through which is determined by the required capacity 'Q' Tonnes per hour with loading efficiency.

$Q = 3600\ h^2\ KC_3\ y'\ V$ tonnes per hr.

Where,

Q = Capacity of the flight *t* conveyor capacity

C_2 = Correction factor to later for inclination as the heap in out of flight will decrease with an increase of the

y' = Through filling factor 0.5 to 0.6 for heavy lump national

B = Width 2.4 to 4

v = 0.1 to 0.5 m/sec

K = B/H

8.17.4 Advantages of Flight Conveyor

The main advantages of flight conveyors may be summarized as follows:

(*a*) Simplicity of design and ruggedness of construction.

(*b*) Ability – to convey loads in two direction on the upper and lower strands.

(*c*) Easy loading and unloading at any point along the conveying run.

8.17.5 The Main Disadvantages of Flight Conveyor

(*a*) The main Disadvantages of flight conveyor are crushing and breaking of material during transportation which for certain materials.

(*b*) Rapid wear of the trough and moving parts especially when handling abrasive material.

(*c*) Considerable power consumption.

8.17.6 Uses of Flight Conveyor

Flight conveyors have found extensive application in local mining, when they are employed for coal handling in concentration plants, chemical industry enterprises etc. They are also used for transportation of hot ash and slag special cable disk conveyors are employed for handling piece goods delivered in mass such as wood, plywood etc.

8.18 PNEUMATIC CONVEYING DRIER

8.18.1 Drying and its Necessity

Drying means the removal of moisture from solid by thermal means. The necessary heat of vaporization is provided from a hot gas by convection, from a solid surface by conduction or radiation.

Mechanical methods for separating moisture from solid are not generally considered drying, although they often precede a drying operation, since, it is less expensive and frequently easier to use mechanical methods than thermal methods.

The purpose of drying are as follows:

(1) To facilitate the handling in further processing.

(2) To permit satisfactory utilization of final product.

(3) To reduce the handling cost.

(4) To increase the capacity of other equipment in processing.

(5) To preserve a product during storage and shipment.

(6) To enhance the value and usefulness of water or by-product.

8.18.2 Description of Pneumatic Conveying Drier

In the pneumatic conveying drier the moisture removal is carried out by dispersing the material to be dried in a hot zone followed by conveying at high velocity.

A pneumatic conveying drier consists of a long tube or duct carrying a gas at high velocities, a fan to propel the gas, a suitable feeder for addition and dispersion of particulate solids in the gas stream and a cyclone collector or other separation equipment for final recovery of solids from the gas.

The solids feeder may be of any type:

Screw feeder, Venture section, high-speed grinders and dispersion mills are employed.

Fans may be induced draft or forced draft type. The former is preferred usually because the system can then be operated under a slight negative pressure. Dust and hot gas will not be blown out of the system through leaks in the equipment.

Cyclone separators are preferred for low investment and ease of cleaning; they are employed usually as primary collectors at least. If maximum recovery of dust or toxic fumes is required, the cyclone may be followed by a wet scrubber or bag collector.

Conveying air is heated by steam, electricity, direct combustion of fuels, or by an indirect – fried air heater.

Pneumatic conveying driers may be single-stage or multistage. The farmer is used for ordinary heating and cooling operations and drying requiring evaporation of small quantities of surface-moisture. Multi-stage installations are used for difficult drying process, *e.g.,* drying heat-sensitive products containing large quantities of moisture and drying materials initially containing internal as well as surface-moisture.

8.18.3 Field of Application of Pneumatic Conveying Drier

Pneumatic conveying driers are suitable for all materials, which can be suitably dispersed in a gas stream and conveyed. The material to be dried remains in contact with hot air for very short time, hence, the temperature sensitive materials can often be dried satisfactorily at relatively high temperature.

This type of driers are suitable for large handling capacities and are usually the cheapest type of equipment for evaporation capacities above 2000 lb of water per hour.

In his type of driers the materials are conveyed in an enclosed pipe, hence these are very useful for chemical products (which are bad smelling, toxic, poisonous, etc.). The flexibility of conveying horizontally, vertically, around bends and risers make it superior to the other driers. They needless number of operators and have less initial cost.

With all these advantages, there are of course, some limitations of the pneumatic conveying driers. They are not suitable for very abrassive materials, or materials which are not broken up or for materials which tend to stick to the wall of ducts. The power consumption is also high, of the order of 4 kWhr per kg. of moisture removed.

We see that the advantages are various limitations only a few. Further the following data comes in support of the wide field and wide range of application of the pneumatic conveying driers.

(*i*) Operating temperature ranges from 300° to 1300°F. The lower temperatures are employed with some sacrifice in heat-economy.

(*ii*) Moisture content: Initial moisture content may range from 3 to 90% (dry basis) or 3 to 90% (wet basis).

(*iii*) The ratio of solid to conveying gas normally ranges from 0.05 to 1.0 kg solid/kg gas, although slightly higher loadings are possible in carefully designed system.

(*iv*) Gas velocity—in actual practice should be about 100 percent greater than the free-falling velocity of the largest particle and may range from 15 to 60 m/sec.

8.19 EN-MASSE CONVEYOR

8.19.1 Introduction

En-masse or continuous flow conveyors with submerged profiled flights differ from general purpose conveyors of solid flights, in that they have a closed trough, partitioned into two sections and that the conveyed material is not supported and conveyed in separate amounts by each flight but moves as a continuous core filling the whole working section of the trough or its larger part. The basic principle is that the friction between the particles on the material must be greater than the friction between the material and the trough surface.

Together with the ability to convey along the most difficult paths, its small space requirements make it the ideal choice for complex installations situated in areas of concentrated machinery and equipment. En-masse conveyor requires less than one-third the space required for a screw conveyor, less than one-fourth the space required for a belt conveyor and less than one-seventh the space required for a bucket elevator.

The conveying element is a series of U-type skeleton links buried in the solid column of material within the conveying run of the En-masse conveyor. These links of flights consist of sturdy, detachable drawbars with skeleton prongs. The joints allow them to move readily around sprockets and corner bends.

Very dry and free-flowing materials like grains, pulverised, flaky, granular and other small lumpy materials can be conveyed and elevated very satisfactorily. The main draw-back of this type of conveyor is that very fine powder cannot be conveyed.

8.19.2 Construction of En-Masse Conveyor

The conveyor consists of two sprockets as head and tail ends. The chain to which rectangular flights are fixed at alternate links. The pulling element is the chain and the carrying elements are the flights. The top and bottom runs are enclosed in open trough as shown in figure 8.86. The advantage of this arrangement is that the material can be conveyed both in the carrying and return runs. The material can be discharged at any point.

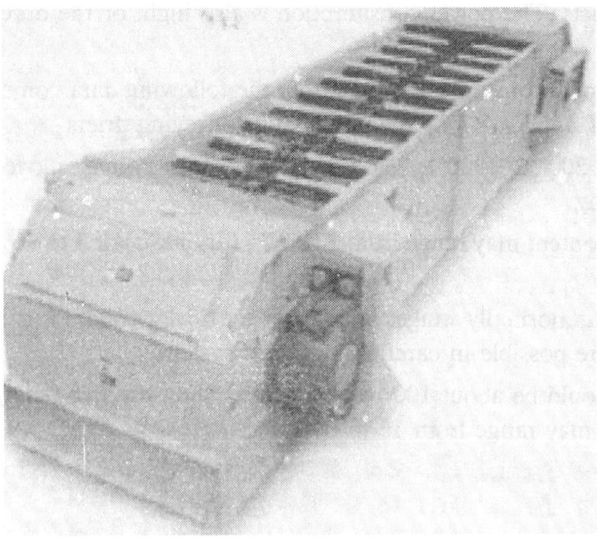

En-Masse Conveyor

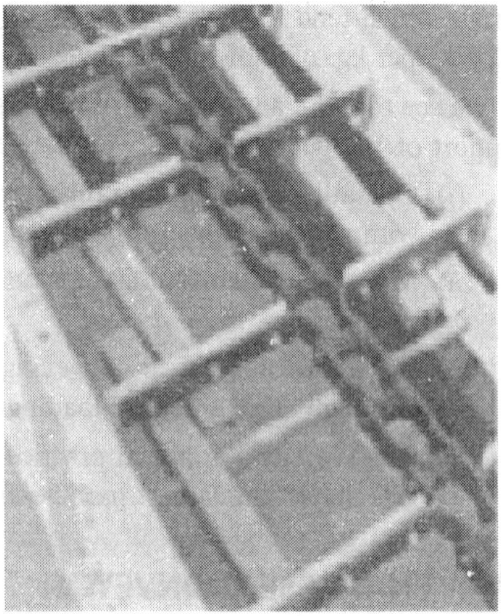

Inside Portion

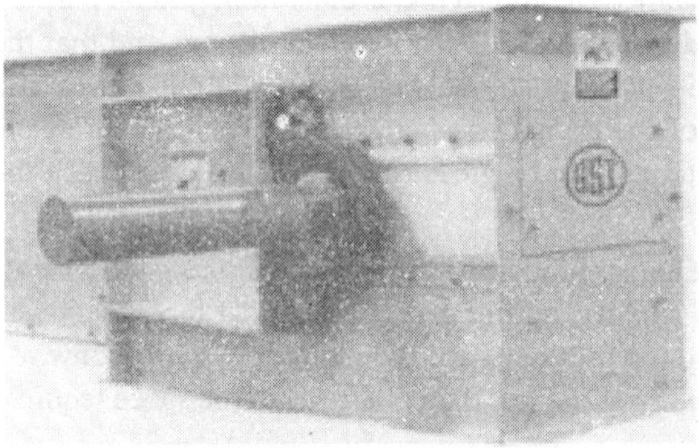

Head End of En-Masse Conveyor

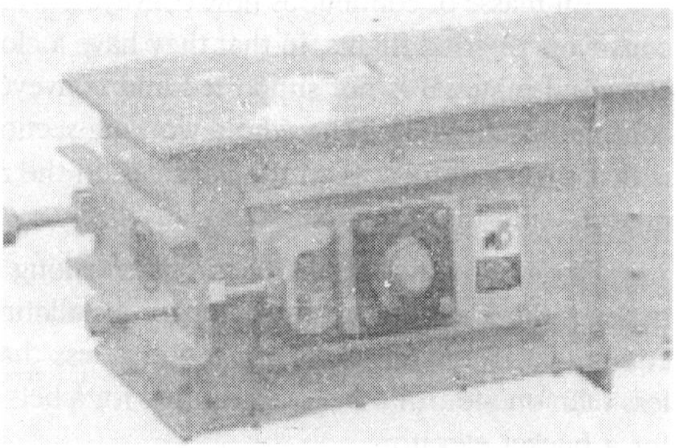

Tail End of En-Masse Conveyor

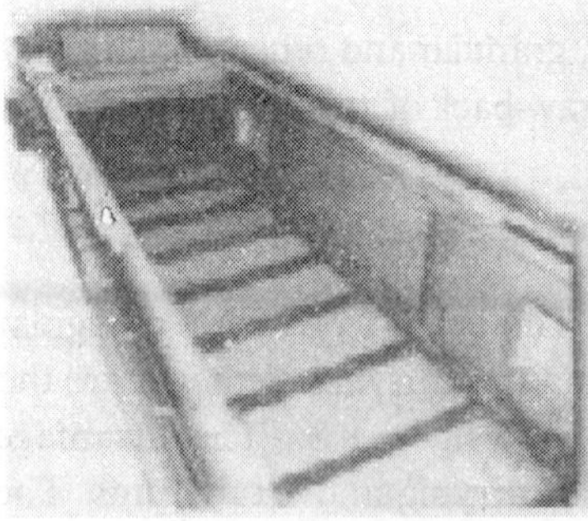

Inside View

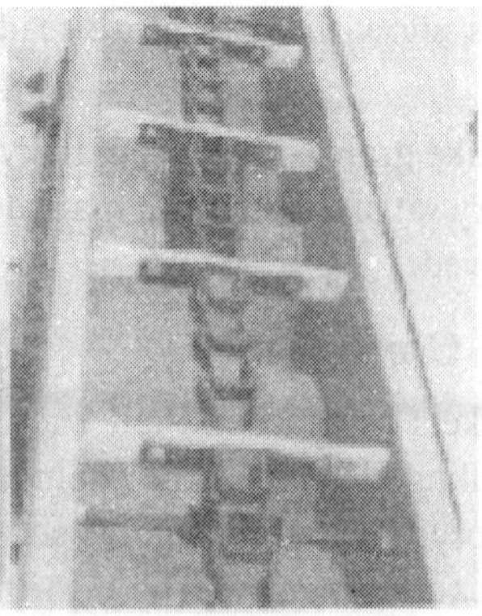

As Saw Dust Conveyor

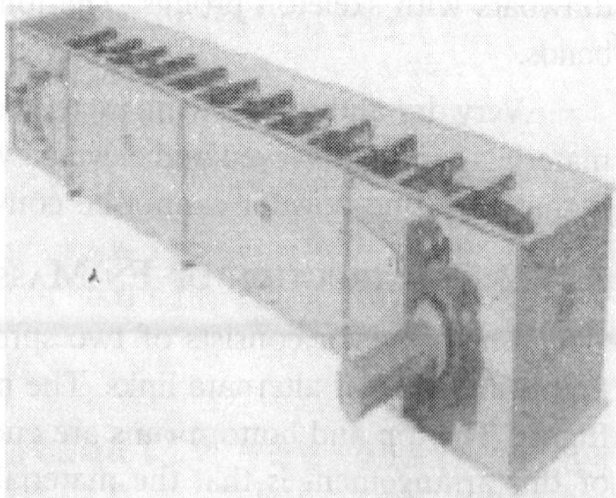

Trough Design

Figure 8.86.

The feed is effected by gravity through a mass flow hopper. The end of the hopper bottom is provided with a box for inserting orifice plates of various diameters, so that the feed rate of the material into the conveyor is varied. The speed of the conveyor is varied by changing the pulley system between the motor and the gear box. Transparent windows are provided at the sides of the trough so that the height of the material column can be measured.

8.19.3 Working of En-Masse Conveyor

The hopper is filled with the material and the opening under the hopper is opened and material is fed into the conveyor. Once steady state is reached the material is collected for a known time to find out the actual discharge. At the same time, the level of the material column in the trough is also measured to find out the theoretical material discharge. This repeated for different feed rates and for different conveying speeds. The ratio of the actual discharge to the theoretical discharge gives the volumetric efficiency of the conveyor. The dependence of the volumetric efficiency on the conveying speed for different flow rates is found out by drawing a graph between volumetric efficiency and the linear speed of the conveyor.

8.19.4 Calculations for Discharge

For a certain chain speed the volume flow is adjusted and the mass discharged for a particular time is noted. This is repeated for several chain speeds.

Q_{Act} = Actual material discharge known from weight measurement

Q_{Th} = Theoretical material discharge

$= n \,.\, b \,.\, h \,. \quad . \, K_1 \,.\, K_2 \,.\, K_3$

Where,

n = Peripheral velocity of the chain

b = Width of the trough

h = average height of the material column in the trough

= Specific weight of sand

K_1 = Factor for the lagging of the material behind the chain speed

K_2 = Geometrical factor accounting for the volume occupied by the chain = 0.95

K_3 = Packing factor for the material

The volumetric efficiency is calculated as the ratio of Q_{Act}/Q_{Th} and its variation with chain speed is drawn as in figure 8.87.

8.19.5 Discussion

When the chain is pulled, theoretically the chain should pull along all the sand. Since, the sand is in a powdered state and in a state analogous to a fluid, only a part of the amount is carried along with the flights.

At higher velocities (chain speeds), the relative velocity comes into play and the En-masse conveyor stops functioning after a certain critical velocity.

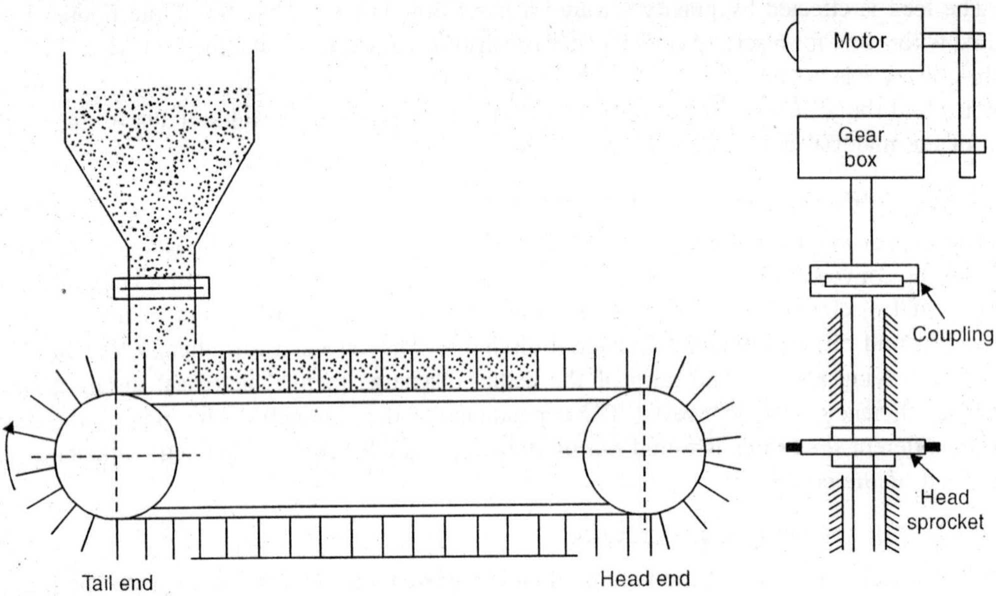

Figure 8.87. *En-Masse conveyor.*

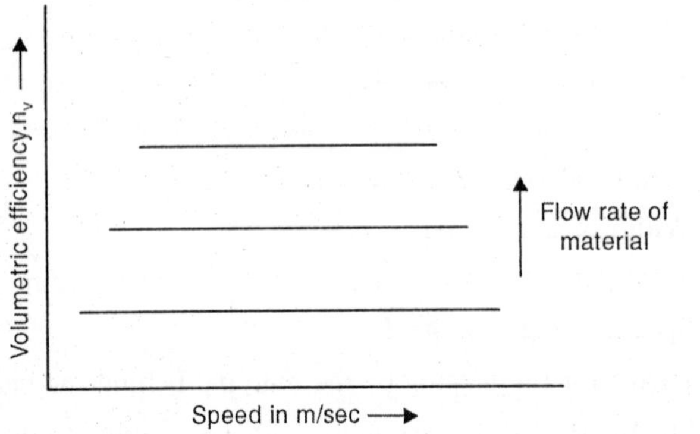

Figure 8.88. *Variation of volumetric efficiency as a function of speed.*

Observations and calculations are recorded in a tabular form as follows:

Table 1 : Observations

Sl. No.	*Orifice size* *mm*	*Chain speed* *mm/s*	*Time* *s*	*Height of the material in the trough, mm*			*Theoretical discharge* Q_{Th} *kg/s*	*Actual discharge* Q_{Act} *kg/s*	*Volumetric efficiency* $\eta = \frac{Q_{Act}}{Q_{Th}}$ %
				max.	*min.*	*mean*			

UNSOLVED QUESTIONS

1. Explain with neat sketch working of belt conveyor.
2. Classify belt conveyors.
3. Explain advantages and limitations of belt conveyor.
4. Explain the function of important parts of belt conveyor.
5. Write short note on indexing conveyors.
6. Write down applications of belt conveyors.
7. Explain with neat sketch working of screw conveyor.
8. Classify screw conveyors.
9. Explain advantages and limitations of screw conveyor.
10. Explain the function of important parts of screw conveyor.
11. Write short note on flights of screw conveyors.
12. Write down applications of screw conveyors.
13. Explain choice of screw conveyors.
14. Explain with neat sketch working of bucket elevator.
15. Classify bucket elevator.
16. Explain advantages and limitations of bucket elevator.
17. Explain the function of important parts of bucket elevator.
18. Write short note on flights of bucket elevator.
19. Write down applications of bucket elevator.
20. Explain choice of bucket elevator.
21. Explain with neat sketch working of roller conveyor.
22. Classify roller conveyor.
23. Explain advantages and limitations of roller conveyor.
24. Explain the function of important parts of roller conveyor.
25. Write short note on types of chains used in roller conveyor.
26. Write down applications of roller conveyor.
27. Explain choice of roller conveyor.
28. Explain with neat sketch working of pneumatic conveyor.
29. Classify pneumatic conveyor.
30. Explain advantages and limitations of pneumatic conveyor.
31. Explain the function of important parts of pneumatic conveyor.
32. Write short note on types of rollers used in pneumatic conveyor.
33. Write down applications of pneumatic conveyor.
34. Explain choice of pneumatic conveyor.
35. Explain with neat sketch working of slat conveyor.
36. Classify slat conveyor.
37. Explain advantages and limitations of slat conveyor.
38. Explain the function of important parts of slat conveyor.
39. Write short note on types of pneumatic used in slat conveyor.
40. Write down applications of slat conveyor.

41. Explain choice of slat conveyor.
42. Explain with neat sketch working of vibrating conveyor.
43. Classify vibrating conveyor.
44. Explain advantages and limitations of vibrating conveyor.
45. Explain the function of important parts of vibrating conveyor.
46. Write short note on factors considered in vibrating conveyor.
47. Write down applications of vibrating conveyor.
48. Explain choice of vibrating conveyor.
49. Explain with neat sketch working of overhead conveyor.
50. Classify overhead conveyor.
51. Explain advantages and limitations of overhead conveyor.
52. Explain the function of important parts of overhead conveyor.
53. Write short note on factors considered in overhead conveyor.
54. Write down applications of overhead conveyor.
55. Explain choice of overhead conveyor.
56. Explain with neat sketch working of chute conveyor.
57. Classify chute conveyor.
58. Explain advantages and limitations of chute conveyor.
59. Explain the function of important parts of wheel conveyor.
60. Write short note on factors considered in wheel conveyor.
61. Write down applications of chute and wheel conveyor.
62. Write short note on vertical conveyor.
63. Give applications of vertical conveyor.
64. Explain with neat sketch indexing conveyors.
65. Explain with neat sketch cart on track conveyors.
66. Explain with neat sketch tow conveyors.

❑❑❑❑

9 Lifts and Elevators

INTRODUCTION

There are so many applications where we need to move the materials or passengers vertically upward in that cases lifts and elevators are used. Lifts/Elevators move goods between different level *e.g.,* between conveyor at different heights or between storey in warehouses and terminals, they can also bridge differences in level which occur between new and old parts of an installation or between loading and ground level. For heavy and/or bulky loads lifts/elevators may be the only realistic solution

9.1 CLASSIFICATION

These are classified in four types :

- **Lifts**

  These are used for intermittent operations, have load carrying cage or platform which moves up and down guided in frames.

- **Paternosters**

  These are the extension of the bucket conveyor (elevator) in which series of load carrying elements were attached to carry the specified unit.

- **Lifting platforms**

  These are used to lift and lower the material mechanically using scissor leg principle.

- **Mechanical Platform less elevator**

  These are elevators in which there is no separate load carrying attachment but the load is carried by clamping or positive pressure. Mechanical platform less elevators are relatively new and were developed originally for vertical handling of mailbags and luggage at airports.

9.2 LIFTS

Every lift consists of a frame, platform or cage, hoist and enclosure. There are many versions from light duty models intended to lift cardboard and wooden boxes, short heights to fully automatic equipment for pallet loads weighing up to several tonnes. These are classified (*i*) according to the purpose or (*ii*) means used for lifting.

(*i*) According to the purpose lifts are classified as

(*a*) Industrial lift used for lifting material from one floor to another floor.

(*b*) Passenger lift used for lifting passengers from one floor to another floor.

(*ii*) According to the means used for lifting lifts are classified as

(*a*) Electric lift

(*b*) Pneumatic lift

(*c*) Hydraulic lift.

Now we will discuss each important part in detail.

A Frame

A frame is used to support and accommodate all important parts such as cage, drive unit, suspension system etc. the main purpose of frame is to guide the cage up and down.

B. Cage or Platform

This is an actual load carrier since load is mounted on it or accommodated in it. Capacity and size depends on the load to be lifted. The door is provided to the cage for loading and unloading.

Whole lift design depend on cage and load to be lifted. Cage is considered as dead or static load and materials/ passengers were considered as live load. Hence the total load is summation of dead and live load. This summation is known as capacity of lift.

C. Drive Unit or Hoist

Lifts are used to move the load up and down by means of some force. There are two approaches which can satisfy this these are :

- Either load is lifted by forcing it to lift from bottom *i.e.,* force can be applied from the bottom.
- Either load is lifted by pulling it to lift from top *i.e.,* force can be applied from the top.

The force which can be used for lifting is provided by

(*a*) Electric means (Electric lift)

(*b*) Hydraulic means (Hydraulic lift)

(*c*) Pneumatic means (Pneumatic lift)

D. Enclosure

It is important part of any lift as it provides safety to the man and material by enclosing all lift parts.

Now it is time to discuss each lift separately.

9.2.1 Electric Lift

Passenger electric lift is comprised of following important parts :

(*a*) Cage or car

(*b*) Guide rail

(*c*) Counter weight

(*d*) Suspension

(*e*) Safety devices.

(*f*) Power source

Electric floor lifts usually have 4 posts between which is a platform supported at the 4 corners by cable which is led by sheaves to a single electric hoist. An electric hoist raises or lowers the lift cage via chains, steel cables or rack and pinion. One end of the steel cables are connected to the top of the cage and other end are attached to the counterweight. The normal location of the hoist is at the top of one corner post.

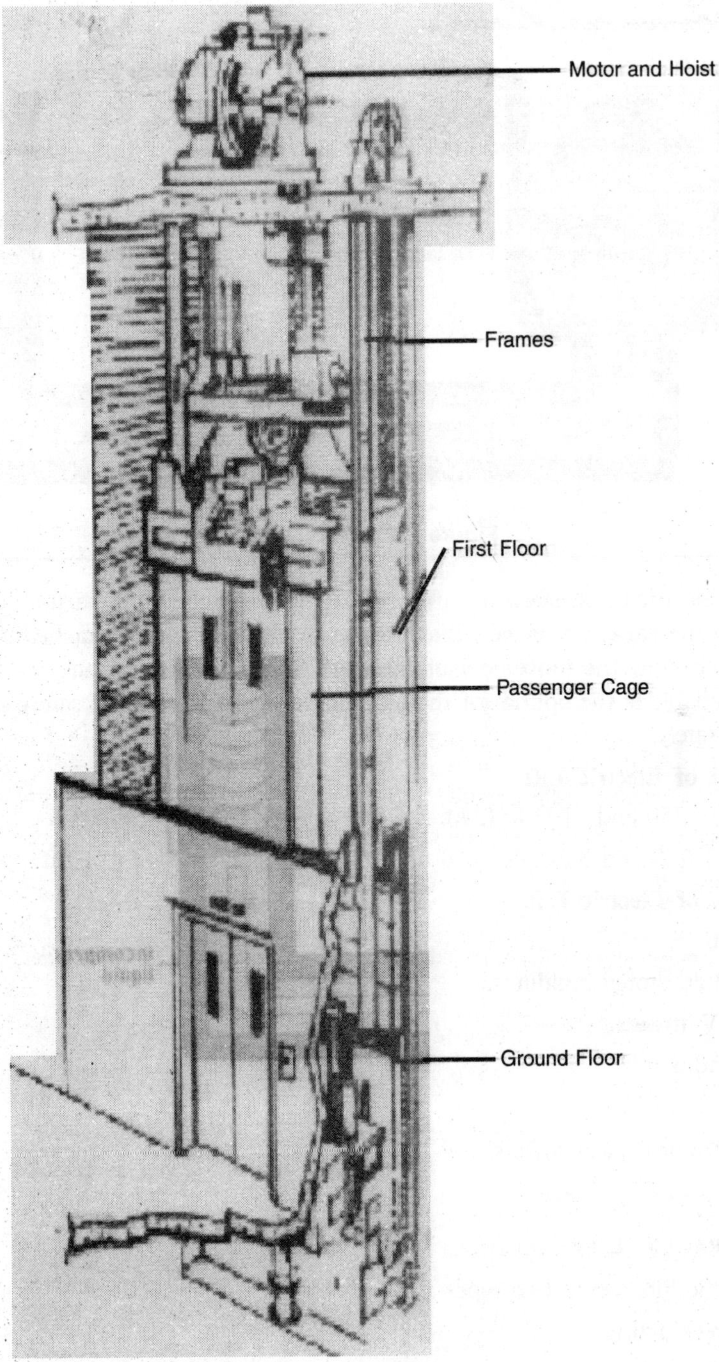

Figure 9.1. *Electric passenger lift.*

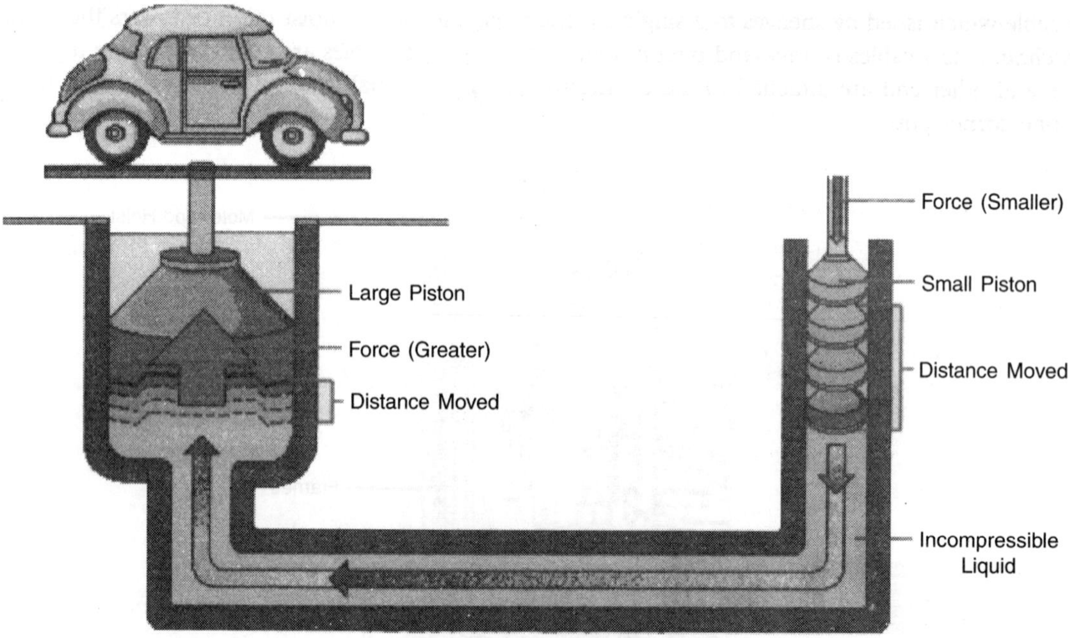

Figure 9.2. *Direct acting hydraulic lift.*

As the electric hoists starts it starts winding up wire rope on the drum due to which pulling force is exerted on the cage and lift gets lifted. Since Counter weight is attached at another end of cage it assists in controlling the movement of the cage. To dampen the shocks of lowering lift, suspension system is available at the bottom of the lift. Lift is provided with the safety devices which enable the lift to work safely.

Specification of Electric Lift

Capacity 750 and 1100 lb (340 and 500 kg).

Speed 2 ft/S and 3.5 ft/S (0.60 to 1.05 m/s).

Applications of Electric Lift

Used in :

(*a*) Multi storey buildings.

(*b*) T.V. towers

(*c*) Radio

(*d*) Relay towers.

(*e*) Meteorological towers.

(*f*) Construction sites.

9.2.2 Hydraulic Lift

Hydraulic lifts are of two types

A. Direct acting

B. Suspended type

Every hydraulic lift consists of ram and cylinder arrangement. Lift is lifted by the force generated by the fluid which exerts pressure on the ram. Since the cage is mounted over the ram, as ram lifts up cage is also get lifted in case of direct acting hydraulic lifts.

In case of suspended type of hydraulic lift one end of cage is attached to the wire rope and another end is attached to the ram-cylinder arrangements. As the fluid exerts the pressure on the movable cylinder the rope is pulled and ultimately lifts get lifted.

Direct Acting and Suspended Type Hydraulic Lift

Hydraulic lift consists of a cylinder and a ram. A platform or a cage is lifted at the top of the ram. Load or persons are carried from one floor to another floor by means of these platform or cages. Water under pressure enters the bottom of the platform and pushes the ram upwards. Thus ram moves up along with a cage in which load or person may be there. In this type of lift (known as direct acting) lift stroke of the ram is equal to the lift of the cage. The cage moves between guides and can be stayed at any level so that one can got in or out of the cage to the floor. For lowering it high pressure water is generally taken out. *Such type of lift is shown in which clotted line indicates the initial position firm line indicates the final position.*

Let,

d = Ram diameter in cm.

P = Pressure of water kg/cm^2

W = Total load of ram

A = Cross section area of ram = $\frac{\pi}{4} d^2$

Then neglecting the friction :

$$W = Pa$$

$$= P \times \frac{\pi}{4} d^2$$

Thus, diameter of ram is designed considering the maximum weight direct acting on lift-required keep well for the ram and cylinder to overcome. This difficulty in modified form of hydraulic lift fitted with jigger is becoming common. By providing jigger these lift requires small ram and cylinder and thus reduces the cost of construction. The cage is suspended with the wire rope and gets it motion by means of jigger in the same way as the hydraulic lift. Such type of lifts is commonly known as suspended type of lift. Hydraulic lifts are useful where electric supply is not available or less frequent.

9.3 ELEVATORS

It is operated with load. There are many different types of drives for such elevators but in general electrical devices (drive) are most common. Hydraulic elevators are used only where it is dangerous to take the chance on electric spark. Elevators requires considerably less space.

Elevator are machines for lifting and lowering loads vertically with intermittent recuperating motion in (1) Multi-floor factories, (2) Warehouses, (3) Stores and Terminals.

By using elevators speed of work enhances considerably.

They consist of a platform to support the load and entire mechanism which actuates the elevator is mounted at the bottom of the elevator.

Types of elevators are :

1. Rope hoist freight elevators.
2. Hydraulic freight elevators.

3. Mast type elevators.
4. Vertical skip hoist elevators.

9.4 LIFTING PLATFORMS OR LIFT TABLES

These are used to lift and lower the material mechanically using scissor leg principle.

Lift tables are used to raise and position work pieces for ergonomic access. They are typically used for material positioning, load positioning, or lifting. They are widely used in construction, automotive and garage service, electrical and power service, telecommunications, manufacturing, inventory management, wire and cable industries, painting and other applications where access to above ground locations is crucial.

These lift tables offered with various specifications such as the lift capacity, vertical lift travel, platform width and length. The lift capacity is the maximum force or load supported by the lift. Vertical lift table describes the difference between fully lowered and fully raised lift positions. The platform width is the narrow dimension of the lift platform. Platform length is the long dimension of the lift platform. These lift tables are actuated by electric motors. These lift tables often equipped with wheels to provide for load transport and positioning. They equipped with brake or stabilizing outriggers.

A lifting platform consists of a base frame and a top platform which is raised or lowered by one or several power operated scissor legs. These can be mounted on the wheels as shown in the figure below.

These are classified:

A. According to the lifting means :
   1. Electrically operated
   2. Pneumatically operated
   3. Hydraulically operated.

B. According to power operated scissor legs
   - Single scissor legs
   - Multiple scissor legs may be horizontally spaced and vertically spaced.

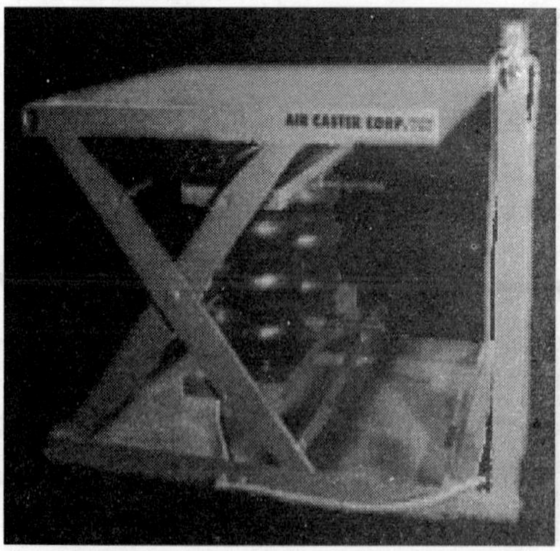

Figure 9.3. *Pneumatically operated.*

Figure 9.4. *Electrically operated.*

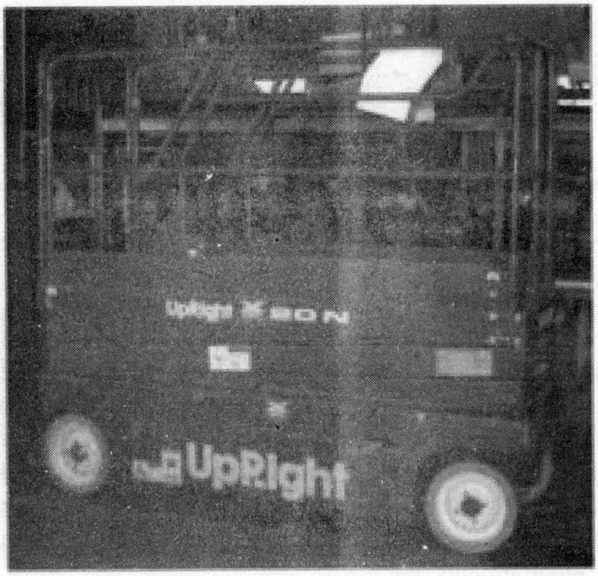

Figure 9.5. *Initial and final position of the lift table.*

UNSOLVED QUESTIONS

1. Explain with neat sketch lift and elevator.
2. Explain hydraulic lift.
3. What is lift ?
4. What is elevator ?
5. Write about electric lift.
6. Write about lift table.

❑❑❑❑

10 Industrial Trucks

INTRODUCTION

Industrial trucks are used to transport materials over short distances. Powered industrial trucks have no route restrictions and are suitable for application where flexibility is important. Trucks are not only often more flexible in certain cases they are the only practical solutions for some cases *e.g.*, for loading and unloading open lorries, containers and rail wagons truck is a mobile, power-driven vehicle used to carry, push, pull, lift or stack material. There are twelve different types of powered industrial trucks and designated types of industrial trucks can be operated safely in varying environments (*e.g.*, flammable areas, dusty environments, etc.).

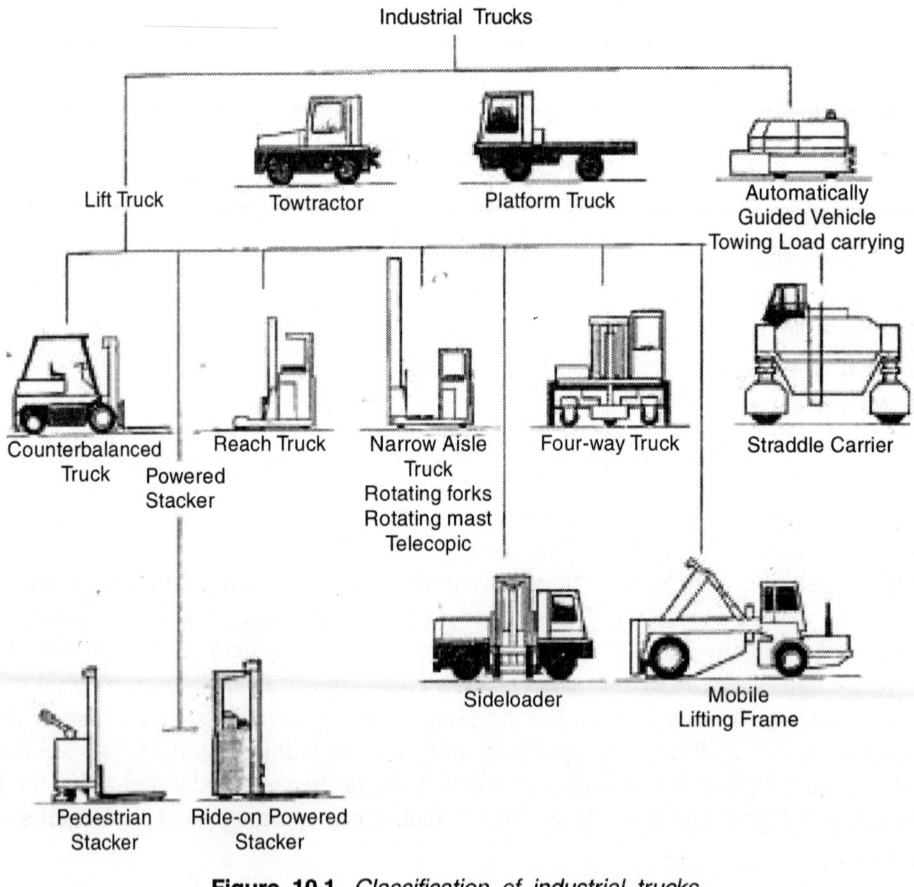

Figure 10.1. *Classification of industrial trucks.*

Powered industrial trucks are a very important part of material handling in many industries. They are also a source of serious accidents in an industry. Most powered industrial trucks injuries are caused by tip over accidents. The primary causes of tip over are excessive speed while turning and raised, unbalanced loads. The best way to avoid tip over is to properly counterbalance your load. All loads must be placed as close to the back of the forks as possible. A powered industrial truck is not a car. Powered industrial trucks are tall and narrow and tip over easily, so operators must drive cautiously. Stooping powered industrial trucks is also not the same as stopping a car. The two small wheels are the braking wheels; powered industrial trucks so do not stop quickly. So a powered industrial truck is a mechanical handling vehicle for load carrying or traction purposes.

10.1 CLASSIFICATION OF TRUCKS

Generally there is no accepted classification of powered trucks. The classification above is an attempt to order on the different types.

10.2 IMPORTANT PARTS

Industrial trucks are generally comprised of following important elements which are listed below.

(i) Chassis

The chassis is self-supporting welded plate and beam structure. It is essential that the chassis is stable and that is designed to give easy access for servicing and to vital components.

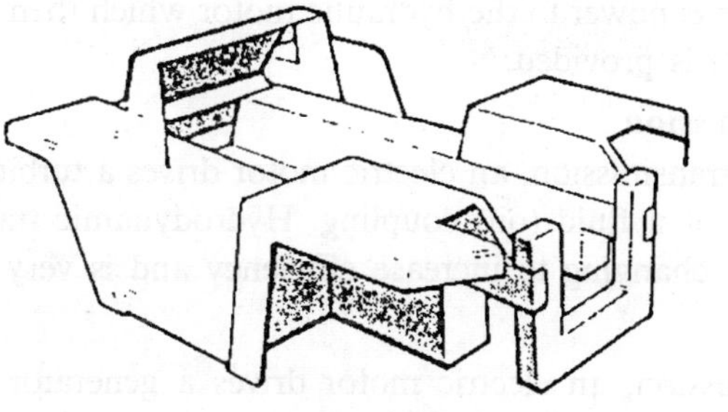

Figure 10.2. *Chassis.*

(ii) Power Sources

A. Battery

Battery sources have great environmental advantages. No exhaust fumes were emitted, noise level is low, and vibration is slight. Consequently, battery power is suitable for indoor operation. Service cost is often slightly lower than for other power sources. Battery storage of electric power is expensive, bulky and heavy. A battery truck is both costly and heavier than an internal combustion engine truck. Its strength is also inferior. Battery capacities are generally adapted to satisfy the requirements of normal operation during an eight hour shift. For multi shift operation, two separate battery sets are used. Breaks for partial recharging of batteries are not recommended as battery life is directly affected by the number of charging cycles. Nor is high-speed charging good for batteries.

B. Diesel Engine

A diesel engine is very powerful in relation to its weight and can therefore use many types of attachment. As it can carry large quantities of attachment. As it can carry large quantities of energy, operating periods between refueling are long. Operating cost are low in comparison with other systems due to low energy consumption with other systems due to low energy costs. This source has also proved itself in other vehicle types.

C. Petrol Engine

Petrol engines provide similar operation advantages to those of diesel but operating cost is some what is higher.

D. LPG

On trucks powered with LPG, the gas is stored under pressure in special cylinders. Refueling, is done by changing a cylinder. Advantage of LP gas is that it has better cold start characteristics than those of a diesel engine. LP gas and diesel engines are similar with respect to ventilation requirements when used indoors.

(*iii*) Transmission

With battery power, the motor is often linked directly to the traction wheels and no special transmission unit is required. With other power sources transmission is normally by one of four methods listed below:

(*a*) Mechanical Transmission

Mechanical transmission in high total efficiency. It generally contains three control elements (accelerator, clutch and gear change) which often demand more work from the driver than do other systems. Due to higher mechanical losses it is now obsolete. It is similar to normal gearbox transmission of the trucks.

(*b*) Hydrostatic Transmission

Electric motor transmits power to the hydraulic motor which then transmits power to the driven wheel. Generally rear drive is provided.

(*c*) Hydrodynamic Transmission

With hydrodynamic transmission, an electric motor drives a turbine wheel which in turn drives another turbine by means of a fluid (oil) coupling. Hydrodynamic transmission is frequently combined with automatic gear changing to increase efficiency and is very common on IC trucks.

(*d*) Electric Transmission

With electric transmission, an electric motor drives a generator which in turn drives one or several other motors linked to the wheels.

(*iv*) Wheels

In industrial trucks wheels are the actual load bearers. They must be able to absorb the deformation and resilience to protect drivers and loads.

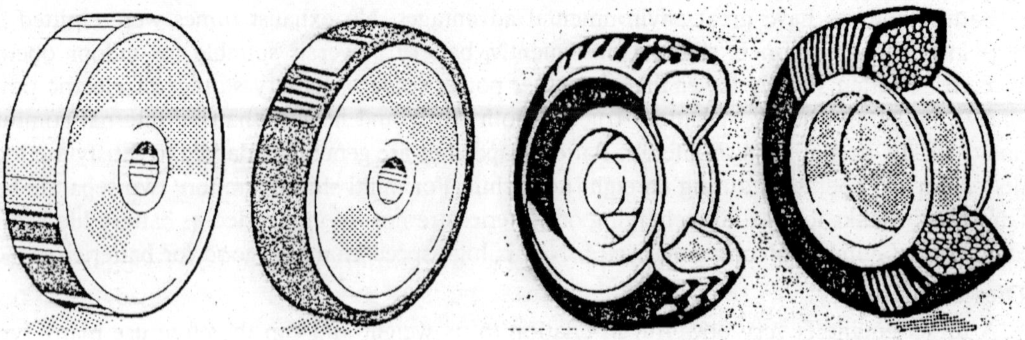

Figure 10.3. *Different types of wheels used In the industrial trucks.*

Solid Wheels

Solid wheels are made from hard material, steel or nylon. They have low rolling resistance. Solid wheels are long lasting and puncture-proof. Steel wheels wear out floor surfaces. It requires even floor surfaces.

Solid Tyres

Solid tyres consist of a natural or synthetic outer layer mounted on a rim. Rolling resistance exceeds that for solid wheels but floor surface need not be so even. Wheel life is shorter in this case.

Pneumatic Tyres

Pneumatic tyres are inflatable (tubeless) and are mounted on rim. Higher rolling resistance (and thus greater energy consumption) is experienced. Since pneumatic tyre adjusts according the shape of the floor, there is no need of even surface. There are chances of hazards due to bursting of tyres.

Semi-Pneumatic (Cushion) Tyres

Semi-Pneumatic (Cushion) Tyres contain a air filled cells or cavities. Flexibility is less than that for pneumatic tyres but life is greater. It is costlier than that of fully pneumatic tyres.

(*v*) Loading Attachments

These are the attachments which directly come in the contact with the load and depends on the shape and the size of the load to be lifted.A loading attachment may be used simply to lift a load or to manipulate it *e.g.,* to rotate a drum etc. The commonest attachment is a fork which consist of tines made of flat or sectional bar, bent at right angles and tapered ends. To increase productivity or to enable trucks to accommodate loads which could not be managed with forks, a special auxiliary attachment may be fitted.

Side-Shifts

Side-shifts are the most common auxiliary attachment since they enable the driver to save time re-positioning the forks precisely in front of the load rather than moving the whole truck. Side-shift should permit driver visibility; otherwise the load may be damaged.

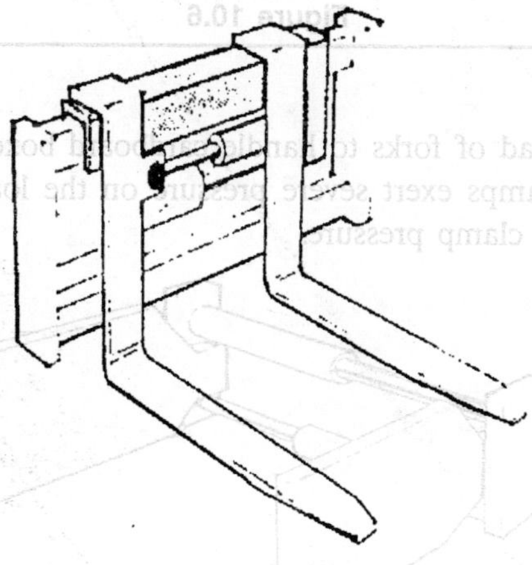

Figure 10.4

Fork-Spreaders

Fork-Spreaders change the distance between adjacent fork and often incorporate side shifting. Fork-spreaders are needed when handling different types of pallet or extra wide loads. Special attachments are used mainly on counterbalanced trucks.

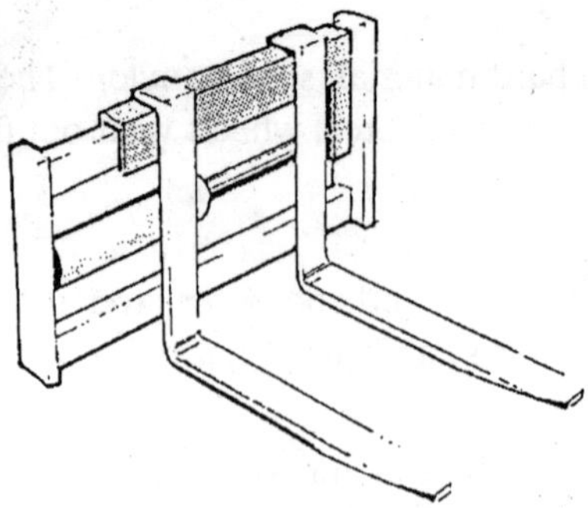

Figure 10.5

Push-Pull Attachment

An alternative to forks is the push-pull attachments. Materials are positioned on a slip-sheet. Slip-sheets take up the little space and are also suitable for closed systems involving bulky materials. A drawback is that handling operations generally take longer time.

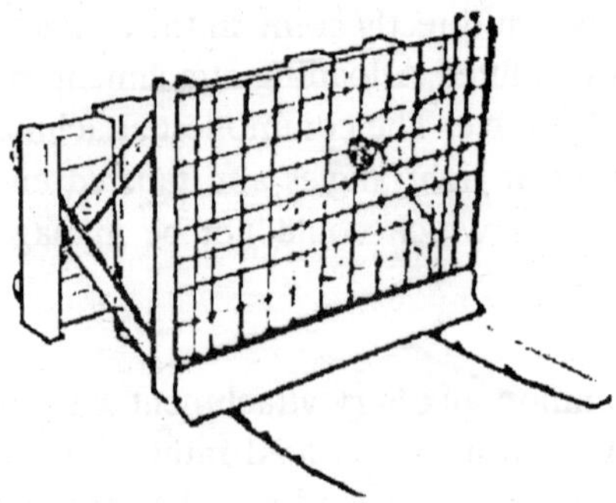

Figure 10.6

Clamps

Clamps can be used instead of forks to handle cardboard boxes, paper bales, paper reels and other similar materials. The clamps exert severe pressure on the load or material so it must have sufficient strength to withstand clamp pressure.

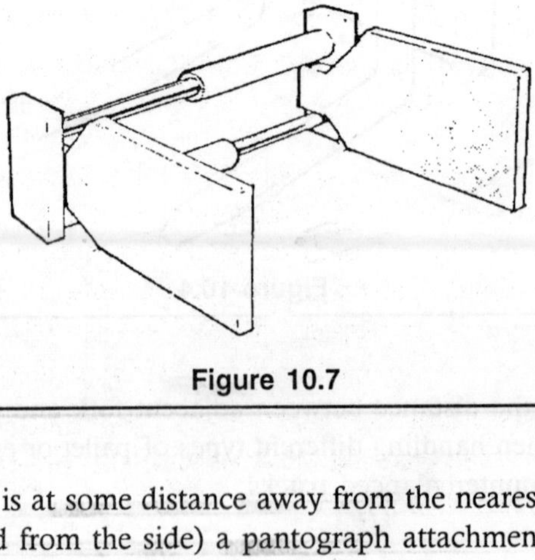

Figure 10.7

Pantograph Attachment

When the load position is at some distance away from the nearest approach point for the truck (*e.g.*, as with a lorry unloaded from the side) a pantograph attachment is one answer.

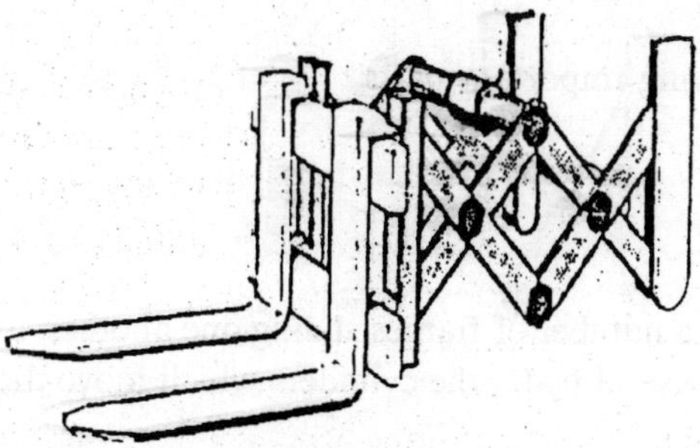

Figure 10.8

Drum Gripper

Specialized units may improve handling rates and/or eliminate the need for a load carrier. In the diagram, two drums are being lifted by a Drum-Gripper; no pallet is needed.

Utilization is improved if the truck serves as an all purpose vehicle *i.e.,* can be fitted with snow buckets, refuse shovels etc.

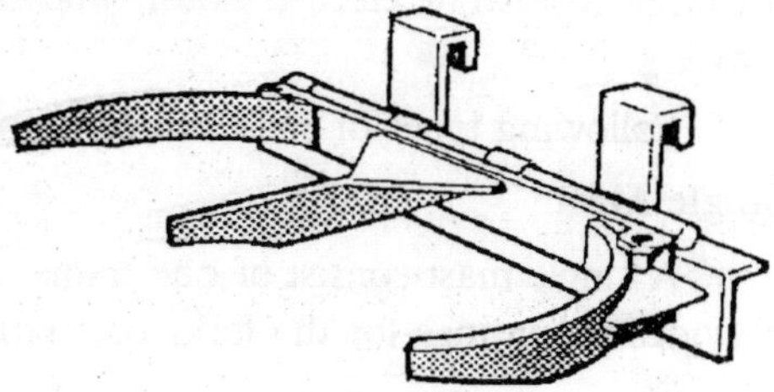

Figure 10.9

Boom

Lengthy and/or cylindrical loads such as coils of wire and rolled carpets can be accommodated on a boom attachment. A boom offers increased handling capacity when compared with forks.

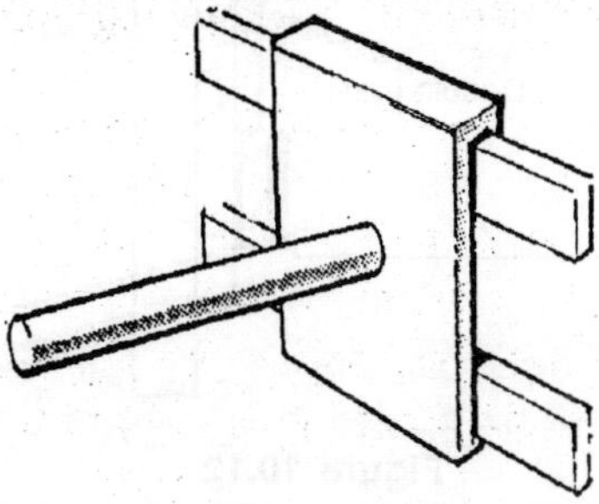

Figure 10.10

Rotating Clamp

Some loads need to be rotated, as when a drum is emptied or a paper reel is removed from a paper making machine. Instead of using a separate manipulator, the truck can be fitted with a rotating clamp.

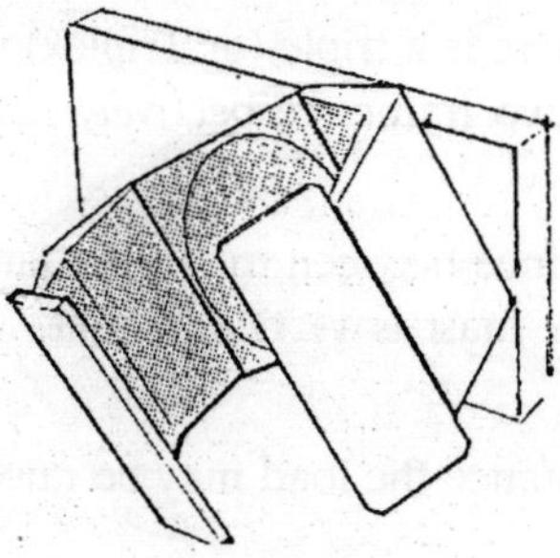

Figure 10.11

(*vi*) Lifting Arrangements

It consists of the following important parts :

1. Sliding frames (mast).
2. Hydraulic cylinder.
3. Pulling chain.

A truck's lift consists of a number of frames sliding one another, usually on roller bearing. Lifting power is provided by one or several hydraulic cylinders which move the sliding frame with or without the aid of chains.

Oil for the truck hydraulics is delivered by a pump. In battery powered trucks, this pump is driven by a separate electric motor, whereas in IC engine trucks there is a direct drive from the prime mover.

Following types of mast are commonly employed:

Single Mast

A single mast consist of one frame. The fork carriage, with fork brackets, runs in the frame and provides mountings for the forks or auxiliary attachment.

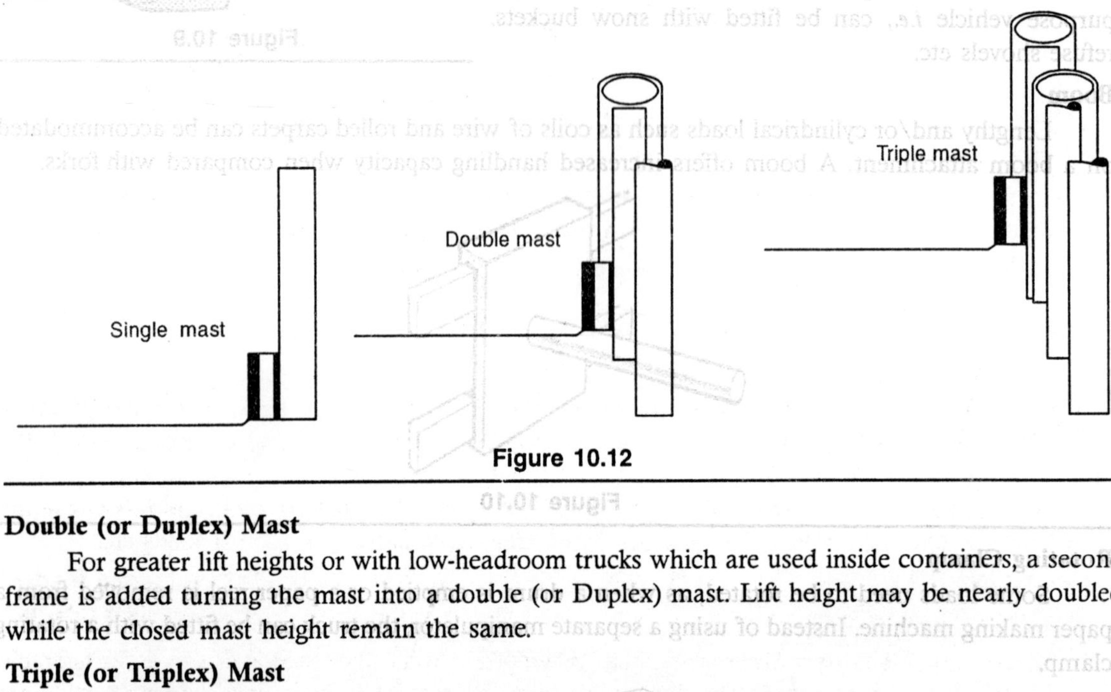

Figure 10.12

Double (or Duplex) Mast

For greater lift heights or with low-headroom trucks which are used inside containers, a second frame is added turning the mast into a double (or Duplex) mast. Lift height may be nearly doubled while the closed mast height remain the same.

Triple (or Triplex) Mast

For even greater lift heights, there is a triple (or Triplex) mast. Quadruple masts have been used but are rare. In these cases one or two frames respectively added.

(*vii*) Performance

Lift height (A) means the distance between the upper faces of the forks and ground level when the floor is horizontal and the truck mast is vertical height.

Free Lift Height

Free lift height refers to the distance the load may be raised with no consequent increase in total truck height.

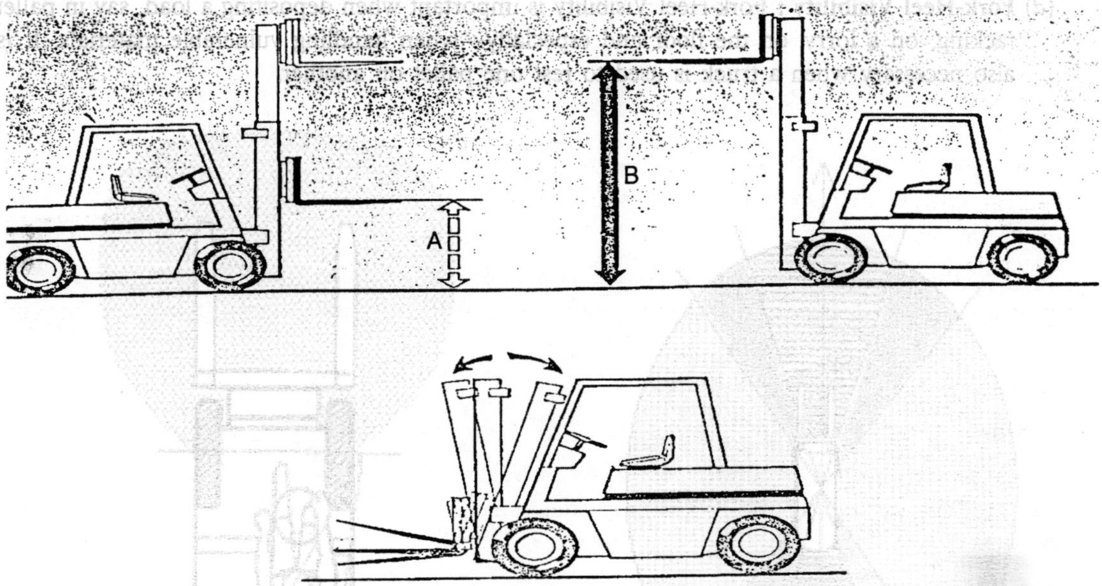

Figure 10.13

Fork Tilt

To prevent a load from sliding off the forks, most trucks forks may be titled backwards. On counterbalanced truck may be tilted backwards. On counterbalanced trucks tilting amount to 10-12°. Trucks used indoors especially reach trucks, travel more slowly and thus require only about 4° tilt.

(*viii*) Turning Radius and Aisle Width Requirements

When a truck revolves, it revolves at some centre. If we draw circle through this point and touching to all corners of the truck it forms the circle. Radius of this circle is known as turning radius. The aisle width requirement is calculated according to as the width of the aisle through which the truck is able to reverse in a straight line with an 800 mm wide and 120 mm long load and then turn through a right-angle (*i.e.,* to deposit or pick up load) maintaining a margin of 100 mm between the truck (and/or its load) and fixed points such as racking, building supports, other loads etc.

(*ix*) Visibility

Visibility is divided into four types :

(*a*) **Driving Visibility :** Driving visibility is important when the truck is driven a considerable distance at high speed. The driving direction considered to be normal varies for different truck types. Driving visibility includes side visibility of importance when the truck reaches a crossing. It may be acceptable to have the driver more quite considerable to achieve the necessary degree of side visibility.

(*b*) **Close Visibility :** Close visibility is essential when the truck is picking up or depositing loads in confined spaces, etc. As such work frequently involves turning, good visibility will be required on the front and the rear "blind" quarters and in other sections close to the truck.

(*c*) **Good-Fork Tip Visibility :** Good-fork tip visibility enables the driver to remove or deposit loads efficiently and safely. Where the truck is lifted not with forks but some other attachment, the visibility requirements for that attachment must be assessed.

(*d*) **Fork-Heel Visibility :** Fork-Heel Visibility is important when depositing a load, say in pallet racking, on a lorry, on the floor and especially – when handling vulnerable materials. It is also necessary when a truck is used to rest one pallet on another.

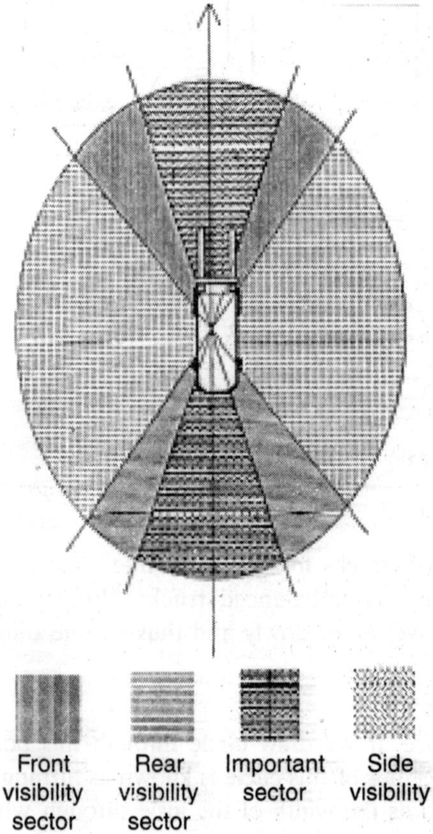

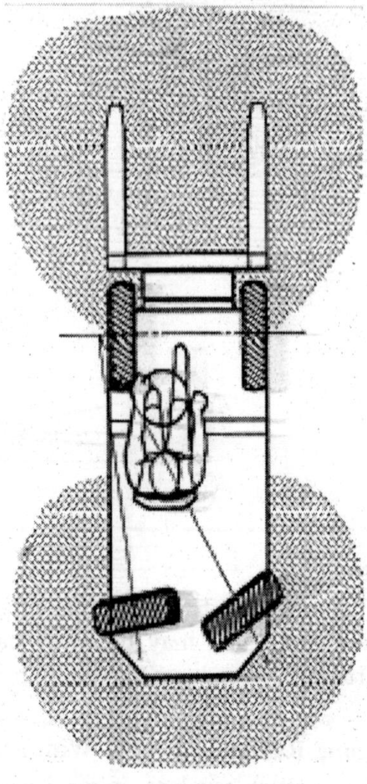

Importance of different sectors for close visibility

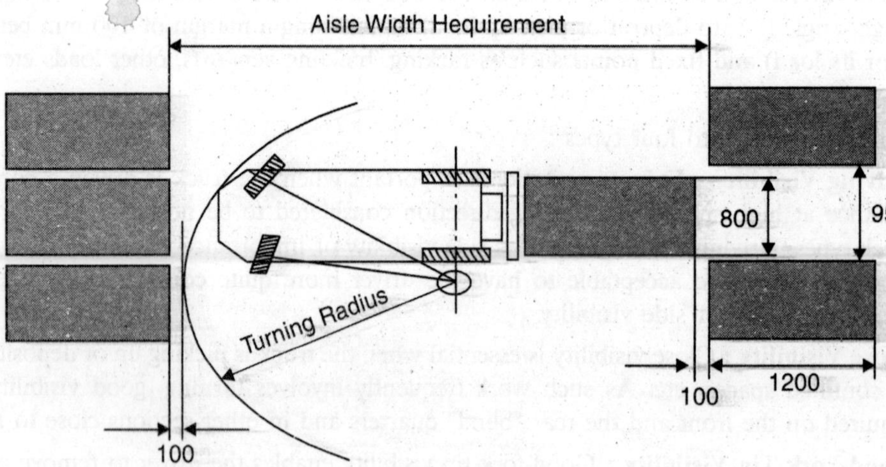

Turning radius and required aisle width for a four wheel truck.

Figure 10.14

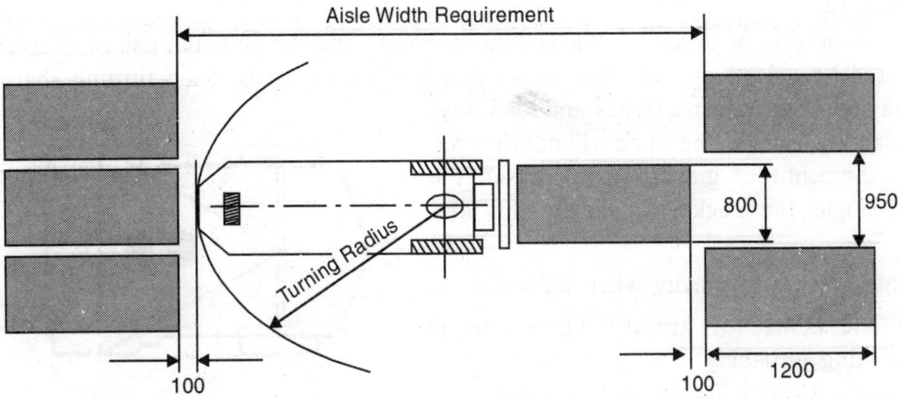

Turning radius and required aisle width for a three wheel truck.

Figure 10.15

(*x*) Stability of Industrial Trucks

Stability is one of the main limiting factors for any industrial truck. Stability can be further broken down into the following components :

Load Centre

The Load Centre (LC) is measured from the vertical face of the forks to the centre of gravity of the load. The manufacturer's rating plate provides safe working load capacities measured from the vertical face of the forks and are at set dimensions of 400, 500 and 600 mm centers for small to medium size forklift trucks,. If the load centre is outside the 500 or 600 mm shown on the rating plate, the operator would have to carry out a small calculation to see if the load is safe to lift, and if not, what reduction must be made to the maximum load capacity.

Note : Calculations are not necessary if the manufacturer has provided a graph showing capacities at different load centers.

It can be seen from the diagram that Load A is within the capacity of the industrial truck (rated capacity 4000 kg at 600 mm load centre) :

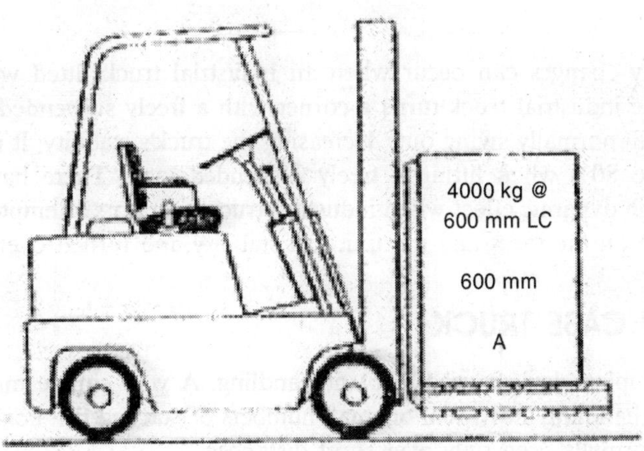

Figure 10.16

Stability Triangle

Operators need to be aware of "the stability triangle" and the potential risk of overturning that may be caused by a load on the forks or additional dynamic loads when turning sharply. If the centre of gravity of the industrial truck and load stays within the stability triangle, the truck will not tip over. However, if the centre of gravity moves outside the imaginary triangle, the truck will lose stability and turn over.

1. Vehicle centre of gravity when unloaded.
2. Vehicle centre of gravity when unit is becoming unstable.
3. Vehicle centre of gravity when loaded.

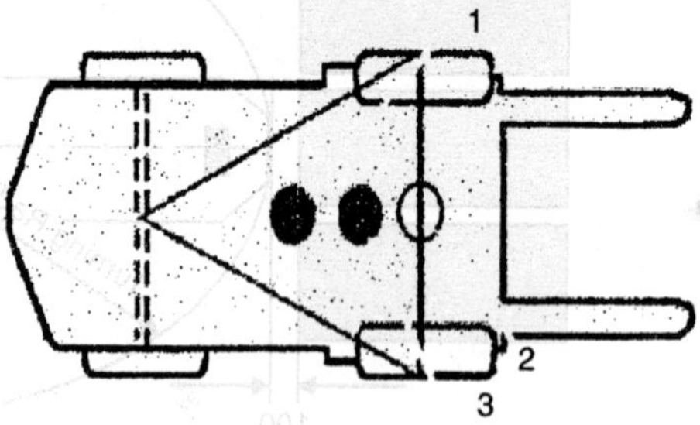

Figure 10.17

Lateral Stability

Lateral Stability is affected by three main factors :

1. Elevated loads.
2. Support surface *e.g.,* potholes.
3. Dynamic effects.

When the load is fully elevated it has a pendulum effect on the industrial truck, especially when the operator is turning the truck. If the operator turns too sharply, this will introduce lateral loading through the centre of gravity of the load and unless the stability of the industrial truck is satisfactory, the truck and load will topple over.

Longitudinal Stability

Longitudinal stability can be best explained as a see saw where the front axle acts as a fulcrum point. As the load is being raised it produces a longitudinal overturning moment. At the same time the industrial truck is applying a downward moment in the opposite direction due to the truck's counter weight and body mass.

A forklift is a mechanical lever. A lever works on the principle of applying a moment to overcome the moment created by the object to be lifted. A moment is the product of force and distance.

Dynamic Stability

Dynamic stability changes can occur when an industrial truck fitted with a jib lifts a freely suspended load. As the industrial truck turns a corner with a freely suspended load hanging from a jib section, the load will normally swing out, decreasing the truck's stability. It requires that the rated capacity be reduced to 80% when lifting a freely suspended load. There have been a number of fatalities caused through dynamic effect when industrial truck operators attempted to carryout a sharp *"U"* turn with no load on the forks and the unit lost stability and turned over.

10.3 SACK OR CASE TRUCK

These are the simplest devices used in plant handling. A very simple manually operated two-wheel unit suitable for handling individual or small numbers of stacked fish boxes of limited capacity, cartoons, boxes and normally used only over short distances.

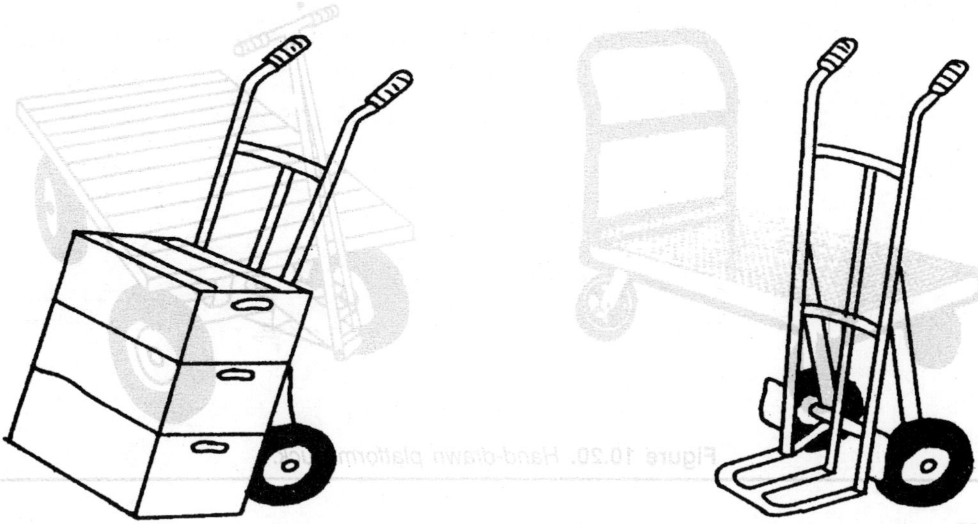

Figure 10.18. *Sack or case truck.*

A rigid platform with wheels at one end which can be maneuvered manually by the use of a towing unit that lifts and supports the other end. Restricted to use on more or less level floor.

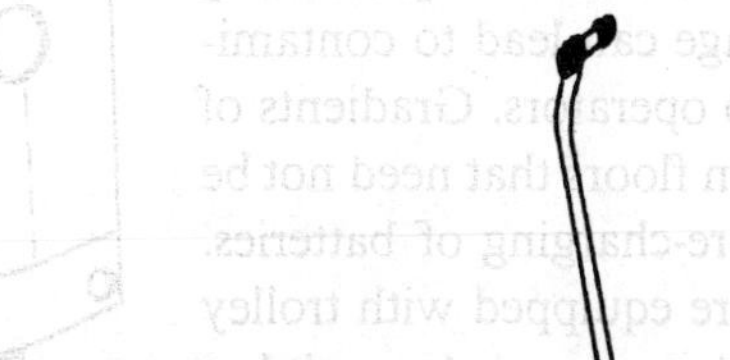

Figure 10.19. *Jak-tug.*

10.4 HAND-DRAWN TRUCK

A four-wheeled manually operated truck suitable for loads of up to 1 tonne. Steering is either by means of a turntable assembly or by a central pivot connected to the steering wheels by track rods. The track rod type of truck is preferred as the wheels remain under and give support to their respective corners on turning. The turntable type has a tendency to tip when negotiating sharp turns if unevenly loaded.

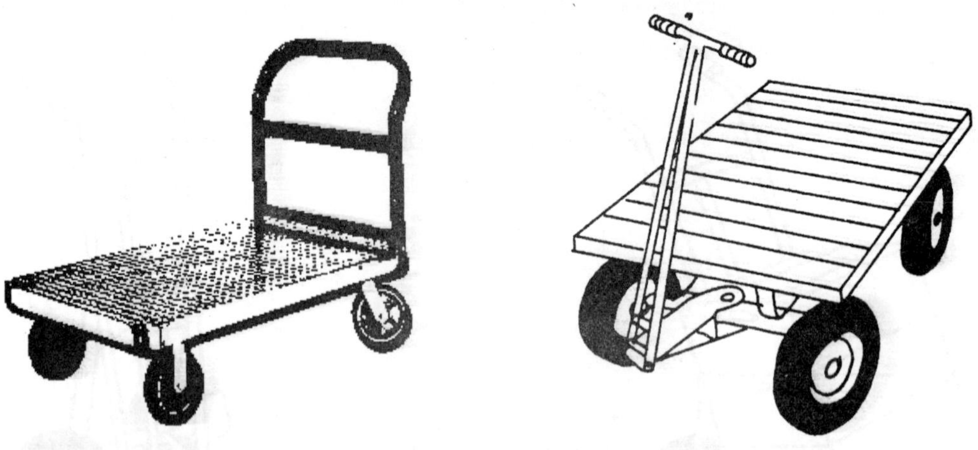

Figure 10.20. *Hand-drawn platform truck.*

10.5 ELECTRIC TOW TRUCKS

They are basically battery-operated versions of the manual equivalent, which are better suited for handling greater loads over longer distances. Diesel versions are also available but should not be used in markets or processing factories where fumes and oil spillage can lead to contamination of product or be a hazard to operators. Gradients of 1:10 can be negotiated under load on floors that need not be smooth. Facilities are required for re-charging of batteries. These are basically tuggers. They are equipped with trolley to transport the material. The load is not carried over it but it is pulled by the truck.

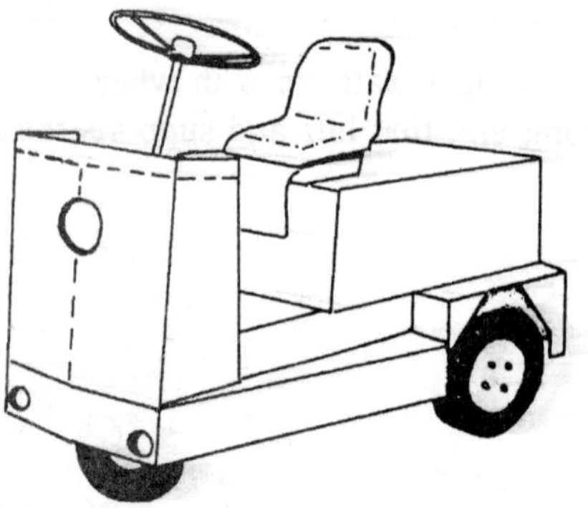

Figure 10.21. *Electric tow tractor unit.*

10.6 HAND PALLET TRUCK

A manually operated truck for use with palletized loads. It consists of two forks and a steering unit. The forks are entered into the pallet which is raised by a hand-operated hydraulic pump, until the load is clear of the ground. Because of the relatively small wheels required for entry into the pallet the floor needs to be smooth and fairly level. Simple modification by means of a frame fitted over the forks makes it suitable for transport of stack/nest types of fish box. It is particularly useful in box, pool or factory operations and eliminates wear caused by dragging or pushing boxes over concrete surfaces.

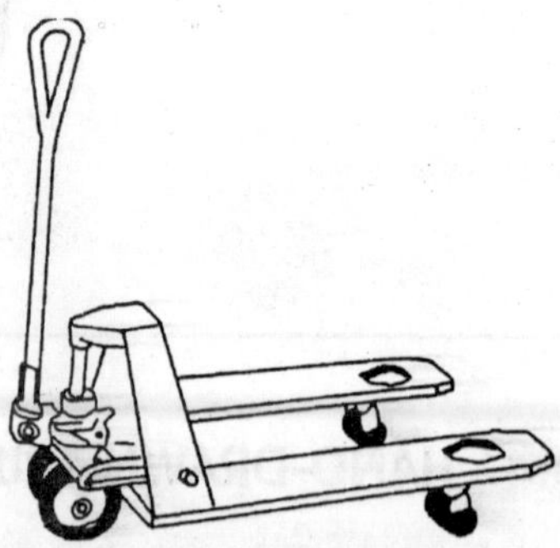

Figure 10.22. *Hand pallet truck.*

10.7 POWERED PALLET TRUCK

An electrically powered truck similar to the hand pallet truck which can be pedestrian or rider-operated depending on design. Quicker in operation than the hand pallet truck and more suited to greater loads and distances inclines. Can be operated on fairly uneven surfaces and inclines.

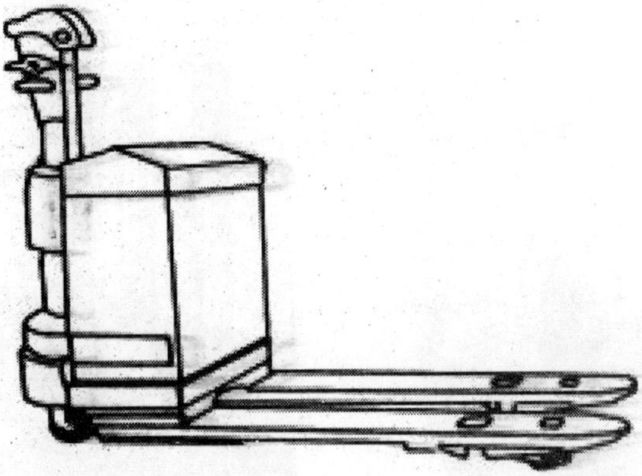

Figure 10.23

Similar to the pallet truck but with the capacity of lifting the load is much higher. The forks do not contain the wheels and the load is balanced by the weight of the truck. They can be pedestrian or rider-operated depending on design, and are particularly useful for loading and unloading of road transport, stacking within chill stores and movement of processing machinery.

10.8 FORKLIFT TRUCKS OR COUNTERBALANCED TRUCKS

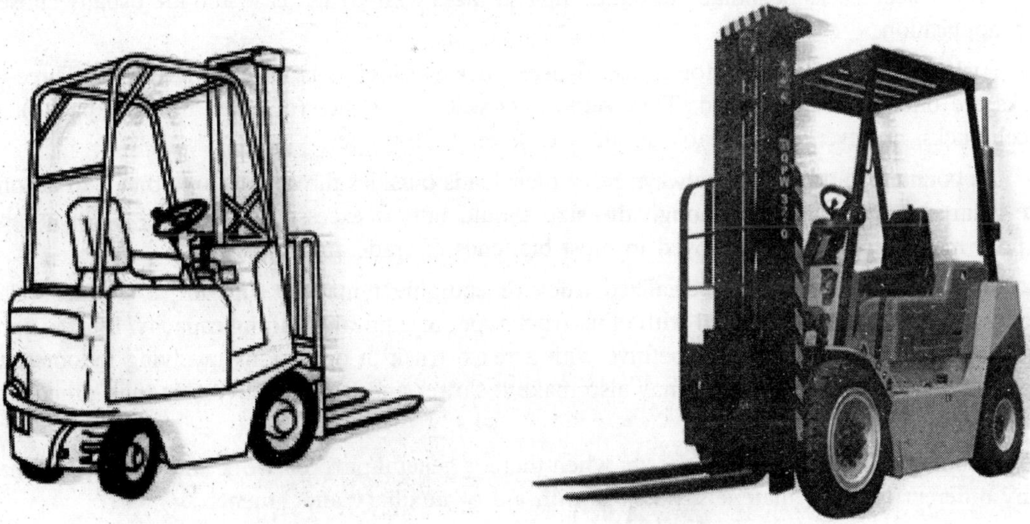

Figure 10.24. *Rider electric fork truck.*

Counterbalanced trucks carry their loads outside the stability polygon-an area limited by assumed lines drawn between the trucks points of contact with the floor/ground.

The loads tendency to tip the truck forward is concentrated by the trucks counter-weight. To increase the load capacity additional ballast is normally mounted at the rear of the trucks. Counterbalanced trucks are the most common truck type and can be used for handling most forms of materials.

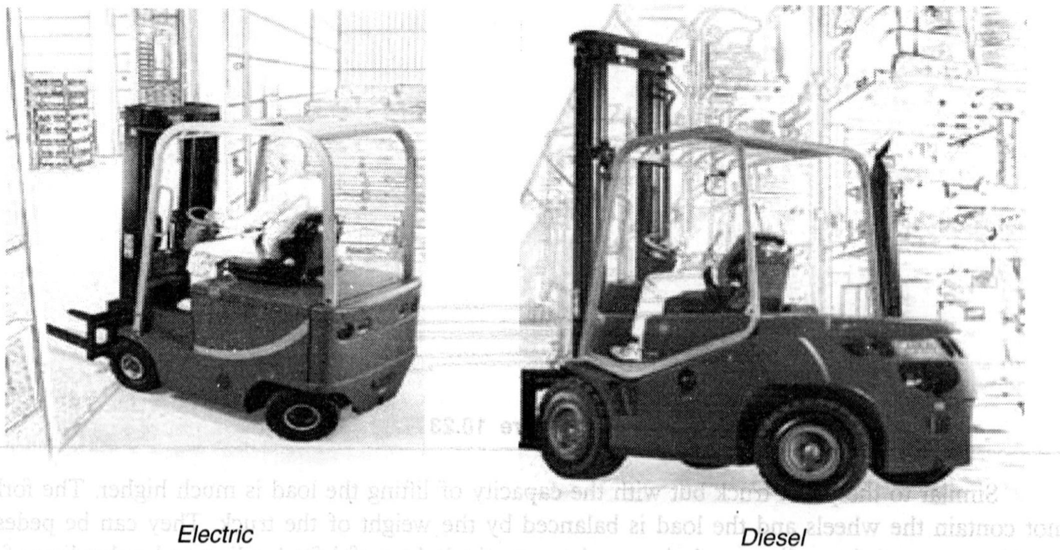

Electric Diesel

Figure 10.25. *Counterbalanced trucks.*

The stability polygons are equal for both forms of trucks. This is because four-wheel trucks are equipped with a pivoted steering. Axle, *i.e.,* the axle pivots at a central point on the truck.

Four wheel trucks negotiate uneven driving surfaces with greater ease and are usually chosen for such application.

Aisle width requirements for a three-wheel truck are less as is its turning radius. Three-wheel trucks are often battery powered. They seldom exceed 1–2 tonnes (of rated lifting capacity). Four-wheel trucks, however, are available in all sizes from 1–30 tonnes.

As counterbalanced trucks always carry their loads outsides the chassis area they can accommodate a large range of items although the size should not be excessive. They are useful as general purpose machines and are employed in most branches of trade and industry.

In some instances more specialized truck or equipment may be superior, a counter balanced truck is large and heavy compared with other truck types of equivalent lifting capacity. It takes up more space and may possibly be uncompetitive with a reach truck in operations involving indoor stacking in pallet racks. Its weight and size may also make it slower and less maneuverable than other, smaller truck types.

Counterbalanced trucks are chosen when there is general-purpose work and the need to manage many different types of materials, possibly with aid of auxiliary attachments.

The work is out of doors under cold conditions or over rough surfaces. Counterbalanced trucks are less suitable : For materials handling in confined spaces, *e.g.,* in door warehouse stacking with stillages.

10.8.1 Advantages of Counterbalanced Forklift Truck

The counterbalanced forklift truck is the most common piece of equipment used. This is because the counterbalanced forklift truck offers the following advantages :

(*a*) It is extremely flexible.

(*b*) It is highly robust.

(*c*) It is readily available off shelf from many different suppliers.

(*d*) It is available in a large range of capacities.

(*e*) It is simple to operate.

(*f*) It can be used in a large range of environments.

(*g*) It can handle a wide variety of products.

(*h*) It can perform a large variety of functions.

10.8.2 The Basic Principle of Counterbalanced Forklift Truck

The basic principle of the counter balanced forklift truck is that it is similar to that of the common see-saw. The concept is that by providing sufficient counter balance at the back of the machine one can off set the weight carried by the machine and therefore ensure that the machine remains stable.

10.8.3 Various Capacities, Shapes and Sizes of Counterbalanced Forklift Truck

Counter balanced forklift trucks are available in various types, capacities, shapes, sizes and applications. The largest counter balanced forklift truck would typically pick up 40 tonnes and may be used for handling large sea freight containers. Typically theses machines will be diesel powered (internal combustion engine) and may cost several lacs.

The smallest capacity counter balanced forklift truck can handle a load of between 500 kg to 1000 kg with a load centre of 500 mm.The small machines are typically battery (electric) powered and will operate in a high quality warehouse environment (where floors are compatible with the small tyres used for these small machines).There is a full range of counter balanced forklift trucks from the smallest 500 kg machine and the largest 40 tonne machine.

10.8.4 Off Road Forklift Truck

This picture shows a counter balanced forklift truck loading a vehicle. Such counter balanced forklift trucks can operate in rough terrains and are typically used in lumber yards, buildings sites and environments where good flat floors are not available.

10.8.5 Empty Container Handling Forklift Truck

This large counter balanced forklift truck is designed specifically to handle empty containers. Such vehicles can stack containers up to 7 levels high.

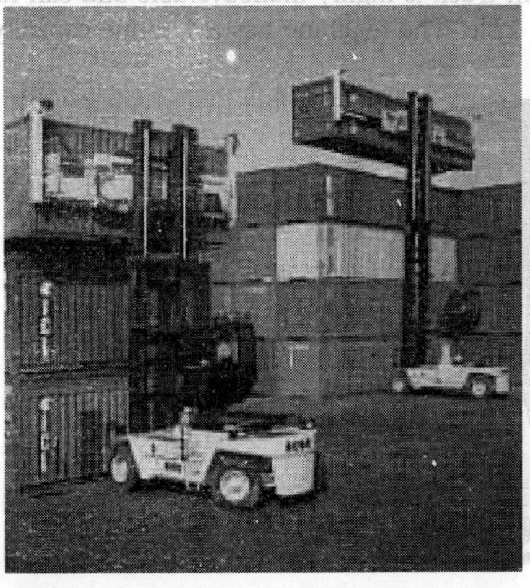

Figure 10.26. *Empty Container Handling.*

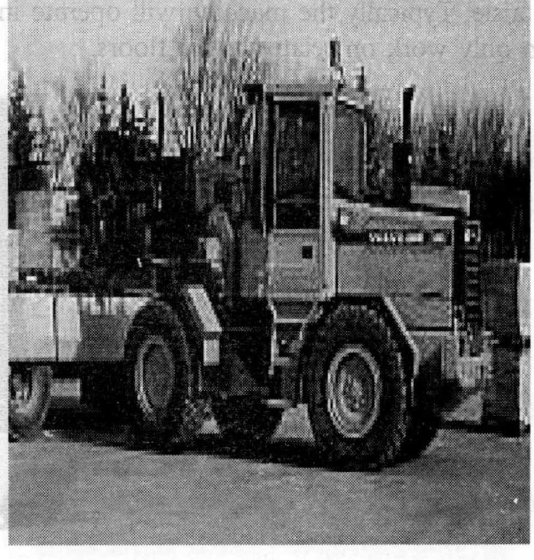

Figure 10.27. *Forklift Truck Off Road FLT.*

10.8.6. Electric Powered Counter Balanced Forklift Trucks

Many counter balanced forklift trucks use battery power. Battery powered forklifts have significant advantages. These include :

(*i*) The battery can be used as the counter balance—reducing the need for heavy metal counter balances.

(*ii*) They are generally far more compact than internal combustion machines.

(*iii*) They are highly efficient.

(*iv*) The cost of running electric machines can be significantly lower than internal combustion machines.

(*v*) They do not pollute the environment and are thus ideal for applications such as food factories.

Generally, electric powered counter balanced forklift trucks are used inside buildings where floors are fairly good. Most electric powered counter balanced forklift trucks are in the capacity range of 1000 kg to 3000 kg. Larger machines are available – but are not common.

One of the main advantages of the electric powered counter balanced forklift truck is its compactness. For example, a well-designed 1 tonne electric counter balanced forklift truck may be able to handle a standard pallet (at 600 mm load centre) in an aisle of between 3 m and 3.2 m. A similar capacity diesel powered counter balanced forklift truck would typically require an aisle of between 4 m to 4.5 m.Typically electric counter balanced forklift trucks used in warehouses have maximum lift heights of 4 m to 6 m (although higher lifts can be obtained under exceptional circumstances).

The electric powered counter balance forklift truck is ideal for use of general functions within warehouses such as :

(*a*) Loading and unloading of trucks.

(*b*) Loading and unloading of containers (with special low mast heights).

(*c*) General moving of goods to racking.

(*d*) General transport within the warehouse.

(*e*) Storage where heights and volume utilisation are not critical.

(*f*) Multitude of other functions within operations.

(*a*) Electric Forklift Truck

This electrically powered counter balanced forklift truck is highly maneuverable and can turn in an aisle. Typically the machine will operate in a 3 m aisle. The machine has a 1 tonne capacity and can only work on relatively flat floors.

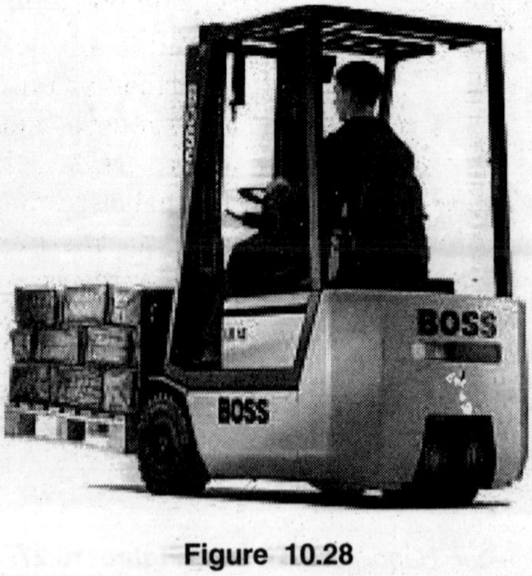

Figure 10.28

(*b*) Walky

This counter balanced forklift truck does not have a seated operator. The operator walks behind the lift truck and maneuvers the lift truck accordingly.The lift truck is electrically powered and is typically used in environments where minimum space is available.

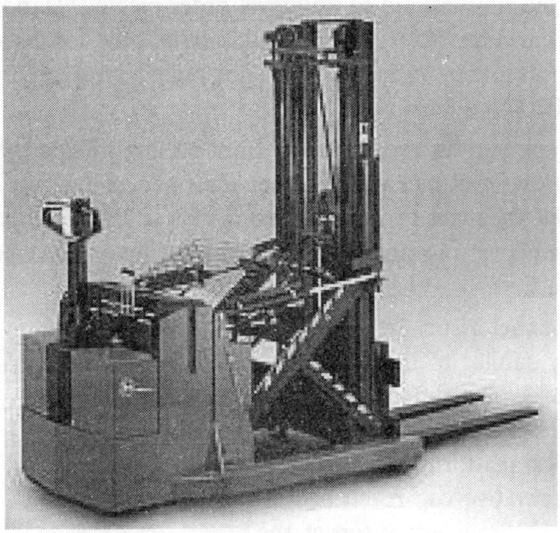

Figure 10.29

10.8.7 ENGINE

The Internal Combustion Powered Counter Balanced Forklift Truck

An alternative to electric power for a counter balanced forklift truck is the use of an internal combustion engine. These might be petrol powered, diesel powered or LP gas powered. The IC engine counter balanced forklift truck is available in capacities from 1 tonne to 40 tonnes.

Many varieties of IC engine forklift trucks are available. These may include :

(*a*) Off road vehicles for outside use (*i.e.*, building sites).

(*b*) Machines to be used in large container yards.

(*c*) Machines used for loading and unloading trucks.

(*d*) Machines used in warehouses.

(*e*) Container Forklift Truck.

This counter balanced forklift truck is handling full 12m containers in a large container yard.

Heavy counter balanced forklifts are expensive, but handle large unit loads extremely efficiently.

Attachment For Forklift Truck

This counter balanced forklift is using a clamp attachment for handling paper rolls. The counter balanced lift truck can be efficiently used for staking rolls up to 5 m high. Numerous other types of attachments for handling hundreds of products are available to facilitate easy handling.

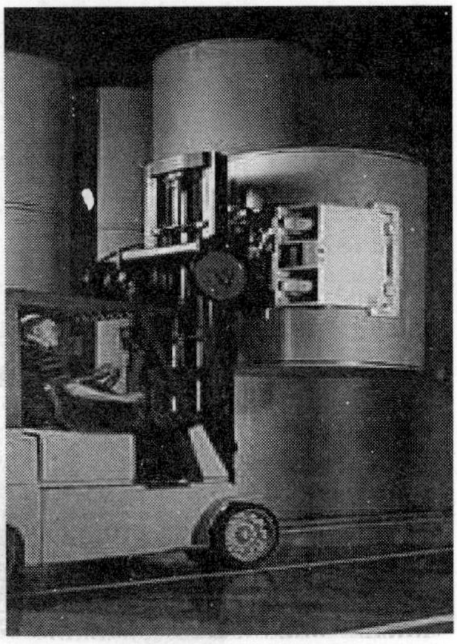

Figure 10.30

10.9 ORDER PICKERS

Order picking refers to retrieval of individual items from storage and normally involves taking goods from one unit load and adding it to another. There are three basic methods.

Hand pallet trucks, the simplest aid to low level order picking. Hand pallet trucks and stackers. A loaded hand pallet truck is heavy to pull and should not be used for distances over 30 metres or for more than 30–40 pallet movements a day. The weight pulled on a hour pallet truck when picking should not exceed 200–300 kg.

A powered pallet truck relieves the operative from pulling the load. Pedestrian guided powered pallet trucks are used for low level picking at higher rates and commonly handle euro pallets or roll pallet. With extended forks, they can carry two euro pallets or three roll pallets although such trucks are long and difficult to maneuver in some situations. The design of powered pallet trucks is dealt with in the section-powered stackers.

The performance characteristics of low level order pickers are basically similar to those of powered pedestrian pallet trucks. The difference is that the order picker is modifies for its task. The driver stands on a spacious ride-on platform, which is easily boarded. So that the operative can reach higher, a skidpro step on platform is provided on top of the battery compartment. This is also easy to climb on to on this upper platform; the driver can pick goods at levels up to 2.5 metres. Low-level order pickers have the disadvantage of being longer than pedestrian pallet trucks and thus difficult to maneuver in cramped surroundings.

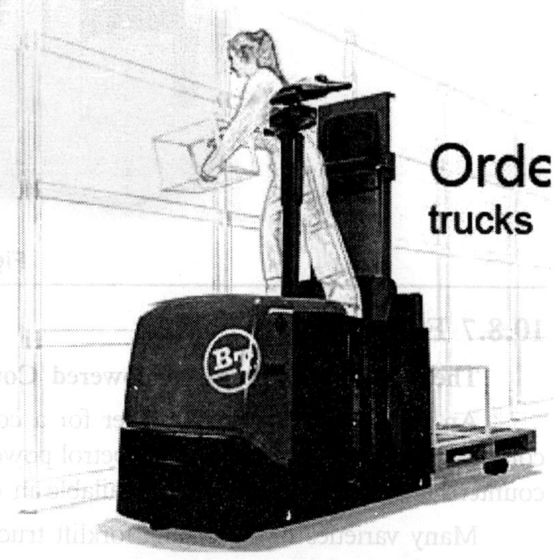

Figure 10.31

Low level order pickers and pedestrian pallet trucks with extended forks have high capacities and good travel speeds. They are therefore suitable for warehouses with high rates of turnover and for bulky goods.

Where there are many product lines and minimal buffer stock requirements for each line high level picking systems are appropriate. There are two main types.

Lofting and lowering speeds are rather slow on driver elevating order pickers. As the driver has to be lowered to the load carrier after nearly every picking operation, his picking rate too is fairly slow, such trucks are not normally used to pick at heights above the second or third level of pallet racking and then only for very low picking rates. Where controls can be operated from the driver's platform, efficiency is improved, but the driver still often has to bring the goods down.

10.9.1 The Man Down Order Picker (Ground Level Order Picking)

Here the concept is that the operator remains at floor level and picks all goods accordingly. Generally, picking at ground level is the most efficient form of manual picking as it allows high-speed access to numerous products.

It has the following benefits :

(*a*) It is highly productive.

(*b*) It is safe.

(*c*) Equipment costs are far lower.

(*d*) It is much easier to carry the load.

(*e*) It is generally easier to carry heavy products.

10.9.2 The Slightly Raised Picker (Medium Level Picking)

Here the picker is raised to just under a 1m height. This allows one to raise the picker to second or third level picking without expensive safety equipment. Generally, it is assumed that if the picker operates below a 1m level, one does not have to have expensive safety devices to ensure the safety of the picker. Furthermore, picking below a 1m height, can be done on a manual or hydraulic step up process—reducing the complexity of the equipment.

10.9.3 The Raised Picker (High Level Picking)

Should one required access to many picking faces it is often necessary to raise the picker above first and second level. This is largely to provide access to a higher number of stock keeping units. In its simplest form, the cheapest equipment for picking at higher levels is a simple mobile stepladder. However, in most warehousing applications one uses more sophisticated equipment to provide access at various levels safely and efficiently. Off self equipment is available to provide access at levels from 1m high to 9 m high. However, most equipment is limited to a lift of approximately 6.5 m.

Numerous types of manual type machines are available for order picking at various levels. Typically these machines are limited in capacity to approximately 500 kg to 1000 kg and are designed to manually access the pick faces.

Many machines are guided in the warehouse aisles (rail guidance or wire guidance) popularly known as AGV.

Again a multitude of different designs are available for various picking applications—whether one wishes to access small light components or large components (such as furniture).

10.9.4 Order Picker–Step Up

This low level order picker is used to pick goods into role containers. The picking machine has a step up function to 1m and can be used for ground and first level picking. (Fig. 10.32)

10.9.5 Order Picker–Low Lift

This picking machine allows hydraulic lift to 1m (second and possibly third level picking) and carriers 3 role containers at once. (Fig. 10.33)

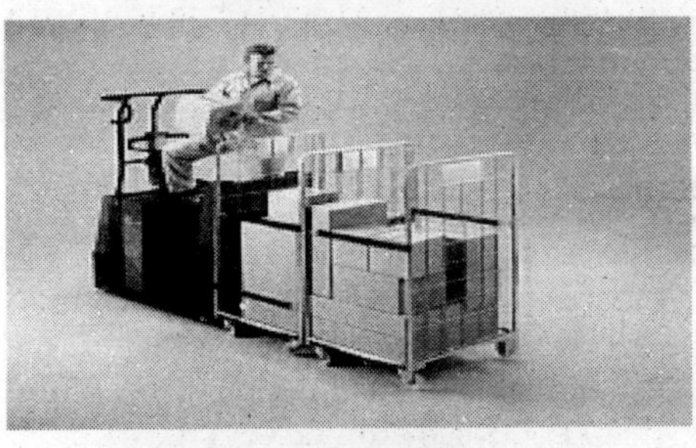

Figure 10.32

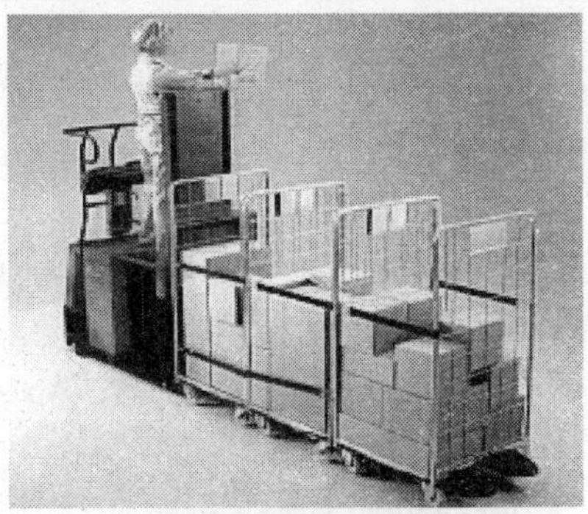

Figure 10.33

10.9.6 Order Picker in Aisle with Reach Truck

Here one can see a reach truck operating in the same aisle as an order picker. The reach truck operates in a wide 3 m aisle and the order picker passes the reach truck. Typically the reach truck is used to let down pallets to bottom level pick faces for the order picker. (Fig. 10.34)

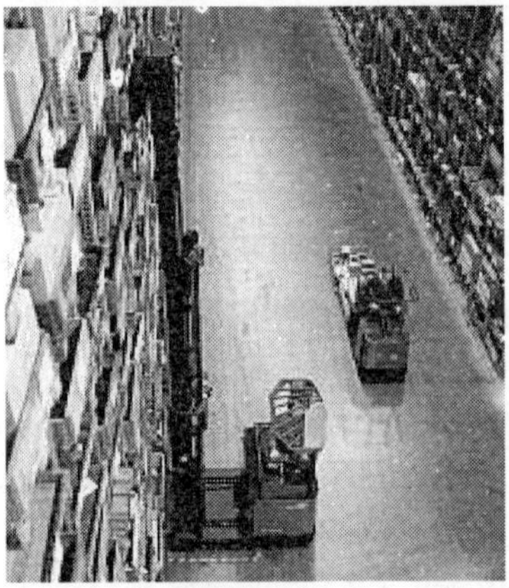

Figure 10.34

10.9.7 Powered Pallet Truck

This powered pallet truck is usually used for loading and unloading trucks. However for low volume throughputs and for relatively small loads, the powered pallet truck can also be used as an order picking machine. (Fig. 10.35)

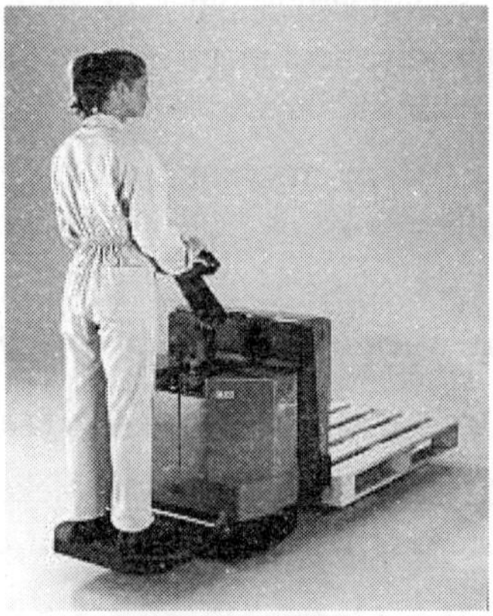

Figure 10.35

10.9.8 Order Picker-Man Up

This man up order picker raises the picker to the workplace. Here the picker can move heights of 6 m (and possibly up to 9 m).The picker thus moves up and down in the machine and can pick at multiple levels. (Fig. 10.36)

10.9.9 Turret–Truck For Picking

Turret trucks provide an ability to pick at heights up to 12 m. However the turret truck is an extremely expensive picking device. (Fig. 10.37)

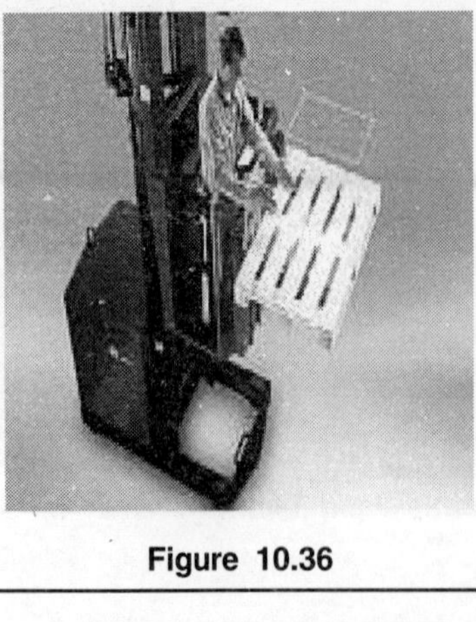

Figure 10.36

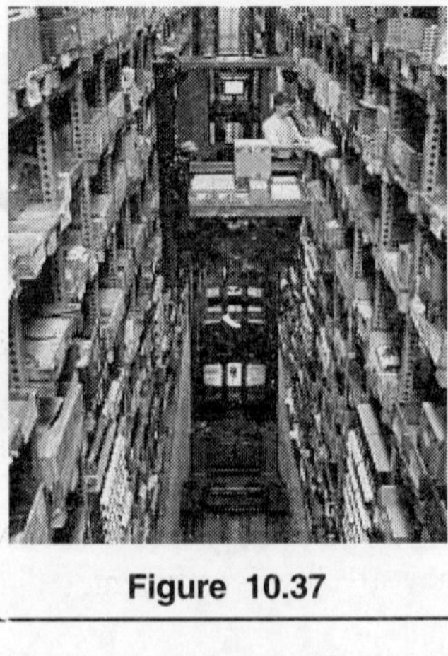

Figure 10.37

The turret truck has the added advantage that can also be used to replenish the pick faces as it can also handle pallets.

Selection

(*a*) The number of stock keeping units one needs to access.

(*b*) The size, shape and mass of product to be picked.

(*c*) The speed at which one needs to pick (number of picks required per hour).

(*d*) The available building heights.

(*e*) The racking design.

(*f*) The aisle design.

(*g*) Product virility and sensitivity.

(*h*) Local regulations and safety requirements.

(*i*) The methods used to replenish the pick faces.

(*j*) The physical limitations (building and racking).

(*k*) Overall customer requirements.

(*l*) Overall service and response needs.

10.10 POWERED STACKERS AND REACH TRUCK

The reach truck is one of the most popular types of Materials Handling equipment for use in warehouses for storage and retrieval of palletsn (Fig. 10.38).

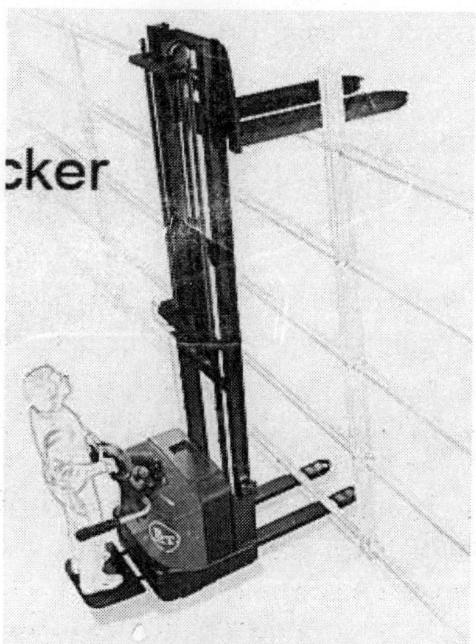

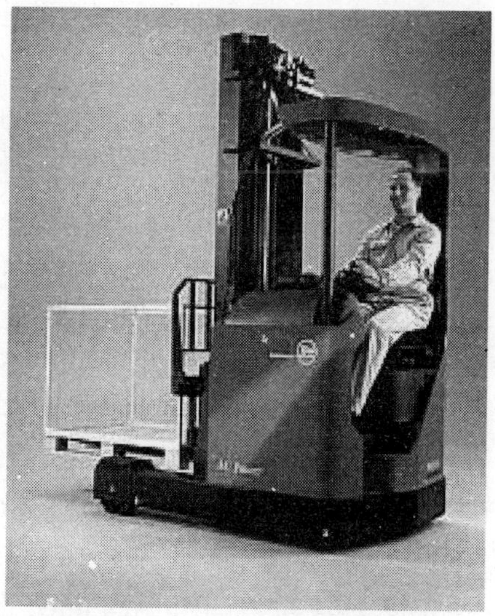

Figure 10.38

The main advantages of a reach trucks are :

(*a*) They are readily available off shelf.

(*b*) They are relatively simple pieces of materials handling equipment.

(*c*) They are highly efficient.

(*d*) They can provide high throughputs.

(*e*) They require relatively low investment.

(*f*) They are easy to hire.

(*g*) They can operate in "conventional" buildings.

(*h*) They are robust and reliable.

(*i*) The types of attachments fitted to the reach truck.

The reach truck carries its load within its wheelbase. This allows the reach truck to maneuver in relatively small aisle. Depending on pallets size and reach truck design, the standard reach truck can operate in an aisle of between 2.7 m and 3.1 m. Modern reach trucks have extensive lift height capabilities. Typically, reach trucks are presently available with lift heights of up to 11.8 m. Most of the traditional reach trucks operate to heights of between 8 m and 9 m. Reach trucks are generally designed to handle pallets and loads of up to 2.5 tonnes.

The modern reach truck is electrically powered using a battery of capacity of between 350Ah and 900Ah and utilizes sophisticated electronics to provide power to the wheels, lifting systems, tilting systems and various add on devices.

In recent years, the development of AC powered reach trucks has further added to the variety and types of equipment available in the market place.

In general, the reach truck is likely to remain the most commonly used storage equipment in warehouses. Its high levels of flexibility and simple operations make it an ideal device for storage operations.

The reach truck can operate in a fairly restricted environment. While it is not as space efficient as the turret truck (lateral stacker), it is significantly more space efficient than the counter balanced forklift truck.

10.10.1 The Reach Truck and Aisle Width Dependability

Typically, the aisle widths for standard reach trucks vary from 2.7 m to 3.1 m. This however depends on :

(*a*) The type of throughputs (higher throughputs, require larger clearances).

The tolerance of installation of racking (the more "tight" the pallet fit in the rack, the larger the aisle one requires).

(*b*) The type of pallet one requires (the larger the pallet, the larger the aisle).

(*c*) The make and design of reach truck.

(*d*) The quality of floors available within the warehouse.

(*e*) The height of lift.

(*f*) The types of attachments fitted to the reach truck.

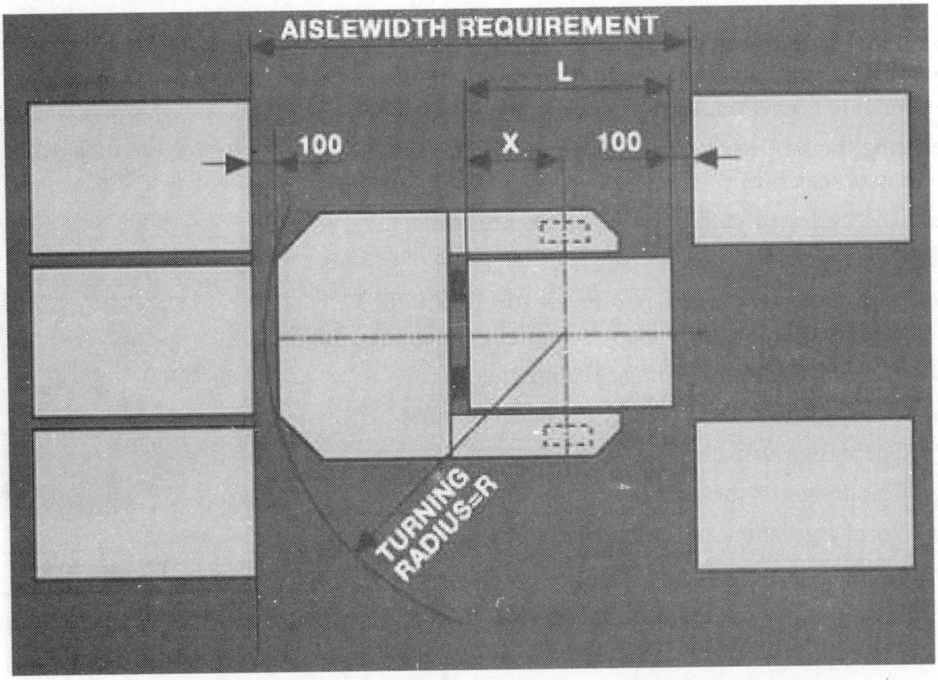

Figure 10.39

The above picture shows typical calculations for aisle width and a reach truck storage environment. However, the tolerances shown should be seen as a guide-line and will need to be varied, depending on the operating environment.

10.10.2 Picking and Storing in the Same Aisle

One of the main advantages of the standard reach truck is that, because it operates in relatively large aisle, one can undertake order picking and storage in the same aisle.

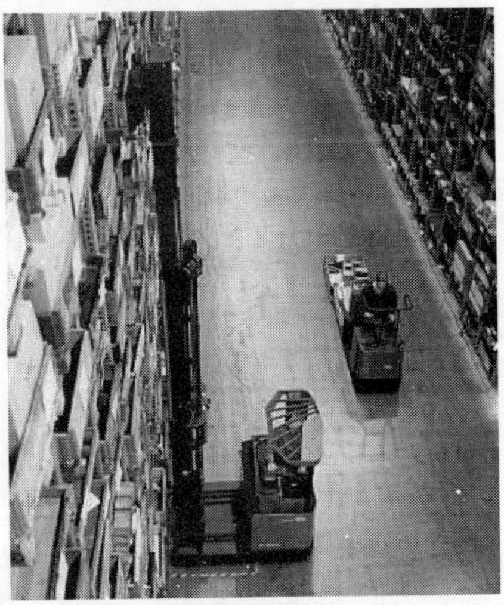

Figure 10.40

In this picture one can see a typical 3m wide aisle, and the reach truck is sharing this aisle with an order-picking machine.The system thus allows one to store at multiple levels with a reach truck and pick at lower levels with standard order picking equipment.

This system is highly flexible and allows high speed picking at ground level, efficient storage retrieval and efficient letdowns to the pick face. This standard system is used in many warehouses and has been found to be highly flexible, highly efficient and relatively easy to manage.

10.10.3 Reach Truck Configurations

Reach trucks are available in various configurations. With some reach trucks the operator sits, while with other reach truck operator stands. Some reach trucks have pantograph forks–where the forks use a scissors action to move forward. In other reach trucks, a sliding carriage moves the pallet forward into storage.

In the picture shown, this reach truck has a fixed base and stores and retrieves a Euro type pallet with no bottom stringers in the direction of fork entry. This allows further aisle reduction. Various configurations are available from many different suppliers. These different configurations each suit specific operational environments.

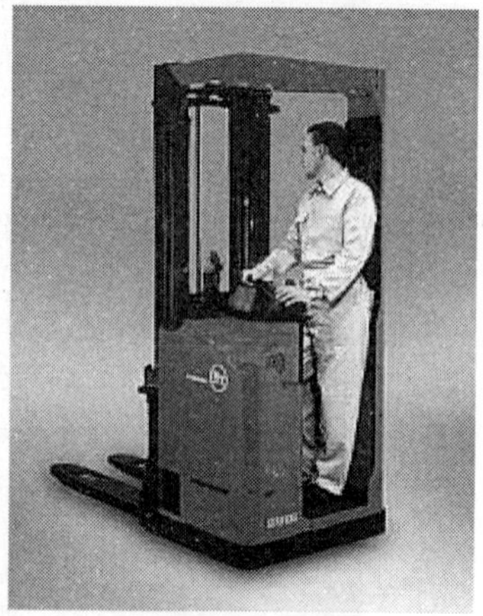

Figure 10.41

Selecting the best reach truck for your operation will depend on issues such as :

(*a*) The cycle rate and throughput requirements.

(*b*) The weight and size of pallet.

(*c*) The number of hours one needs to operate the reach truck.

(*d*) The lift height and storage requirements.

(*e*) The quality of the building (especially floors).

(*f*) The design of pallet.

(*g*) The design of the racking.

(*h*) Local legislation and safety requirements.

(*i*) Available space for asle.

(*j*) Interfaces with other equipment - either feeding the reach truck system or taking product away from the reach truck system.

10.10.4 Reach Trucks Applications

Reach trucks have many applications in warehouses :

They may be used in storage and conventional ambient building or they may also be used in cold stores.When operating in a cold store, one can install a special cab in the reach truck. This allows for comfortable operations at temperatures as low as –30°C. Refrigerated cab reach trucks are used in many cold stores.

10.10.5 The Storage Efficiency of Reach Trucks

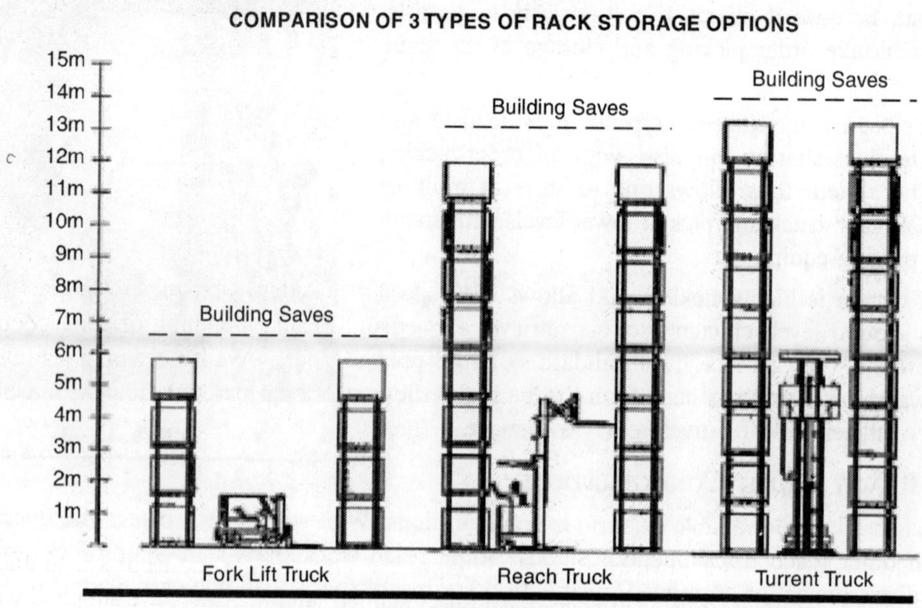

Figure 10.42

In this picture, reach trucks are compared with other types of storage equipment such as : Counter balanced forklift trucks, Turret trucks, Pallet stackers.

The storage density one will achieve in ones own warehouse will depend on many other restrictions such as :

- Available roof height.
- Position of columns.
- Order picking and throughput requirements (which will determine aisle requirements). Other environmental influencing factors.

The picture does however give one an extremely good indication of the relative storage efficiency of reach trucks - as compared to other equipment types.

10.11 SIDE LOADERS

It is a device which carries the load on its sides. These are built to handle specific product and engine operated. These are mostly used for longer and wide products.

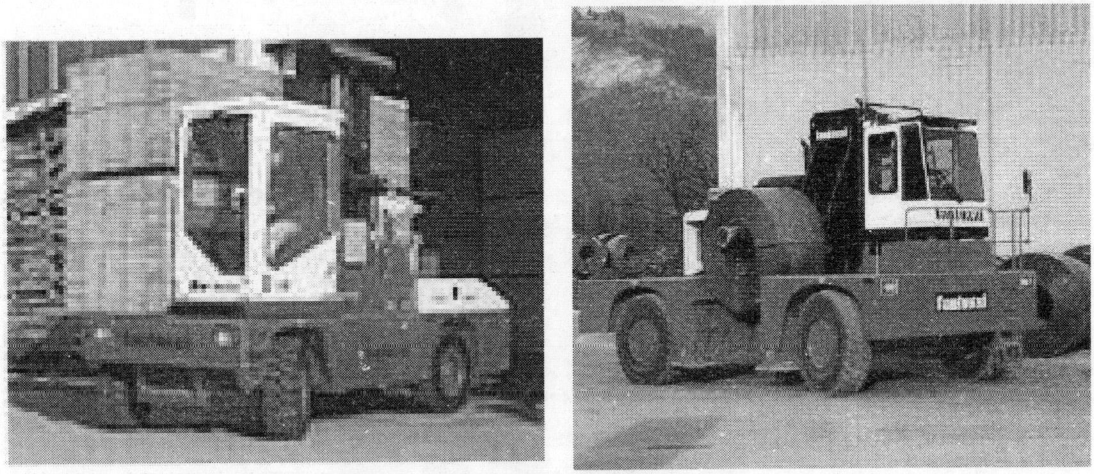

Figure 10.43

10.12 FOUR WAY TRUCK

The four-way truck is a reach truck which can also be driven sideways thanks to the fact that all 3 wheels can swivel. It therefore gives a very space-saving and flexible handling of long goods and other bulky goods such as windows, vinyl siding, shutters and doors. Available with hydraulic fork spread that allows the driver to transition from pallet handling to long load handling with a push of a lever. Provides for complete load lifting security and safety.

10.13 STRADDLE CARRIER

A mobile 4-wheeled vehicle which straddles a container and lifts and transports it to and from the vessel within the container yard. These are used for container handling, these are supplied with the load lifting attachments. These carriers lifts the load and transport it for a shorter distance in shipyard. They can travel on any path. It is manually operated.

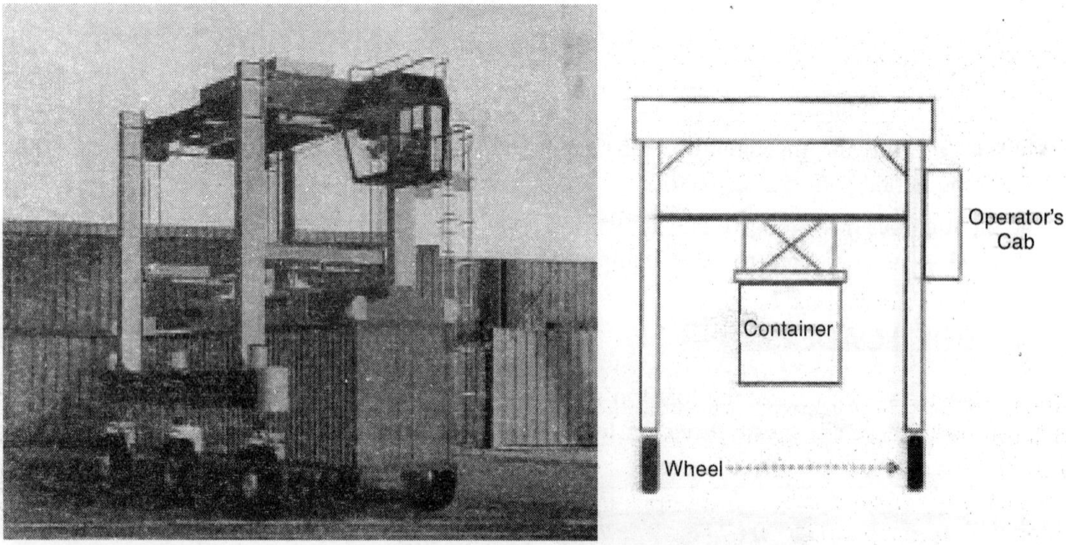

Figure 10.44. *Straddle carrier.*

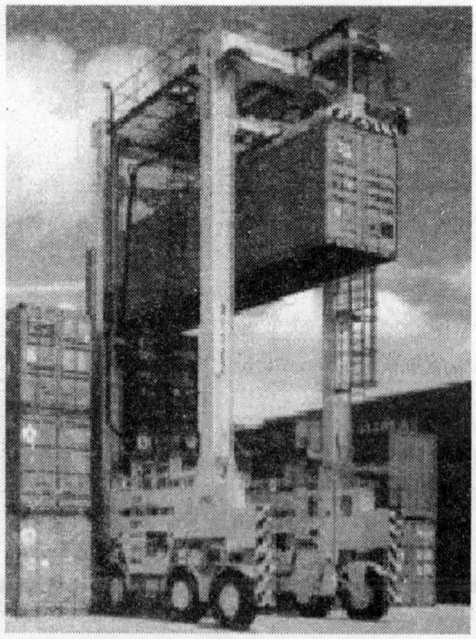

Figure 10.44. *Straddle carrier.*

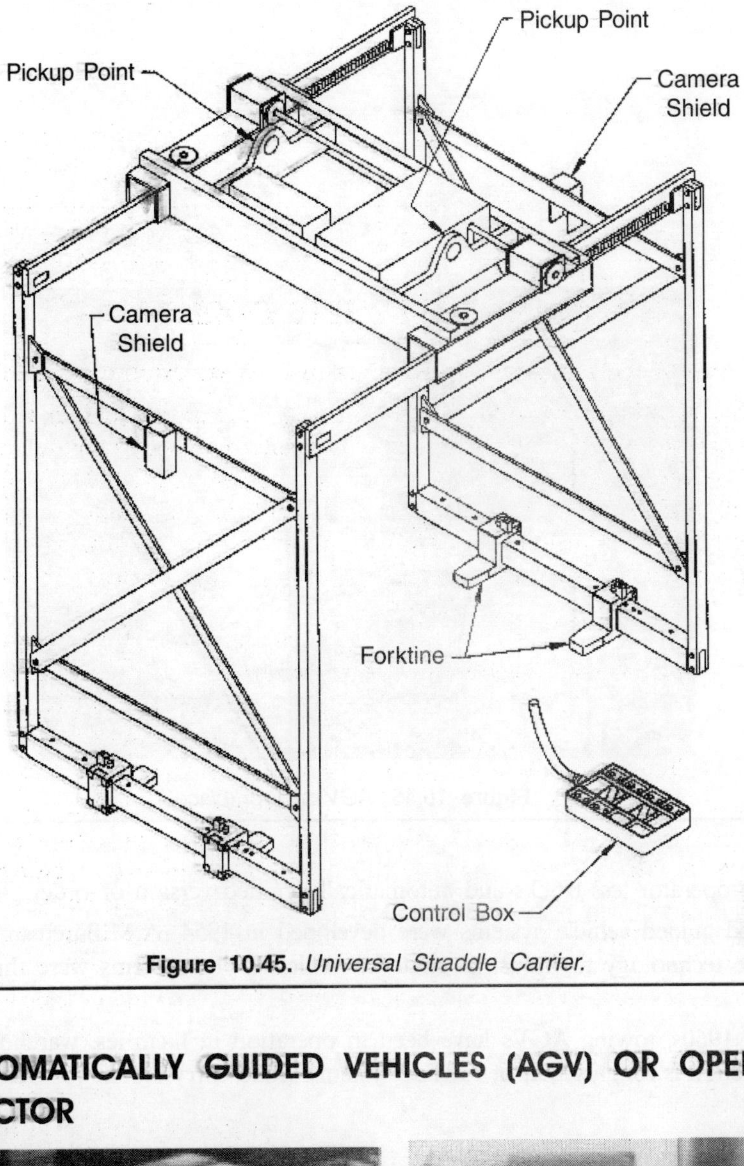

Figure 10.45. *Universal Straddle Carrier.*

10.14 AUTOMATICALLY GUIDED VEHICLES (AGV) OR OPERATOR LESS TRACTOR

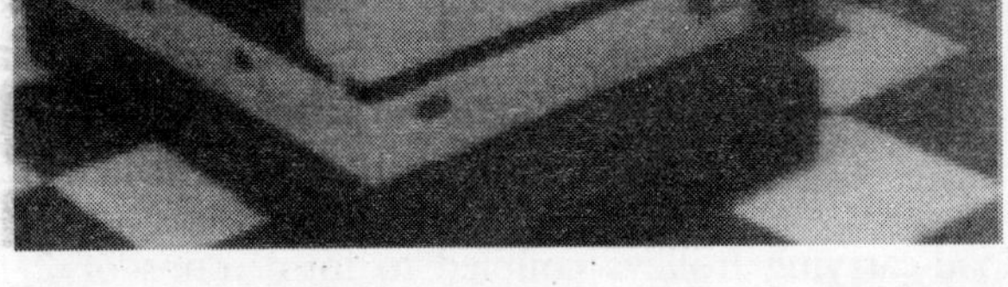

Pallet Carrier AGV

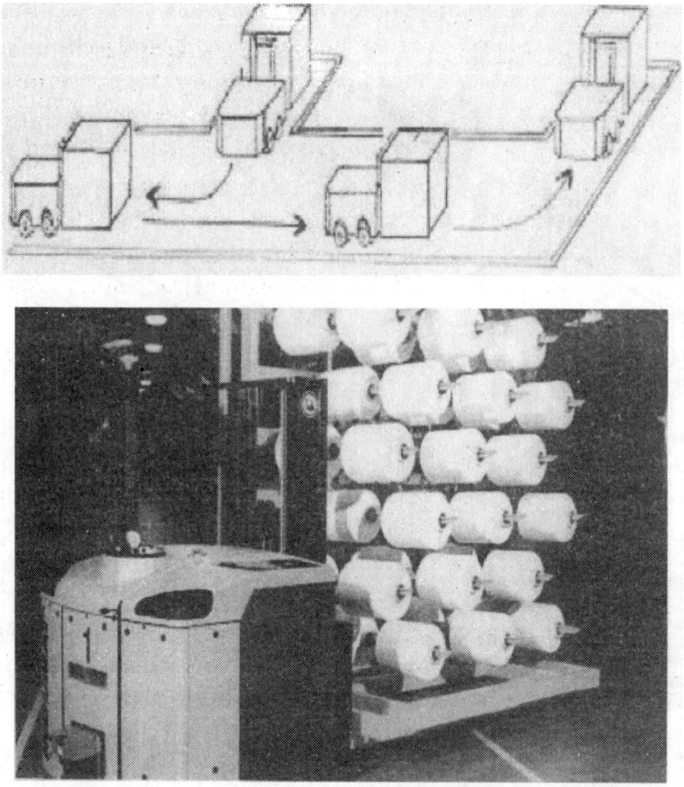

Figure 10.46. *AGV at Workplace.*

Introduction

These are operator less trucks and automatically guided version of order pickers.

Automated guided vehicle systems were developed in 1954. A.M.Barett are the inventor they apply electronic technology to industrial material vehicles. The systems were then called driverless vehicle system.

Since the 1960s, towing AGVs have been in operation in factories, warehouses and terminals where their function is roughly that of towline systems. Load carrying AGVs where developed during the 1970s.

10.4.1 Classification

Automatically Guided Vehicles

- Towing vehicles
- Load Carrying Vehicles
  - Load collection from floor
  - Load collection from shelves or racking

(*a*) Towing vehicles

Towing AGVs consist of a tractor and load-carrying trolleys coupled to form trains of 2–5 vehicles. The trains follow preset truck (100ps), stopping and vehicle uncoupling may be pre-programmed.

The tractor AGV speeds are 60–80 m/min and normal-tractive power is up to a few tonnes, minimum radius of bends is 2 metres approximately, since trolley drawbars enable excellent tracking, the tractors can handle slight gradients and can also be used out-of door for transport between building.

Equipment for automatic opening of doors etc., is available. This is the type of vehicle normally meant when reference is made to an AGV. Such vehicles may handle loads at floor level or from shelving. Space requirements for driverless trains vary, depending upon the length of train, type of steering on the trailers and the width of the aisles they will be operating in. Most driverless trains system applications require side tracks or spurs to spurs to permit moving trains to pass trains parked at stop stations.

(*b*) Load Carrying Vehicles

The load-carrying trolleys have platforms or roller tables conforming to the loading/unloading systems used.

Towing as well as load carrying AGVs may be equipped with guidance system which follow floor marking because such system are vulnerable to mechanical wear and dirt, they are not used extensively.

10.14.2 Safety Features of AGV

- Emergency bumpers to prevent collisions.
- Lights (rotating red lights).
- Guide path monitor to shut down a vehicle if it deviates from path.
- Audio Signal (Continuous).

10.14.3 Various Flow Pattern

(*a*) Standard system with basic controls.

(*b*) Advanced system with microprocessor controls.

(*c*) Microprocessor system with interface to other systems.

(*a*) Standard System with Basic Controls

- Having 3 or 4 stop stations.
- 1 or 2 vehicles plus associated trailer.
- Total length 300 to 600 m.

(*b*) Advanced System with Microprocessor Controls

This system permits more vehicle over longer and complex routing. The individual vehicles may be remotely programmed and automatic loading and unloading may be provided.

(*c*) AGVs have a tremendous range of automated capabilities. Many systems are called upon to go up and down ramps, upon and close automatic doors, cross draw bridges use elevators and perform advanced automatic loading and unloading junctions. These capabilities give a materials handling engineer a powerful tool to use in solving horizontal materials transportation problems.

10.14.4 AGV are Most Efficient When

- A heavy flow of material must be moved between 2 points.
- Several pallet loads or equivalents must be dropped off or picked up at intermediate locations along a route at least 45 m.
- Finished work on pallets is to be transferred to storage.

10.14.5 Advantages of AGV

(*a*) AGV uses existing aisles to transport materials throughout a facility.

(*b*) Any where fork trucks were previously transporting loads, an AGV can work.

(*c*) AGV used to change the function.

(*d*) Installation of AGV system is relatively easy compared with conveyors on in-floor flow lines with relative complexity.

(*e*) The installation time is 2 to 3 week for 600 m.It is portable. Manual operation of AGV equipment is important if the automatic central fails.When the vehicle fails in the automatic mode, it can be driven manually.

(*f*) Not require concrete floors, although concrete is the best conveyor.

(*g*) Real-time color graphics monitoring on board microprocessor to make it's own decision as it travels along the guide path, central computer, automatic summoning with accountability.

10.14.6 Limitations of AGV

(*a*) If the tractor must travel on upon space between 2 buildings, the operation is usually very precarious and hazardous especially weather cold, wet, snow-ice, winter.

(*b*) In dry and fairly climate it is very well in performing their function.

(*c*) Reliability is marginal.

(*d*) Very expensive.

(*e*) Easily damaged.

(*f*) Require extra skill to maintain electronic components.

	Towing version	**Load carrying version**
Speed	60 m/min	60 m/min
Capacity	2.4 trolley each carrying 1000 kg.	Normally heavier loads possible in special cases.
Locating accuracy	–	± 5 mm
Turning Radius	2 m	0–6 m
Acceleration	–	0.5 m/s^2

Table 10.1 : AGV Specifications

10.14.7 Applications of AGV

- In textile Industry : Move sperels of yawn's.
- In paper Industry : Rolls of paper or flat stock.
- In rubber Industry : Green and finished tyres
- In general Manufacturing Industry : Used to move incoming material to wearhouse locations and from wearhouse location to shipping dorks.
- Used to move material from phase of their operation in small moderate movements.

Besides handling bulk movements driverless trains are highly effective for chain movement.

A Chain movement occurs when material is loaded on to a train for dispersal throughout a system.

Among *e.g.*, are assembly AGVs with bodies specially designed for specific workplaces *e.g.*, car engines, car bodies or lorry cabs. Such vehicles are used above all as an aid for new factory layout where assembly operation are carried out by self contained trams (Volvo).

10.15 TOW TRACTOR AND TROLLEY

Introduction

Everybody knows the tractors used in the farms for the farming work, these are no way different from these tractors except the purpose. In short, tractors which are used for farming they can be effectively used for material handling.

Consider a handling of sugarcane from the farm to the sugar industry. In rural India it is done by bullock carts, tractors with one or two trolleys or either by trucks.

Trolleys are the equipment commonly used for short distance transportation of bulk goods or general cargo consisting of small items. These are one of the most common modes of horizontal transportation. They are either moved manually or mechanically. Trolleys are generally coupled with the tow tractors.

10.15.1 Classification of Tractors

(*a*) Tow Tractors can be classified according the source of power :

(*i*) **Engine Operated (Diesel Engine)**

These are conventional tractors which are used for farming used as tow tractors.

(*ii*) **Battery Operated**

Battery operated counterbalanced trucks are used as tow tractors. it has low capacity that conventional tractors. Used for in plant handling. These have very compact construction so that it can be maneuvered easily.

(*b*) These are also classified according to the operator.

(*i*) Operator Controlled Tractor.

(*ii*) Operator less Tractor (AGV used as tuggers).

Operator Controlled Tractor

The operator-controlled tractor called the "tractor ox" is controlled by an operator who carries or his belt a small radio control box. By means of this he is able to steer it to either the right or the left. The operator rides on the tractor and operates all the controls. Steering wheel is provided for the maneuvering (movement and transportation). These are used for in plant as well as outdoor handling. These are very powerful devices with high pulling capacity. Very heavy tyres generally pneumatic sometimes solid tyres were also used.

10.15.2 Trolleys

Trolleys can be classified as :

(*a*) According to place where load is put :

(*i*) **Box Trolleys**

These are characterized by box type of construction. It may be solid or wiremesh construction. Solid trolleys are made up of metal or plastics. These are used to carry bulk loads. Capacity may vary according to the place. Consider a example of bricks trolley, this is a good example of box trolley.

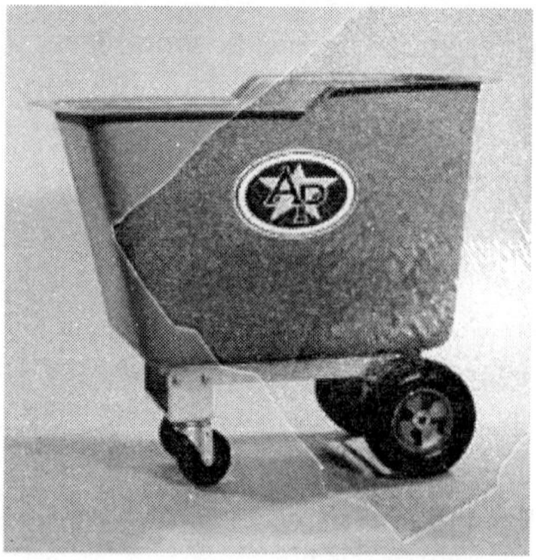
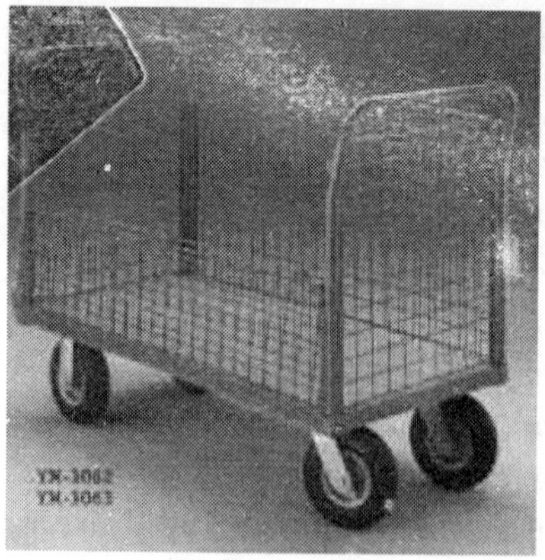

Figure 10.46

(*ii*) **Platform Trolleys**

These are characterized by flat platform used to put the load on it. These are used to carry unit loads.

(*b*) According to the wheels trolleys can be classified as

(*i*) **Two wheeled trolleys** have back wheels and front portion is attached to the towing tractor through coupling or joints.

(*ii*) **Four wheeled trolleys** have two back wheels and two front wheels and front portion is attached to the towing tractor through coupling or joints.

10.15.3 Uses of Tractor

These are used in both indoor and outdoor operation. Important function as a horizontal transportation device.

(*i*) Material transfer by means of trolleys generally known as trailer system.

(*ii*) Move material from one storage and operating area to another. They are available for fairly long runs or congested travelling areas.

(*iii*) Serve a number of load collecting and dispatching location over rectangular route.

(*iv*) Used at airports for goods as well as plane handling.

(*v*) For moving passenger is another important benefit especially in rural areas.

10.15.4 Advantages of Tractor

Great flexibility is achieved as tractors can be used to hauled different types of trailers. Trolleys can be left loaded and can be picked up by different tractors. This system has the advantages of great flexibility plus all the advantages of industrial railway, as there is no investment in laying tracks. This is one of the most important methods of materials handling inside the plant, or work area and from one building to another. Can be coupled any type of attachment made for the purpose.

10.15.5 Disadvantages of Tractor

(*i*) These generally require smooth terrains (ground).

(*ii*) These must be designed as a compact units if used for indoor handling such as warehouses, otherwise there is a problem of maneuvering.

10.16 CARTS

These are same as that of tractor and trolley system except power source. In simple words carts uses human or animal power for maneuvering.

Depending on that they can be classified as :

Human cart : Human carts are used for low capacity short distance transportation.

Animal cart : Animal carts are used for high capacity long distance transportation compared to human cart. Varies according to the animal used to tow the trolley. Generally used animals are bulls, horse, buffalos, ox, yak, and donkey.

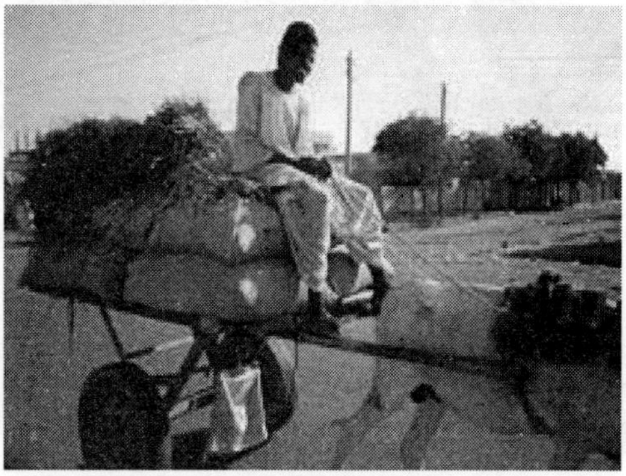

Figure 10.47

UNSLVED QUESTIONS

1. Write about industrial trucks.
2. Classify trucks.
3. What are important parts of truck.
4. Write about sack or case truck.
5. Write about hand driven truck.
6. Write about electric tow truck.
7. Write abvout hand pallet truck.
8. Write about powered pallet truck.
9. Write about forklift truck.
10. Write about counter balanced truck.
11. Write about walky.
12. Write about order picker.
13. Write about powered stacker.
14. Write about reach truck.
15. Write about side loader.
16. Write about four way truck.
17. Write about straddle carrier.
18. Write about automatically guided vehicles.
19. Write about tow tractor.
20. Write about trolleys.
21. Write about carts.

11 Pallets and Containers

INTRODUCTION

In industry variety of material gets in and gets out such as packaged material, bulk material. One cannot handle all these material directly because some materials are fragile, some are brittle, some are softer etc. Geometry of material such as shape and size also affects on selection of material handling. All the material handling equipments cannot handle the material directly; due to this the material is either put on something or contained in the something. To do this job pallets and containers are used.

Load is put on the platform and the platform is moved using material handling equipments are known as load carriers. These include pallets, skids and dolly. These are generally used for unit handling. Load is contained in some form of container and it is handled. These includes containers, drums, barrels, trays, crates, bins, bags, cartons, boxes, jars etc. These are the structures used to hold several items and together to facilitate handling and storage by standardized means. These are designed to be compatible with existing storage and transporting system.

These above equipments works on the unit load concept.

11.1 UNIT LOAD CONCEPT

Everybody needs vegetables and fruits in day today life. Consider a case of buying the one kg of fruits of five varieties say mangoes, jam, apple, grapes and bananas. You purchased from one wholesaler. Now you want to carry all these fruits to home, what you will do? Logical as well as practical solution is obvious *i.e.,* you will ask wholesaler to give some kind of container to carry all the fruits. Let's suppose some how he didn't gave you a container and still you can bring the fruits at home. Now you can easily visualize what will be the situation. You have to make at least five trips to home and to wholesaler.

Its better to carry the items in the bulk in a container rather than to carry them individually this is a concept of unit load. Consider a box containing 5 kg apple will form a unit load, a bag of 5 kg onions will form a unit load. Hence unit load depends on the capacity of the container as well as on the capacity of the material handling equipment and vice versa. Characteristics of a unit load :

(*a*) These should be minimum number of handling.

(*b*) Manual handling should be eliminated.

(*c*) Material assembled into a unit load for economy of handling and storage.

(*d*) Redesign packages, containers etc. for better assembly into unit loads and returns them to prevent product damage.

(*e*) Unit load should be as large as possible within the constraints of building equipment etc.

11.1.1 Advantages of Unit load Concept

(*a*) Possible to handle large loads.
(*b*) Reduce handling cost of one unit.
(*c*) Reduce packaging cost.
(*d*) Maximize space utilization.
(*e*) Faster movement of materials.
(*f*) Reduced loading and unloading time.
(*g*) Reduces pilferage/damage during transit or storage.
(*h*) Safer handling.
(*i*) Reduces cost of transportation.

11.2 PALLETS AND ITS TYPES

These are very important load carrying equipment used in almost all industries to carry unit loads over it. They are square or rectangular in shape. Although standard pallets are available size depends on the load to be carried. These are made up from wood, metal or plastics.

Softwood is offered as two different Pallet Design Types: Stringer and Block. The stringer is standard pallet design while the block is designed for more accessibility for moving the pallet.

11.2.1 Softwood Pallets

Softwood is the most economical platform available. It is the standard limited-use pallet designed for an average of up to nine trips, with an average of five handlings per trip in an average environment.

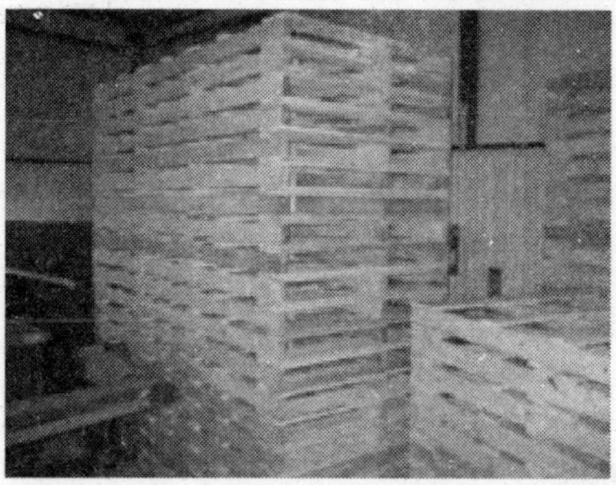

Figure 11.1

(*a*) Pallet Design Type

Figure below shows the two types of the designs of the pallets and important parts of it.

Stringer Design

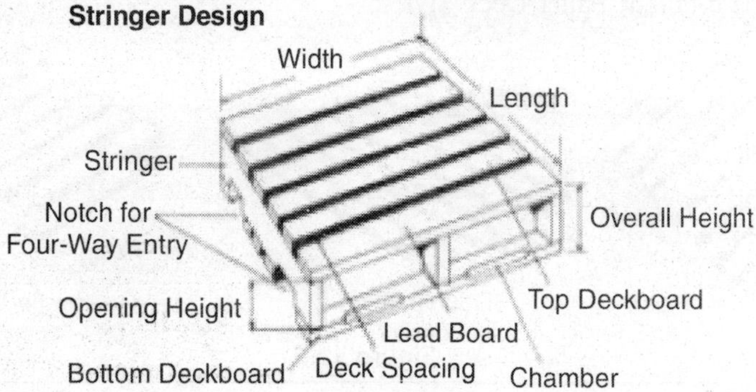

Figure 11.2

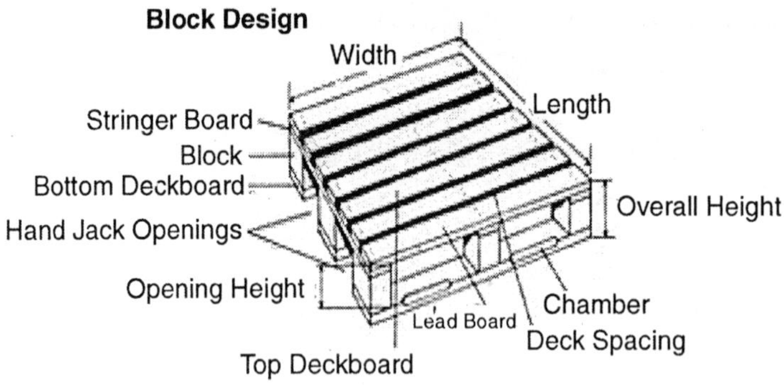

Figure 11.3

(b) Pallet Styles

There are three different available pallet styles:

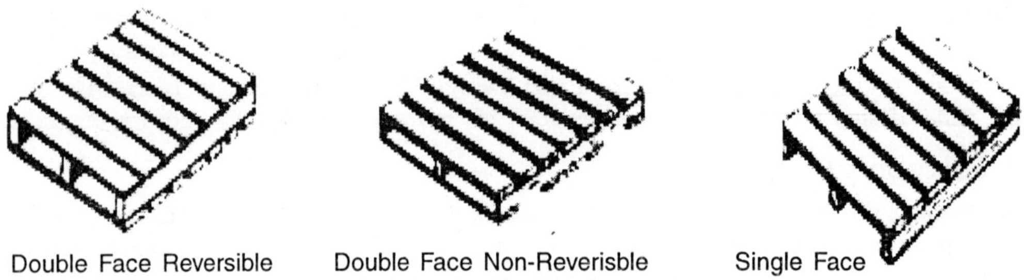

Figure 11.4

1. **Double Face Reversible:** Can be used from both sides.
2. **Double Face Non-Reversible and Single Face:** Can not be used from both sides.
3. **Single Face.**

(c) Entry Types

According to the type of the fork entry. There are two types.

2-Way : A pallet without notches in the stringers. Its un-notched solid stringers allow entry from only the two opposite pallet ends.

4-Way : A pallet with notched solid stringers to allow entry from both ends and both sides. Forklift access is through the notched stringers.

(d) Pallet Deck Style

There are three different Pallet Deck Styles:

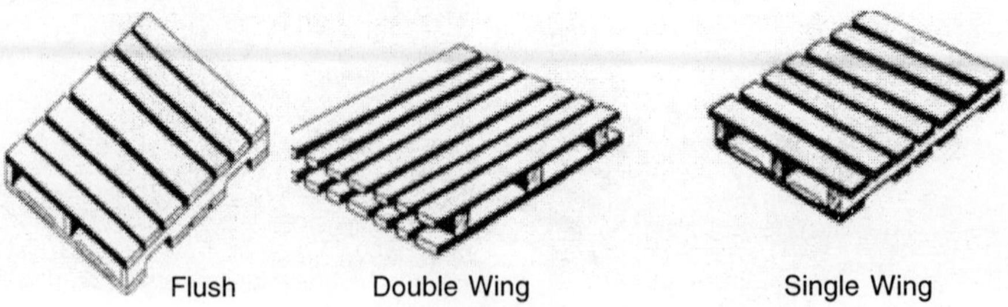

Figure 11.5

Wing pallets in which one (single wing) or two (double wing) decks projects beyond the spacers.

11.2.2 Hardwood Pallets

These are same as that of the softwood pallets except the material is changed one. Hardwood, made from broad-leaved tree species, is the most economical storage platform available. It is the standard multiple-use pallet designed for repeated uses for more than one unit load with an average minimum life-to-first repair of ten trips or more, with an average of five handling per trip in an average handling environment.

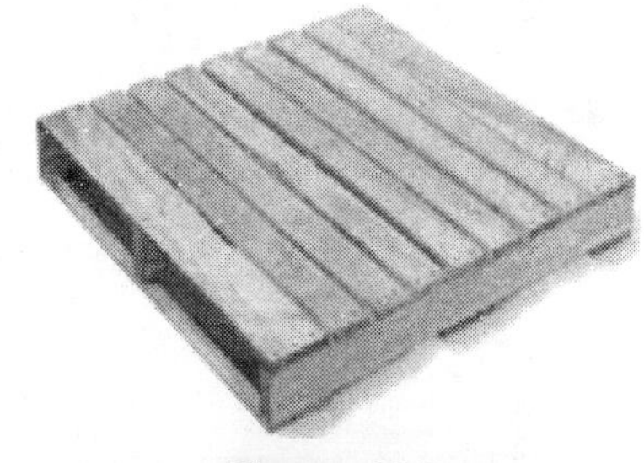

Figure 11.6

11.2.3 Molded Plastic Pallets

Molded plastic is typically a captive/slave pallet intended for use within the confines of a single facility, not intended to be 'exchanged'. It is versatile, custom designed, 100% recyclable, easily stored and assembled, and maintained in a climate range or – 30 °F to +150 °F.

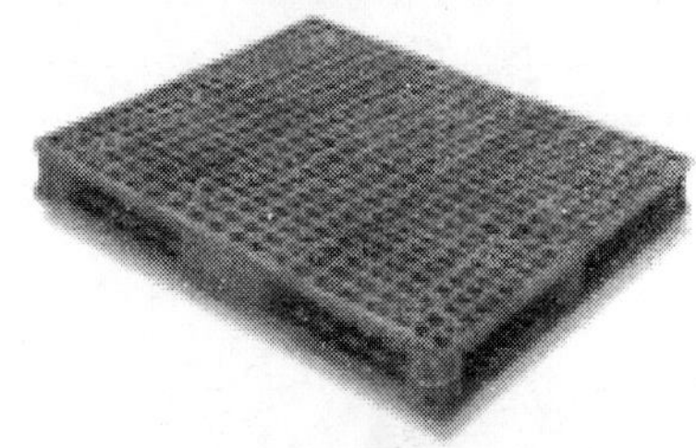

Figure 11.7

Molded Plastic Pallet Styles

Molded plastic is offered in four different pallet styles: Double face Reversible, Double Face Non-Reversible, Single Face and Nestable. Nestable pallets are Shaped or designed to efficiently stack pallets inside one another, saves on storage space and shipping costs and is safer for product.

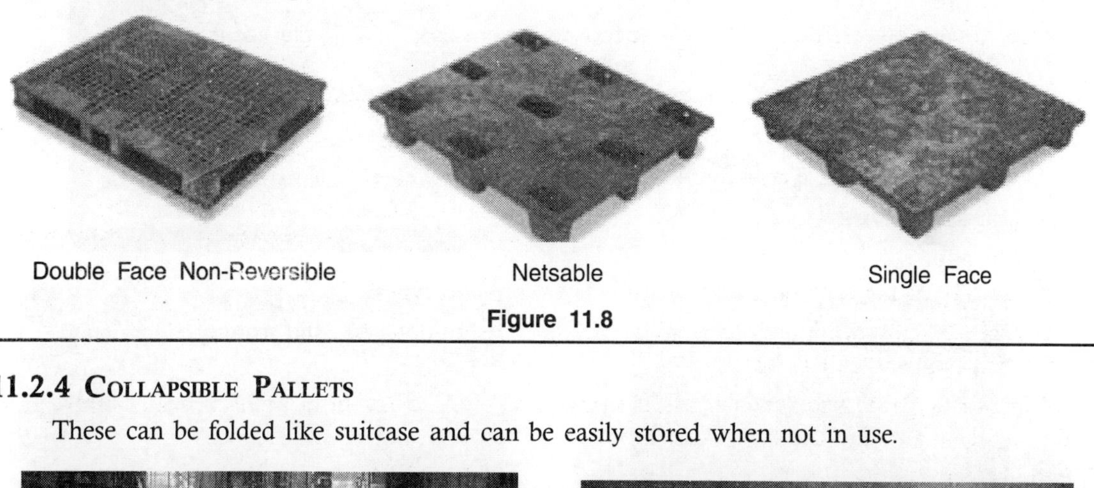

Double Face Non-Reversible | Netsable | Single Face

Figure 11.8

11.2.4 Collapsible Pallets

These can be folded like suitcase and can be easily stored when not in use.

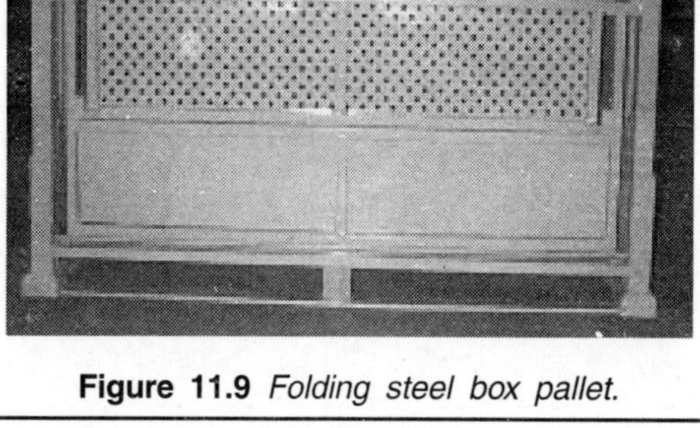

Figure 11.9 *Folding steel box pallet.*

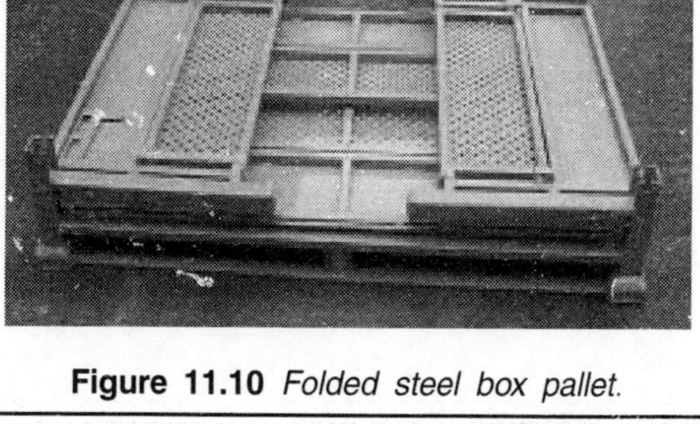

Figure 11.10 *Folded steel box pallet.*

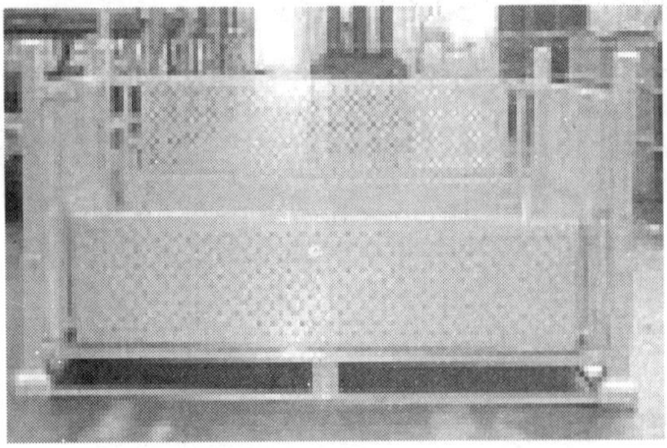

Figure 11.11 *Collapsible side pallet.*

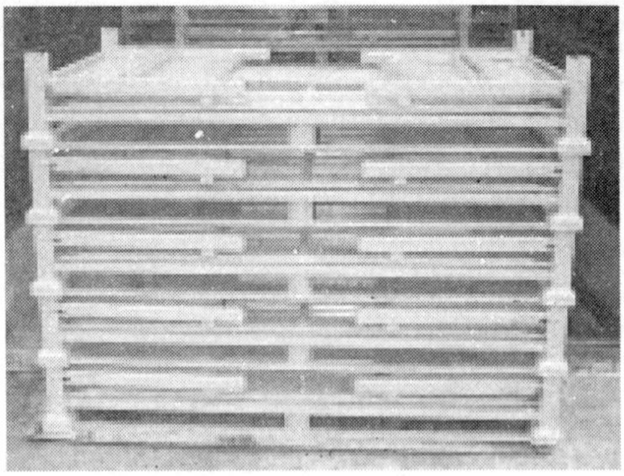

Figure 11.12 *Stacked box pallets.*

11.2.5 PPLICATION EXAMPLES OF PALLETS

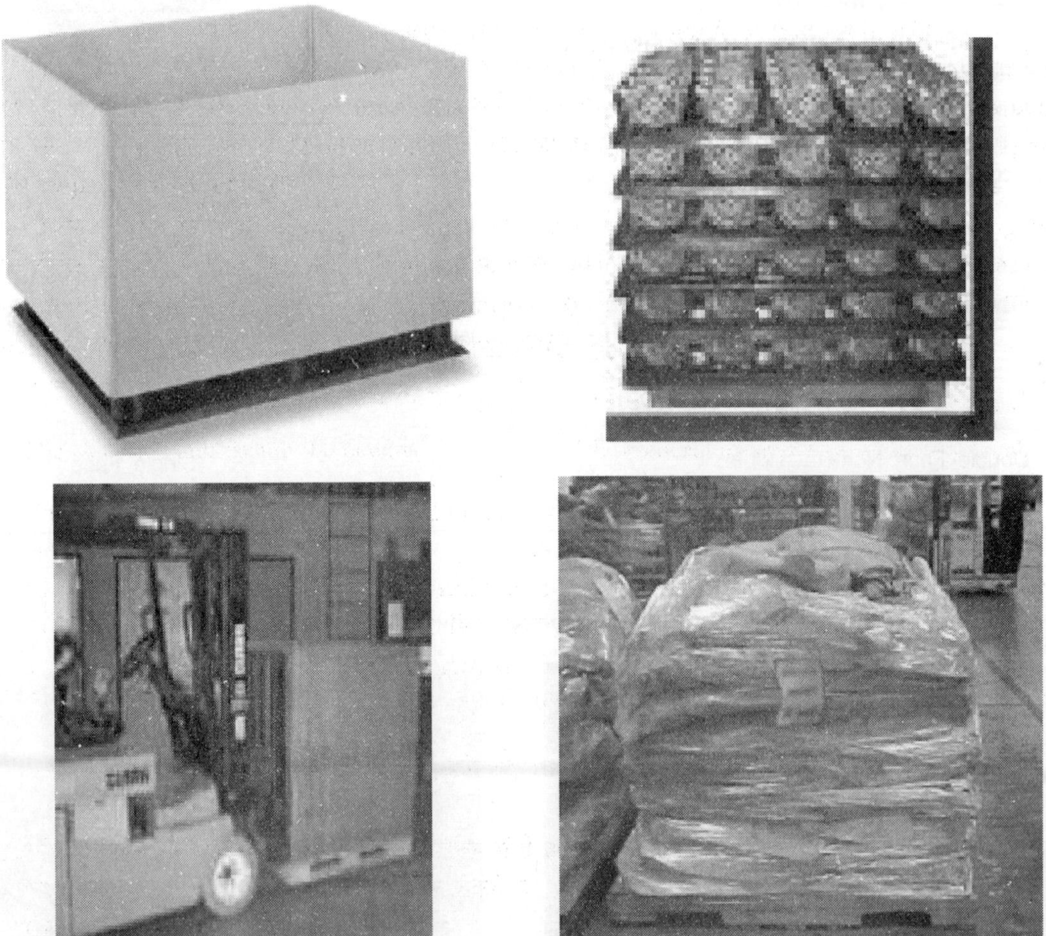

Figure 11.13

11.3 SKIDS

These are used with lift trucks. Goods may be loaded into the skids and then picked up by fork lift trucks. The skids can be loaded as a unit and transferred from position without subsequent loading and unloading. Both skids and pallets raise the load off the supporting surfaces and allow the easy insertion of conveying means. Skids have at least two steel legs.

There are two types of skids as shown in the figure below :

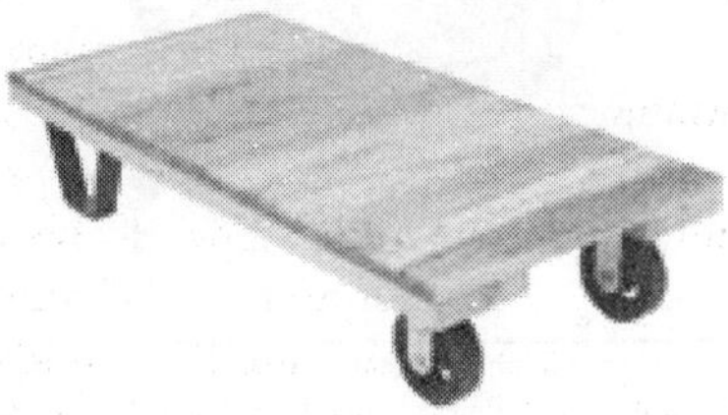

Semi-Live

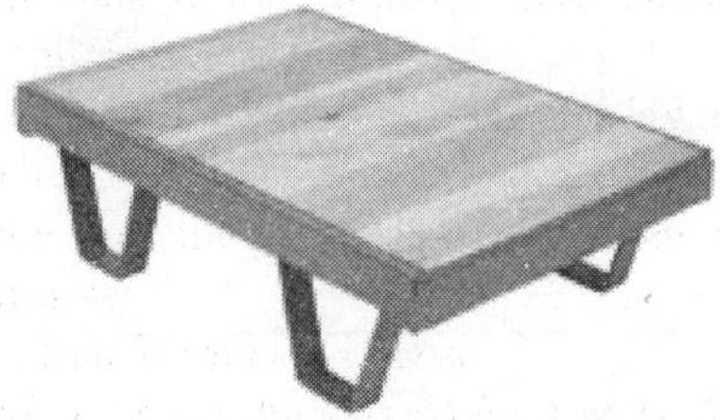

Dead Skids

Figure 11.14

11.4 DOLLY

Some materials are stored on dollies. This can be moved on their casters (wheels). Dollies are specially made for special products. They are made up of wood or some times of metal depending on the load to be carried. Dollies do not have legs they have all four casters (wheels) which can rotate in any direction.

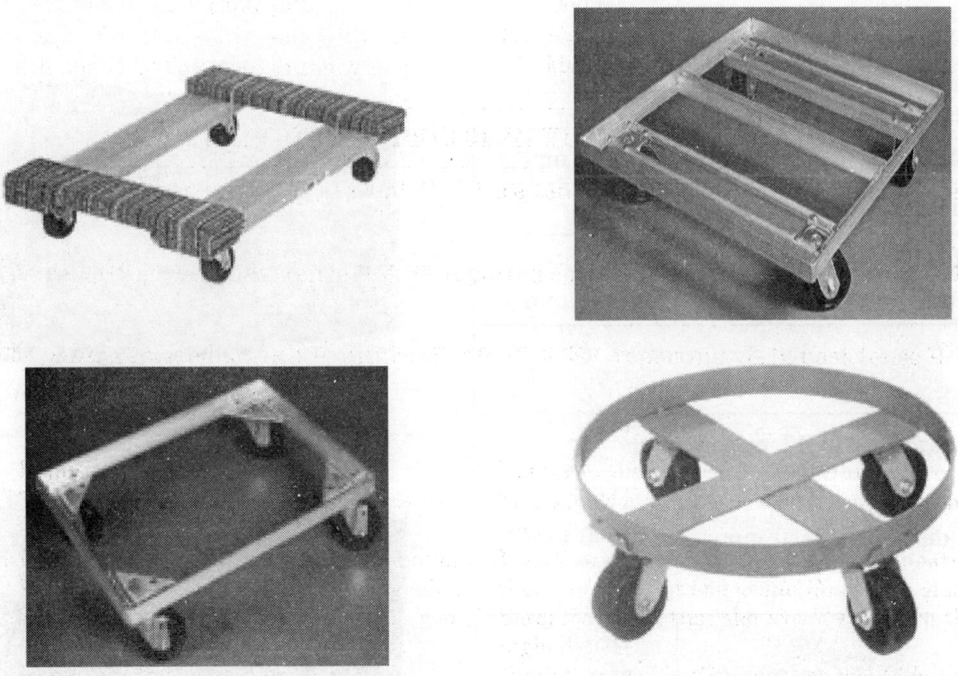

Figure 11.15

Application Examples of Dollies

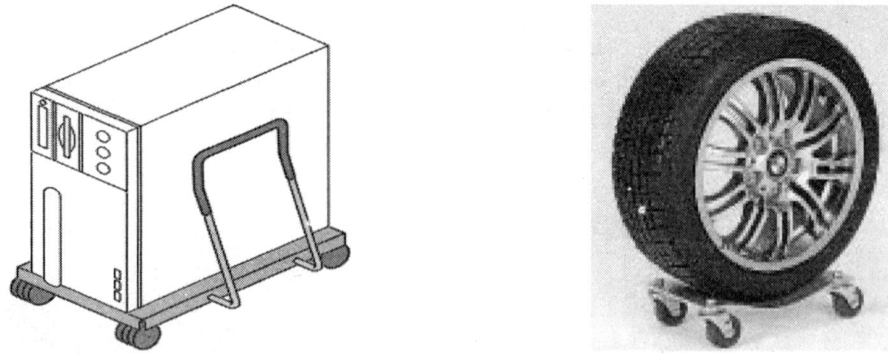

Figure 11.16

11.5 CONTAINER

These use the principle of unit handling. Containers are very useful for the bulk handling of solids and liquids. They are available in various shapes and sizes. These are specifically more useful in shipping and oil industries.

(*a*) Shipping Containers

These containers are equipped with exterior hinged double doors at one end which open completely for the emptying. Generally walls of these containers were made from beaded steel sheets, steel profile frame, wooden floor on steel cross members. This is a container in which 2 forklift pockets were provided on each longitudinal side.

Dimensions of 6 m containers

Inside dimensions		
Length 5898 mm	Width 2350 mm	Height 2390 mm
Outside dimensions		
Length 6058 mm	Width 2438 mm	Height 2591 mm
Weight 2300 kg		

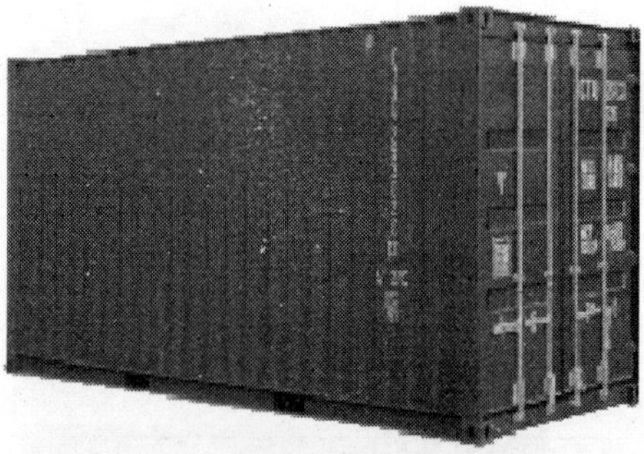

Figure 11.17

Dimensions of 12 m containers

Length 12030 mm	Width 2350 mm	Height 2390 mm
Outside		
Length 12190 mm	Width 2430 mm	Height 2590 mm
Weight 4100 kg		

Figure 11.18

(*b*) Drum Containers

Stainless steel containers used for fluids and solids handling such as food, flavors, fragrances, oil, lubricants etc. Capacity varies from drum to drum. It ranges from 13–250 litre. Different sizes of the drums were available as per the ISO standards. This type of drum containers are useful in oil industries, nuclear waste, chemicals, electronics, pharmaceuticals industry and other various applications at low costs.

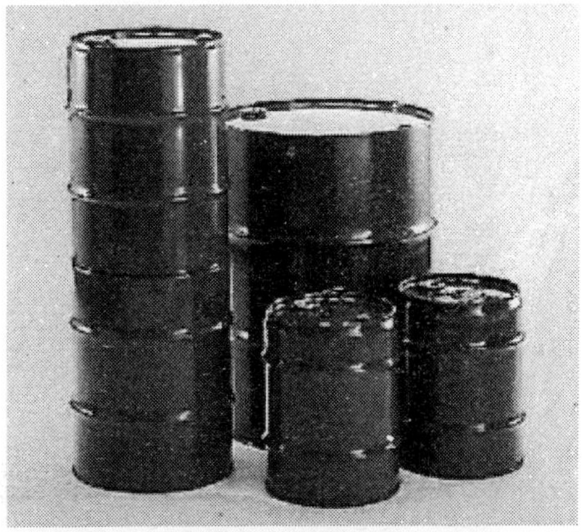

Figure 11.19

11.6 CRATES

Crates are very useful bulk handling equipments which are extensively used for handling bottles, eggs, vegetables, tea, fruits, fish etc. These are available in the various forms, shapes and sizes. These are made up of wood, metal or plastics depending on the material to be transported. These containers provide very high strength and durability for many applications. They are very cost effective because they are cheaper, light in weight and reusable. Design is made such a way that number of crates can be stacked over each other and stored.

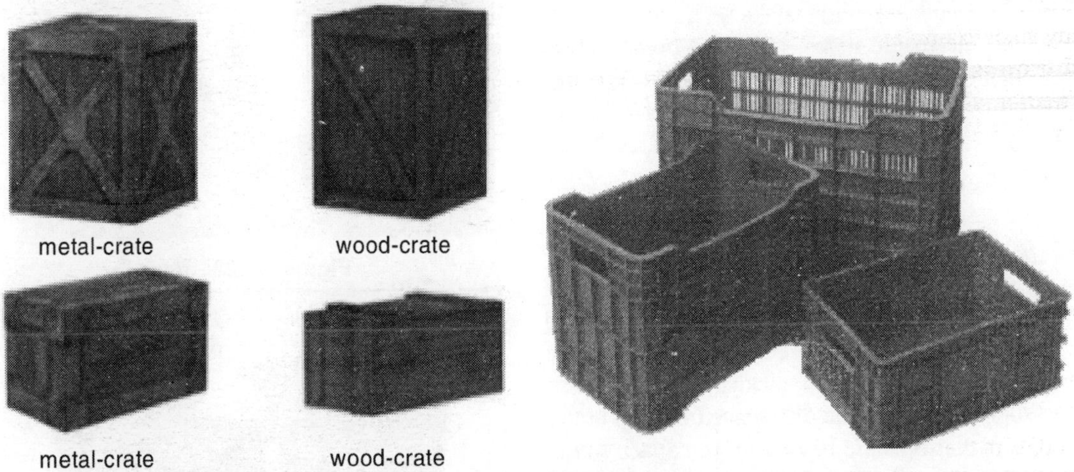

Figure 11.20. *Plastic crates.*

(*a*) Fold and Carry Crates

These are special type of crates which can be folded for easy transportability. It is supported by plastic or wood at the bottom.

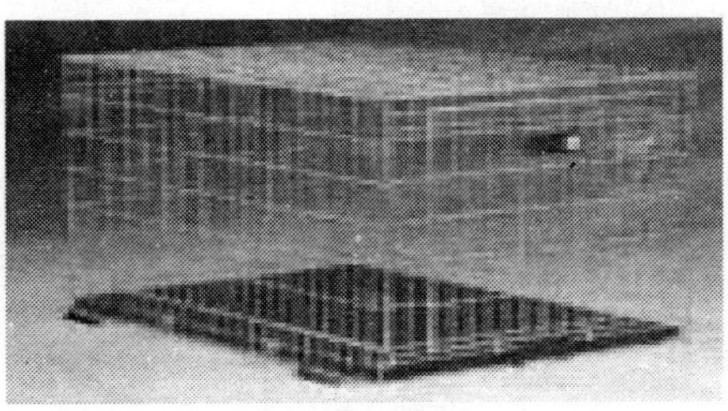

Figure 11.21

(*b*) Application Examples of Crates

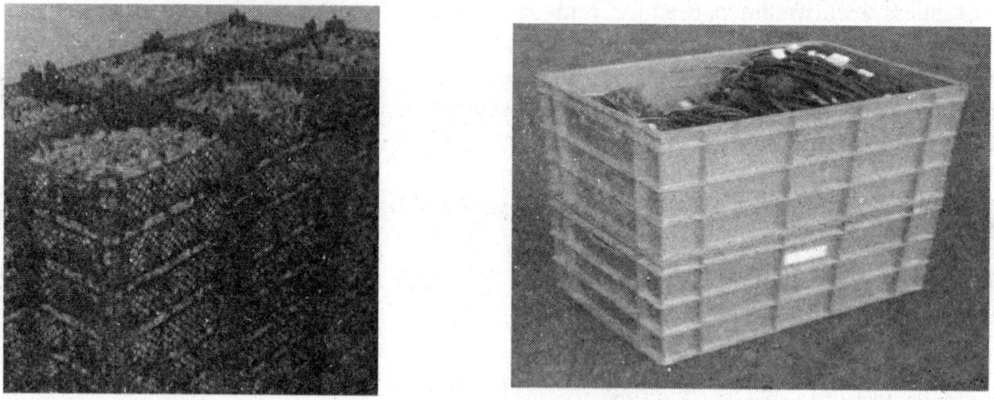

Figure 11.22

11.7 TRAYS

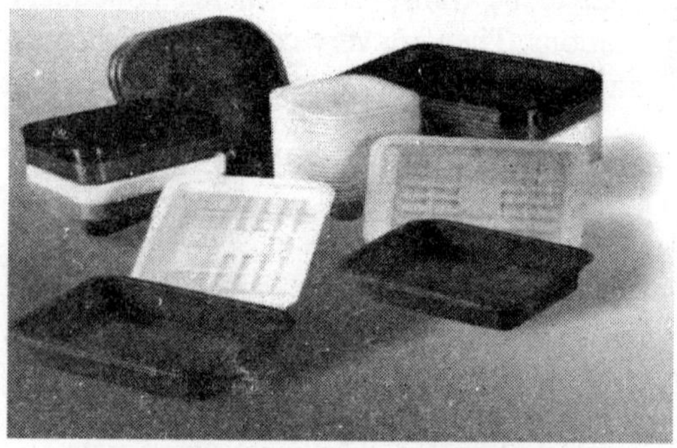

Figure 11.23. *Trays.*

Trays are different types of containers which are used for bulk handling of loose materials such medicines, electronic items, nuts and bolts and so many such examples. The difference between crates and trays is that crates are having more capacity than trays. Also trays are more flat than crates. These are light in weight and easy to clean.

11.8 BINS

Bins are types of containers which are used for bulk handling of loose materials such as waste papers (as a dustbin), as a pickles carrier and so many such examples. The difference between crates and bins is that bins are having more capacity than crates. Also bins have more height than length. These are light in weight and easy to clean.

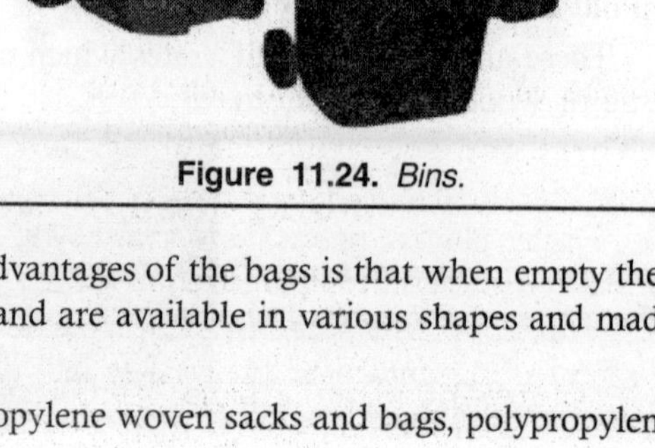

Figure 11.24. *Bins.*

11.9 BAGS

Bags are types of containers which are used for bulk handling of loose materials such as powders (cement, soil, wheat etc), grains, sand, house hold appliances and so many applications can be found. The advantages of the bags is that when empty they can be stored very easily. The capacity of the bagsvaries and are available in various shapes and made up of different materials as listed below.

Paper bags, T-shirt bags, plastic T-shirt bags, polypropylene woven sacks and bags, polypropylene woven fabric.

11.9.1 Paper Bags

Paper bags are used for packing of cement, minerals, flour, tea, cattle feed etc. Capacity varies from 1 kg to 50 Kgs.These are available in so many varieties such as with block bottom, top open, top block with valve made out of single ply to 5 ply with or without PE lining and aluminum foil.These are useful for tea and coffee, food packaging paper bags. Open top paper bags for packing of flour, maize, cattle feed wrap sheets and rolls from single ply to 5 ply.

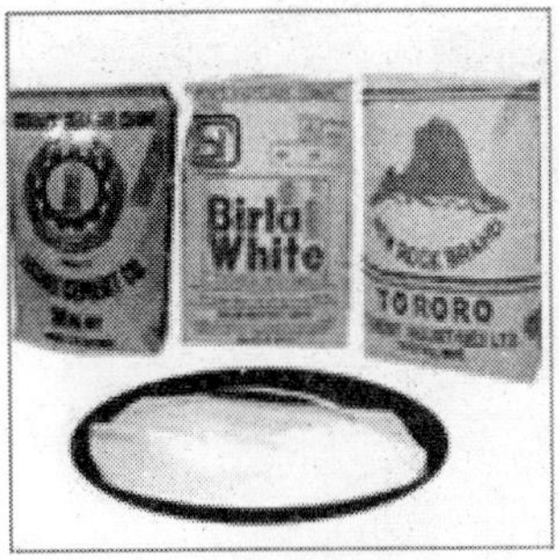

Cement Bags

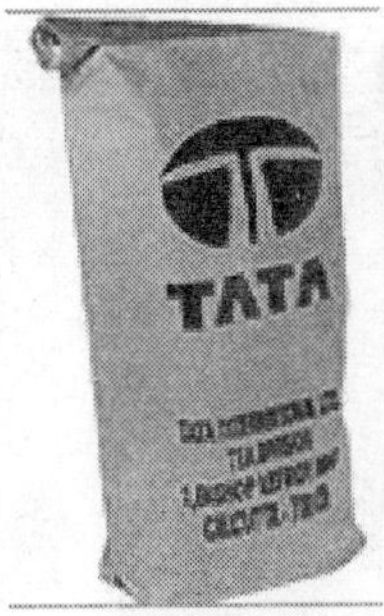

Tea Bags

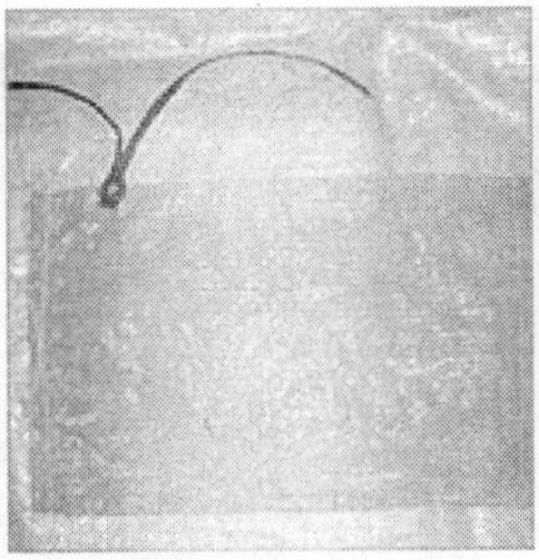

Woven Bags

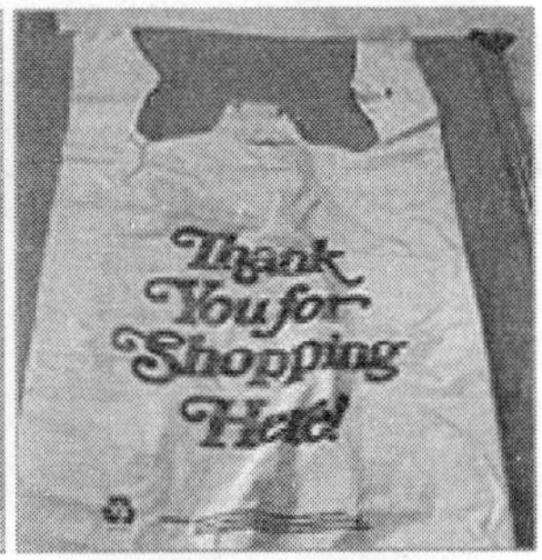

T-Shirt Bags

Figure 11.25. *Bags.*

11.9.2 Application Example of Bags

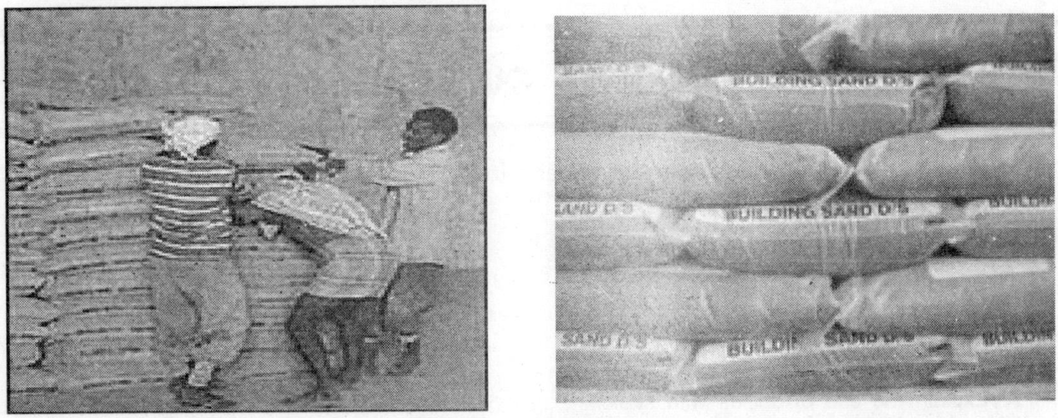

Figure 11.26

11.10 JARS

These are generally useful as bulk container. Standard jars to use in home canning. These reusable jars are produced in a variety of sizes, some with decorative finishes for gifting. They are made to seal properly, to be durable with repeated use and to be used safely in.

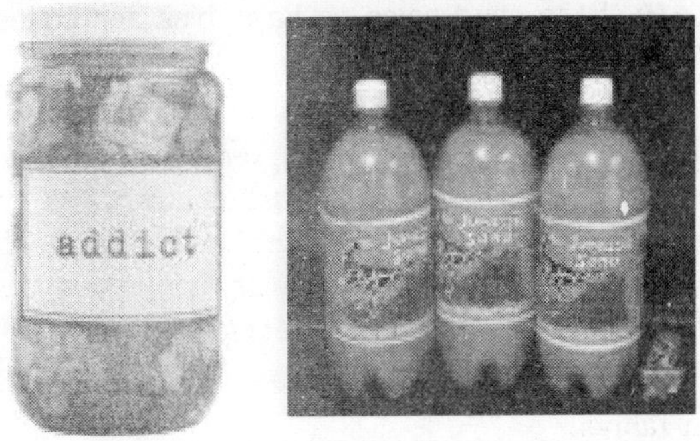

Figure 11.27. *Jars.*

11.11 CARTONS

These are very useful packing and containing device. Made up card boards used in almost all types of applications. These are light in weight and can be folded easily when empty. Volume is calculated by multiplying the length × depth × height. Colorful containers were also used for packaged food.

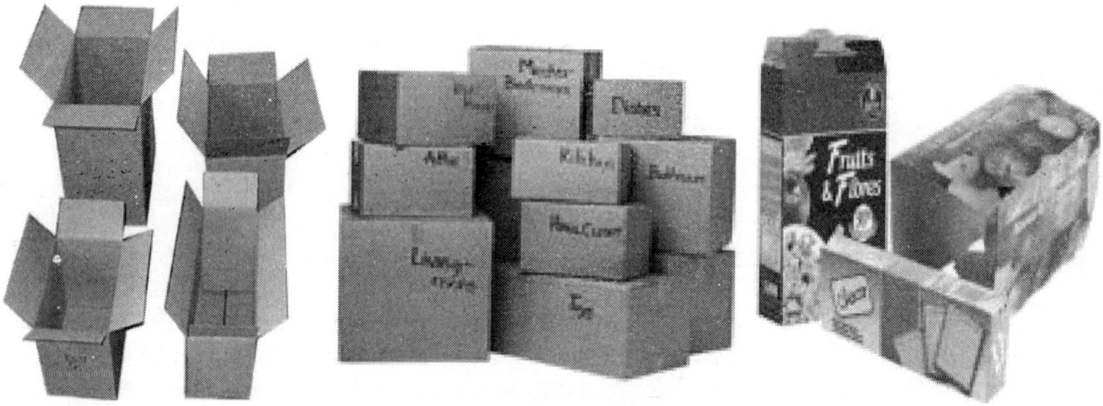

Figure 11.28. *Cartons.*

UNSOLVED QUESTIONS

1. What is meant by palletisation and containerization?
2. Explain the concept of unit load.
3. Classify pallets.
4. Explain pallet design type.
5. Explain different pallet styles.
6. What is meant by skids?
7. Explain the use of dollies with sketch.
8. Write down about ISO ship containers.
9. Why bags are used? What are their different types?
10. Write short notes on trays, bins and crates.

❑❑❑❑

12 Aerial Cableways and Tramways

12.1 HISTORY

The first indication of transport using rope comes from the rugged Asiatic countries of China, Japan, and India. Men used fiber rope to cross the chasms, initially transferring themselves, hand over hand, with the body suspended by a crude harness. The next application was to pull oneself in a basket that also had a few belongings of the traveler. Although Fausto Veranzio of Venice illustrated a bicable passenger ropeway in 1616, the ropeway industry generally credits Wybe Adam, a Dutchman, with erecting the first successful operational system in 1644.

Early ropeway technology and development were lead by the Europeans, particularly Germany and the alpine countries, Austria and Switzerland; later Italy and France. Most rapid development followed introduction of wire rope and later, the electric drive. Many innovations were introduced by extensive use of military tramways in the ferocious mountain warfare between Italy and Austria in World War I.

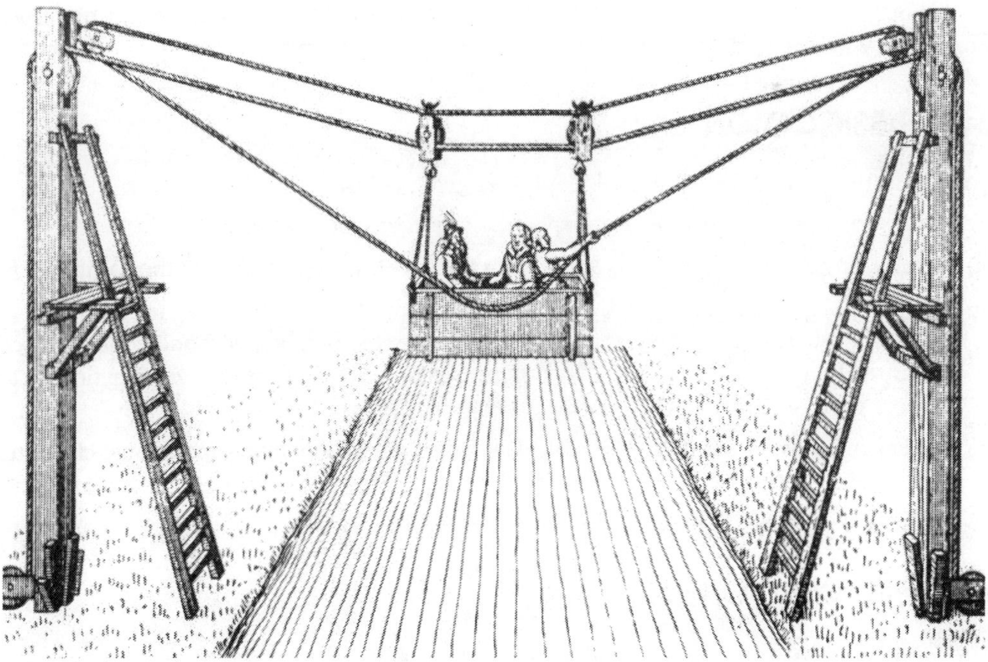

Figure 12.1

Ropeway technology and use of wire rope has a long history in North America dating back to the last century when wire rope was used for transportation applications, such as canal transport or the famous San Francisco cable cars. During the past years, however, the major growth of ropeway systems has been generated within the ski/recreational industry and most recently for people movers, an urban transportation application.

Ropeway trolleys, for passenger traffic and material handling, and steel and cable suspended bridges, are growing in popularity in India, and not just in the north and north-east. Ropeway trolleys are, of course, widely used in the inaccessible terrains of the Himalayas and in the north-east region. The movement of people, goods and other material across rivers and ravines in most remote villages and townships located in the higher tracts of the mountain range, in Himachal Pradesh and Uttaranchal, is achieved through ropeway trolleys.

There are some useful ropeways such as Village ropeways to transport goods to and from remote villages; Logging ropeways to transport timber from remote forests; and, River ropeways to transport people across the river.

Two prominent ropeway companies in India are Damodar Ropeways and Construction Company Pvt. Ltd and Usha Breco Ltd.

12.2 INTRODUCTION

A ropeway is a system for transporting materials and/or passengers in carriers suspended from or controlled by ropes. A system would qualify when a rope propels or controls carrier motion in transit on rail or other fixed support (funicular).

It transports a load for a short distance, in a single carrier travelling back and forth on a single carrier travelling basis and forth on a single cable or on multiple parallel cables, a hoisting operation being combined with the transfer of load; the operation is intermittent.

12.3 CLASSIFICATION

The ropeways is classified in different ways as given below

According to what is intended to transport : Passengers or Materials.

(*a*) Materials ropeways or tramways are seldom further classified other than monocable or bicable systems.

In a monocable (single cable) system, one rope serves to both support and control the carriers in transit. In a bicable (double cable) system, separate ropes serve the two functions : a static support rope or "track cable" and a moving "haul rope."

(*b*) Passenger ropeways are more extensively classified depending upon operational characteristics and carriers. The following are most common : aerial tramway, gondola, funicular, detachable chair, fixed grip chair, surface lift, tows, and unique types.

12.4 BASIC COMPONENTS

Every ropeway has some basic components.

1. Carriers

Carriers are an essential and distinguishing component of every ropeway. On materials ropeways they are generally buckets. On passenger ropeways they may be large cabins, usually described by capacity (*e.g.,* two 80 passenger cabins on a double reversible system). They are almost always totally enclosed. There is usually standing room only at full capacity; otherwise, seating is optional.

- The carrier for a funicular is large and, generally, seating is provided. It can also be called a car. It is most often enclosed. Wheels below the carrier support it on fixed steel rails; carrier motion is controlled by wire rope.
- Surface lift towing units are designated : disc type when a single circular disc is inserted between the legs, J-bar (single) or T-bar (for 2 or more) when between and behind passengers
- Gondolas are usually totally enclosed for year-round use or (less often) partially enclosed as in amusement parks. The term is most often preceded by capacity, for example, 6 passenger gondola.
- Chairs are described by capacity such as single, double, triple, quadruple, 6 passengers, 8 passengers etc. Chairs may have a foot rest and restraining bar (do not call restraining bar a safety bar).

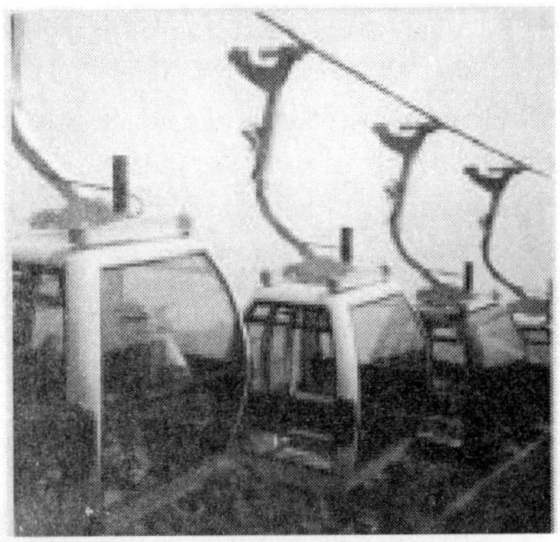

Figure 12.2

2. Terminals

Virtually all ropeways have two terminals. If a vertical change takes place, the terminals are upper and lower terminals. For ski lifts, the two terminals are referred to as loading and unloading (or discharge) terminals. Always, there is a drive terminal and a return (idler) terminal. Either may be a tension terminal where counter- weights or a tensioning device is located. On material ropeways an intermediate structure could be used to change the otherwise straight alignment of the ropeway. This called an angle station.

Every terminal station must be designed for 3 distinct functions :

(*a*) To secure track cables.

(*b*) To operate traction rope.

(*c*) To transfer carriers from cables to rail, lead them to loading or discharging point and return them to the cable.

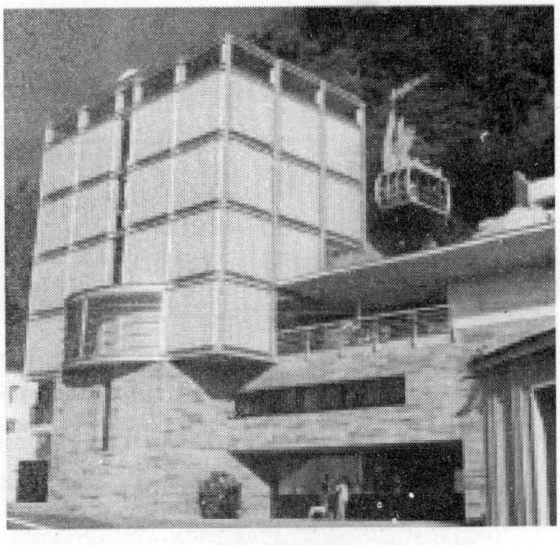

Figure 12.3

3. Towers

Intermediate structures support carrying and haulage ropes between terminals. These are most often called towers although occasionally they can be a large pressure frame. Seldom is a pylon used. Towers carry line sheaves for moving ropes and/or "saddles" for stationary track ropes.

Towers at each end support, the entire equipment and must be high enough for loads to be hoisted to the elevation desired at any point. Stationary towers are often 45.72 m and movable towers 30.48 m high; usually of steel. Generally steel, with bolted connections is used, so the towers can be taken down and re-used. Intermediate stations are needed to apply working tension to track cables which are too long for terminal apparatus to be throughout.

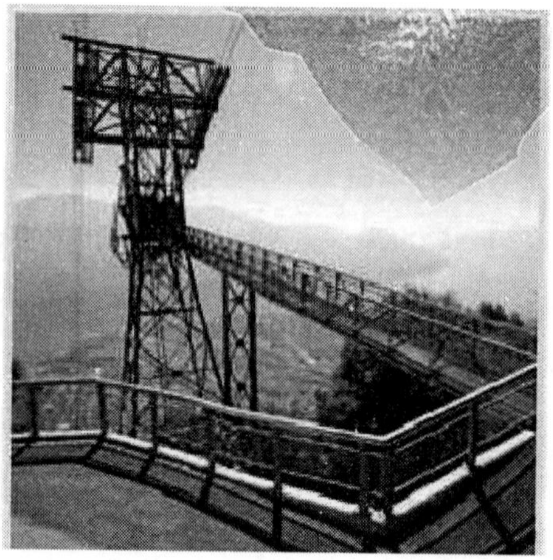

Figure 12.4

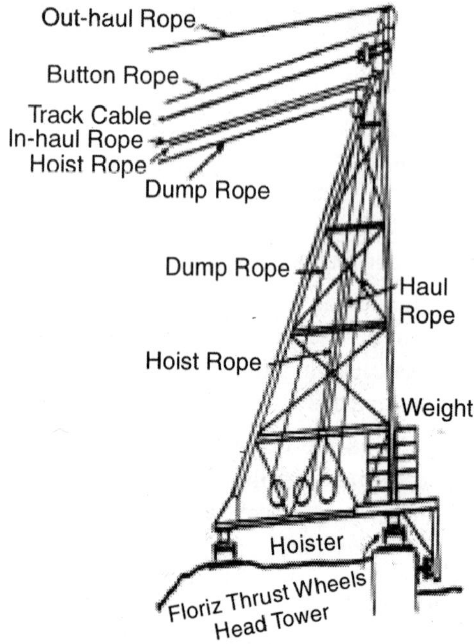

Figure 12.5

Towers on stationery cableways are designed chiefly to resist the vertical resultant of the cable tensions. Towers must also designed to resist a moderate Bide pull, if cable system is out of line, and to resist wind press acting on the tower

Stresses in towers have 3 sources :

(*a*) Downward press on saddles, due to vertical component of tension in track cables, weight of loaded carriers and vertical component of tension in traction rope.

(*b*) Racking motion parallel to line of tramway, due to changes in direction of press from the cable when a load approaches and leaves the tower.

(*c*) Side press, due to wind.

4. The rope

The rope is the heart of the ropeway. The rope (almost always a wire rope) is formed by inter-twining individual wire to form a strand and then the strands to form a rope. There are many variations of the processes used in manufacturing rope and in choosing the rope for the application. This is a major concern for rope manufacturers and ropeway designers. Of more impact to those of us would be to determine if there are both a haulage rope and track rope (bicable) or that one rope support both functions (monocable).

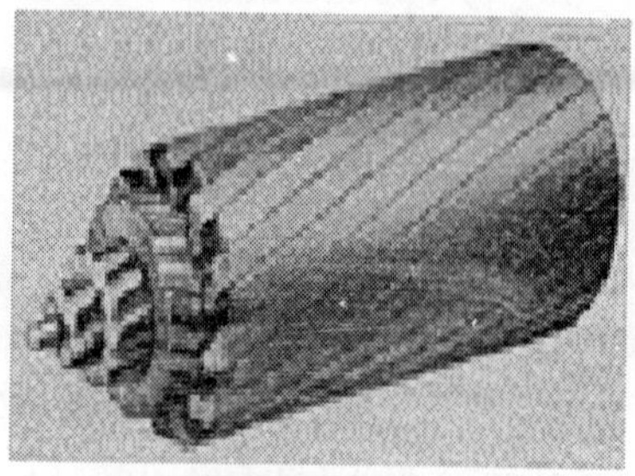

Figure 12.6

Ropes are generally described by their outside diameter in cm. Common usage would be a haul rope and a track rope for a bicable system or a rope for a monocable system. Ropes are made endless by using a "long splice" or they are terminated by attaching an end to a carrier, anchor, or counterweight by use of end connections such as sockets.

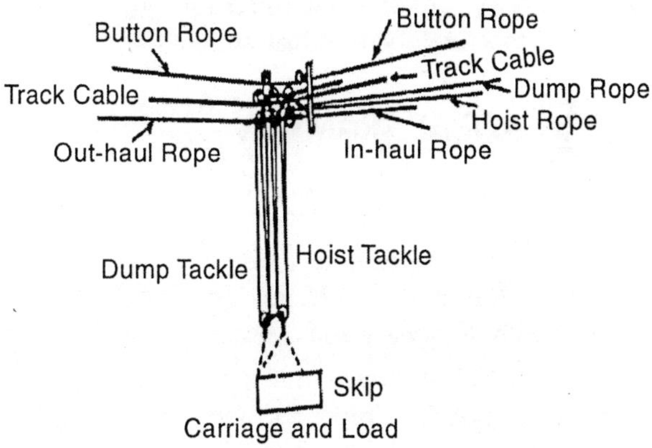

Figure 12.7

Hauling rope has one end linked to near end of carriage, passes to sheave on head tower, then down to make several turns on a concave-faced drum on the hoister; returns to head tower sheave, passes above the track cable to top of tail tower, thence around a sheave and back to the carriage. The rope is usually the same size as the hoisting rope, but has less bending stress, and may be of larger wires, to withstand surface wear due to slippage on drum.

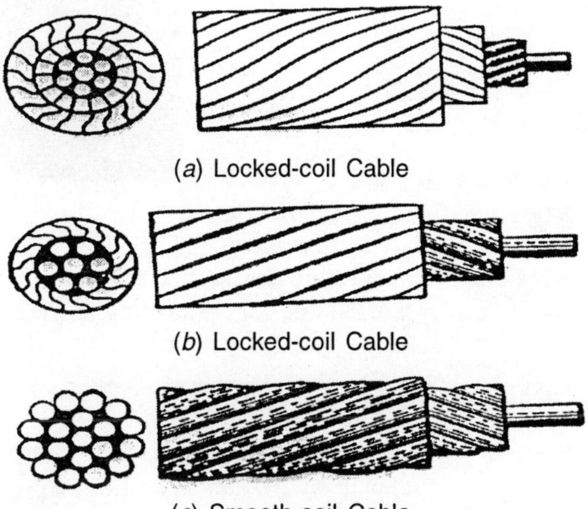

(*a*) Locked-coil Cable

(*b*) Locked-coil Cable

(*c*) Smooth-coil Cable

Figure 12.8

5. Evacuation and Rescue System

Aerial ropeways have provision for an auxiliary drive in the event of electric power failure, usually gasoline or diesel driven. Large reversibles have a rescue system which send an independently powered, small carrier out to remove passengers from a stranded carrier. Most aerial systems have provision for evacuation of stranded carriers using hardness and rope to lower individual passengers to the ground below.

6. Hoister

Hoister comprises : (*a*) hoisting drum, holding the rope in one layer; (*b*) hauling drum, having wide enough face for the rope, with several laps, to travel and fro across it (or light work, a capstan-like drum is used); (*c*) dumping drum; (*d*) small auxiliary drum for moving towers. Hoisters are electrically driven; controlled like electric hoisting engines. Carriers, skips are used for excavated material; chain or wire-rope slings for large rocks, quarried stone, or heavy equipment; bottom-dump buckets, sometimes, for depositing concrete.

12.5 AERIAL TRAMWAY

An aerial tramway uses large carriers or cabins and travels high above ground. The carrier(s)—one or single reversible, two or double reversible—travel between terminals, stop, reverse direction and travel back on the same, usually stationary and counter weighted, track rope.

An aerial tramway transports loads in carriers suspended from wire ropes forming the tracks, between fixed points, usually a long distance apart. Tramways are divided into 3 class : (1) bi-cable (formerly called double rope), (*a*) continuous, (*b*) reversible. (2) twin-cable; (*a*) continuous; (*b*) reversible. (3) mono-cable (formerly called single-rope), continuous.

Carriers are said to reciprocate between terminals. These systems are commonly referred to as "reversibles".

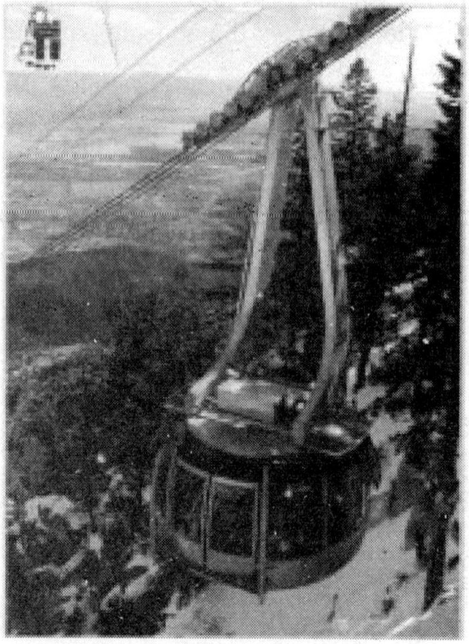

Figure 12.9

12.6 GONDOLAS

Gondolas are small carriers set at regularly-spaced close intervals. The systems are continuously circulating with carriers passing around terminal bull-wheels. Carriers detach from hauling rope in terminals, are decelerated and carried through the unloading and reloading areas at very slow speed, then accelerated for reattaching to the haulage rope for high speed travel "on the line" between terminals.

These systems are popular in ski areas and amusement parks.

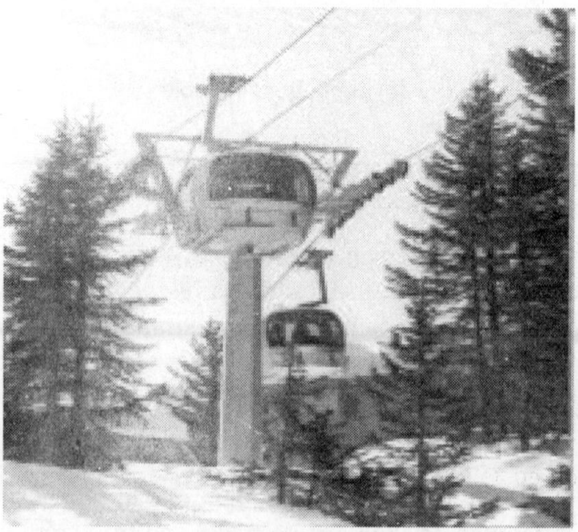

Figure 12.10

12.7 FUNICULAR

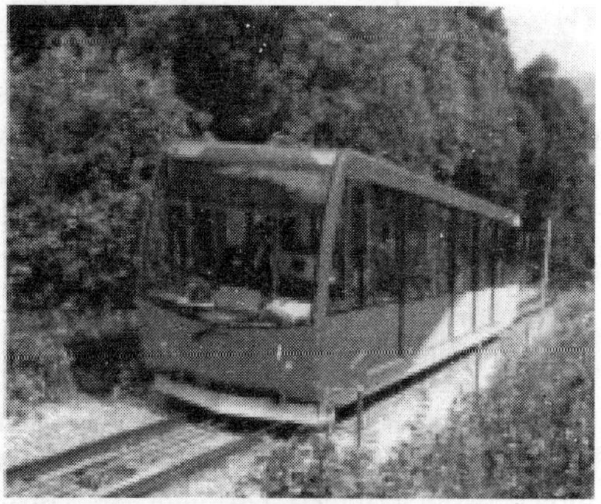

Figure 12.11

The funicular is controversial among rope-ways. It qualifies in that a wire rope controls the carriers even though a funicular may travel at ground level or on structurally supported steel tracks. The installation at Royal Gorge, Colorado, is classic. It was built by Otis Elevator in 1929 and is considered by many as an inclined elevator; but it is a ropeway.

12.8 DETACHABLE CHAIR LIFTS

Detachable chair lifts are virtually the same as gondolas, but the carrier is a multi-passenger open chair (usually 4 or more) with restraining bar and footrest. Extremely high capacity is common for skiers using special arrangements to facilitate loading at a comfortably slow carrier speed and travelling at a high line speed.

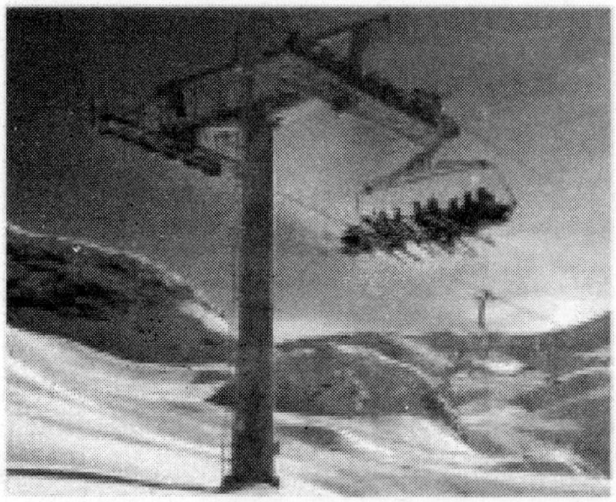

Figure 12.12

Fixed-Grip Chair Lift

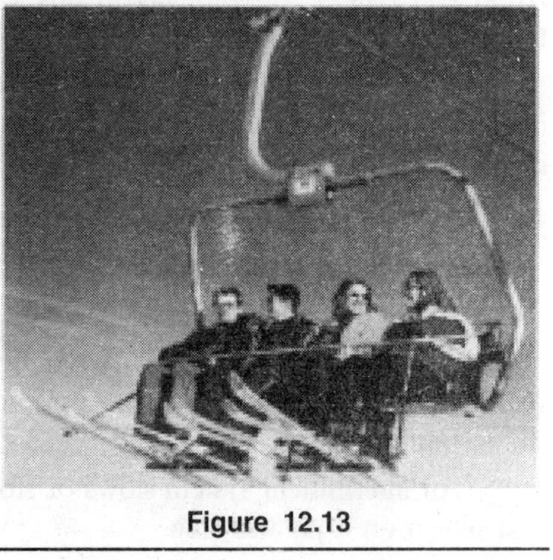

Figure 12.13

Fixed-grip chair lifts were "standards of the industry" from the early 1940s to the mid 1980s. Multi-passenger carriers circulated between and around terminals at a constant speed. The drawback was, the carrier speed comfortable for loading and unloading was slow "on the line" between terminals. Systems were popularly designated by carrier capacity; *e.g.*, single-chair, double-chair, triple, quadruple etc.

12.9 SURFACE LIFTS

Surface lifts are largely used in sky areas to move skiers along the snow surface by means of an overhead haulage rope with attached towing devices. Further designation was by carrier type; *e.g.*, disc, J-bar, T r, etc. Loading and unloading is accomplished between terminals with empty carriers circulating around the terminal bull-wheels at constant line spe

12.10 TOWS

Tows are a circulating rope that a skier grasps directly on a fiber rope or uses a "handle" on a wire rope. No intermediate support is used on the upcoming rope; the returning, empty rope may be carried by line sheaves or supports.

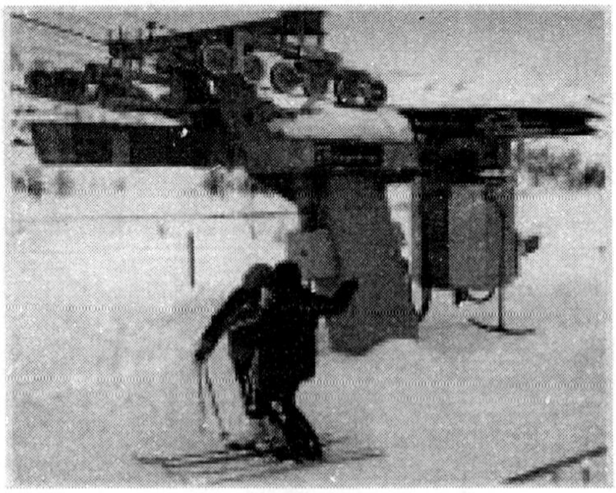

Figure 12.14 *Surface lift.*

12.11 SKI LIFT

Length of a ski lift is considered the slope length between terminals. Vertical rise is the change in elevation between terminals. Downhill skiers love knowing how much "vertical" they will ski each time they run between ski terminals. The skiing industry was largely responsible for introducing a new term to ropeway terminology : vtmh—vertical transport m per hour. This is a measure of lift capacity in passengers per hour multiplied times the vertical rise in m.

Vtmh influences the horsepower requirement of the drive and therefore, the cost of the lift. It determines the common classification of large or small.

12.11.1 Line Speed and Loading Speed

Carrier speed is measured on the line between terminals. If working with a detachable gondola or chair, the speed at the load or discharge point is loading speed. Speed "on the line" determines time in transit and speed "at loading and discharge points" facilitates loading and unloading, increasing comfort and safety.

Line speed for fixed grip quad-chairs serving skiers is 135 m per minute; the same carrier as a detachable system can run 345 m/minute. Carrier speed of loading (or unloading) of a fixed grip system is the same as line speed; for the detachable it could be 45 m per minute.

Standards governing passenger ropeways include The American National Standard for Passenger Ropeways–Aerial Tramways, Aerial Lifts, Surface Lifts, Tows and Conveyers–Safety Requirements ANSI B77.1-1999. Additionally in this state, Colorado Passenger Tramway Safety Board Rules and Regulations must be followed.

In addition to the classifications as discussed above, ropeways are described and compared by : operational characteristics, capacity, length and vertical rise, line speed, and loading speed. Each are briefly discussed here.

12.11.2 Operational Characteristics

Operational characteristics indicate whether the system operation is : continuous circulating, intermittent circulating (stop to load), or reversible.

In continuous circulating, carriers move around the terminals and maintain their order on the return line; most commonly, a fixed grip chairlift with carrier speed constant on the line and through the terminals.

An intermittent system slows or stops periodically but carriers eventually go around a terminal and return on a parallel line.

For a reversible, the carriers stop at each terminal, reverse direction of travel and return on the same line.

12.11.3 Capacity

Capacity of a ropeway is stated in passengers per hour. Capacity is a function of carrier capacity, carrier spacing, and the speed of the line at the loading point. A double chair ski lift with carriers at 18 m spacing and travelling at 165 m per minute has a capacity of 1100 passengers per hour.

12.12 CABLEWAY

Cableway is a misnomer in discussing ropeways. A cableway is a system for hauling materials on a hook or is a carrier supported and controlled by wire rope(s), but it has the unique ability to raise or lower the load in transit. Most often a cableway is used for construction as on a dam, bridge, pipeline, or similar projects. It can also be used for dredging, moving materials, stock piling, etc. Cableways have their own national standard : Safety Standards for Cableways, Cranes, Derricks, Hoists, Hooks, Jacks and Slings (ASME/ANS/ B30.19).

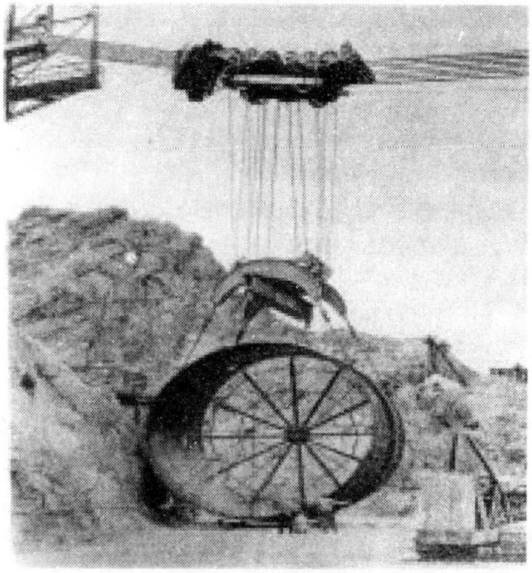

Figure 12.15

A passenger ropeway otherwise qualifying and used for transporting large numbers of foot passengers, particularly in an urban environment, is classified a "people mover."

An elevator is a mechanical device with a moving platform or cage for conveying persons or goods from one level to another as in a building. The carrier most often raises and lowers vertically and operates in a guided shaft. Confusion may arise between an inclined elevator and the funicular which qualifies as a ropeway.

Applications of Cableways

(*a*) Light cableways are used for sewers, subways and similar long, narrow, shallow excavations.

(*b*) Inclined cableways have a high head tower at discharge end and a low one at other end.

(*c*) Excavating cableways are adapted to cases requiring mechanical excavation before transport

UNSOLVED QUESTIONS

1. Write down classification of aerial ropeways.
2. Write and explain basic components of aerial ropeways.
3. Explain the function of terminals.
4. What is a function of towers ? What type of stresses arises in it ?
5. Explain rope construction and its importance in the aerial ropeways.
6. What is meant by aerial tramway ? Distinguish between cableways and aerial tramway.
7. Write about gondolas.
8. What is meant by funicular ?
9. What are the different types of chair lifts ?
10. Explain working and applications of cableway.

❑❑❑❑

13 Sorting Systems

13.1 INTRODUCTION

Materials have to be identified, labelled and sorted both in production and distribution. With manual handling this process is fairly simple as sorting staff can handle and inspect goods, decide on their destination and place then in corresponding compartments. With mechanized handling, on the other hand, each unit must be labelled with an item or address code for machine reading into a control unit which in turn generates signals to mechanical switching devices.

A range of devices are in use today, mainly in installations with large materials flow, *e.g.,* in central depots for multiple retail stores, spare parts stores, mail-order firms, goods terminals and baggage handling installation at airports.

Sortation systems may include shoe sorters, cross-belt sorters, pop-wheel and roller sorters, a variety of conveyors and other automated sortation systems. These systems are designed for either low or high volume order sizes, to promote shipping accuracy, and the ability to ship more frequently. They also offer increased productivity, labour savings, cost reduction, and vastly improved customer satisfaction.

13.2 SORTING SYSTEM SELECTION

A suitable sorting system (*i.e.,* with reference to type and capacity) may be chosen of these factors :

(*a*) Physical characteristics of materials, such as size, weight, packaging, durability etc. So, for instance, small unpacked units require handling by belt conveyors and metal chutes while goods in cardboard boxes can also be conveyed on wheel or roller conveyors.

(*b*) Rate of flow, *i.e.,* goods through out per unit of time. Small rates of flow may be sorted by hand but where the rate exceeds 5–10 units per minute, mechanical/automatic handling may be better.

(*c*) Number of destinations. One or two operators can possibly manage a small number of destinations. Large number may justify more advanced solutions because of space requirements and transport distances.

(*d*) Goods subdivisions required, *e.g.,* each article by itself; a given number of units by themselves (*e.g.,* in accumulation for automatic pallet loading); each order (destination) by itself.

13.3 STORAGE SYSTEMS

Generally involve sorting of out-going goods. These may be retrieved by one of two methods : Order Picking or Batch Picking. In the former case, goods are placed either directly on designated load carrier (Pick and place method) with no sorting equipment required, or transported to a sorting installation subsequently to be divided between the respective orders. In connection with picking operation, goods have to be identified by some means, *e.g.,* labels which are turn off the picking list and struck on the goods.

13.4 PRODUCTION SYSTEMS

In production systems sorting is performed between different production stages whether some items are to be selected from a production accumulator for further processing or whether rejected components are to be remove dafter a sub-operation. In the former case, goods are stored in the transport system in the later case sorting is usually a stage in production.

13.5 DESIGN OF SORTING SYSTEMS

A sorting system comprises :

(*a*) Arrivals Area

This is an area intended for accumulation and collection of goods waiting to proceed to sorting, in its simplest form, floor space is allocated for the reception of units prior to processing. Normally, however, it is desirable to save and facilitate further handling by organizing the material, for instance, using chutes.

(*b*) Feed Station

Here, each unit is separated for identification, destination and labeling.

(*c*) Separation

Separation can be hand *i.e.,* the operator takes one unit, or mechanically, as when goods mechanical, as when goods on a conveyor are transferred to another, faster conveyor line. To release one item at a time special arrangements can be incorporated on the basis of "Stop and Separation".

(*d*) Identification

Identification in its simplest form is the attachment of labels which can be read by humans. To facilitate reading and to produce the risk of error, label information should be uniform and of sufficient size and contrast. Only short words or group of figures should be used. Colour coded labels aid checking. Identification can also be automatic if reflecting tape, bar codes (UPC or EAN codes), etc. are employed. Codes are then read by photoelectric sensors or by a light pen.

Address allocations means that identified units are given destinations. An address may be the same as the identity of the item or a conveyor line.

(*e*) Labelling

Labelling means that the product-or the part of the conveyor which carries the product—is provided with the identification. The methods are based on the type of the control system used.

In a control system based on routing information which accompanies the item, adjustable pins, printed squares, bar codes or magnetized spots endure the appropriate action at switching points.

Where each item is monitored by a central system, pulses are generated to operate the required switches. Destination can be entered by hand through a keyboard by voice labelling or automatically. Voice labeling is suitable where operators must use their hands to manipulate goods in order to see label texts.

(*f*) Conveyors and Switching Elements

These include roller, wheel, chain, slat and overhead rail conveyors. Switching points incorporate switches with lifting or tilting sections, deflectors, pushers, etc.

(*g*) Collection Areas

Sorted goods normally need an accumulation system upstream from packing and dispatch. Systems with limited flow use containers, larger systems use chutes or wheel/roller conveyors.

(*h*) Sorting Installations

Roller and/or chain conveyor are used to sort pallet loads. Where there are many destinations, however, this solution becomes expensive and systems incorporating roller conveyors and transfer trolleys are then preferred.

13.6 SORTATION CONVEYOR

Sortation Conveyor are one of the more common and effective sortation systems. These conveyors are designed to provide intelligent sortation. Sortation systems are designed to meet the growing demand for smaller order sizes, high shipping accuracy, and more frequent shipments while improving efficiency in your distribution operations.

Type of product handled—Unit load.

Location of the conveyor system—On Floor/Overhead.

Sortation conveyors are used for merging, identifying, inducting and separating products to be conveyed to specific destinations.

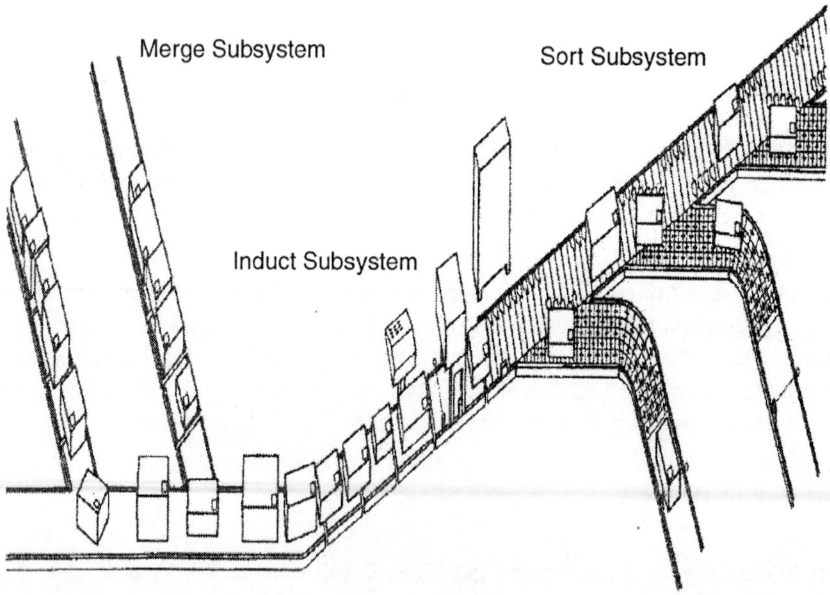

Figure 13.1

(*a*) Sortation Conveyor : Diverter

Stationary or movable arms that deflect, push, or pull a product to desired destination. Since they do not come in contact with the contact with the conveyor, they can be used with almost any flat surface conveyor. Usually hydraulically or pneumatically operated, but also can be driven. Simple and low cost.

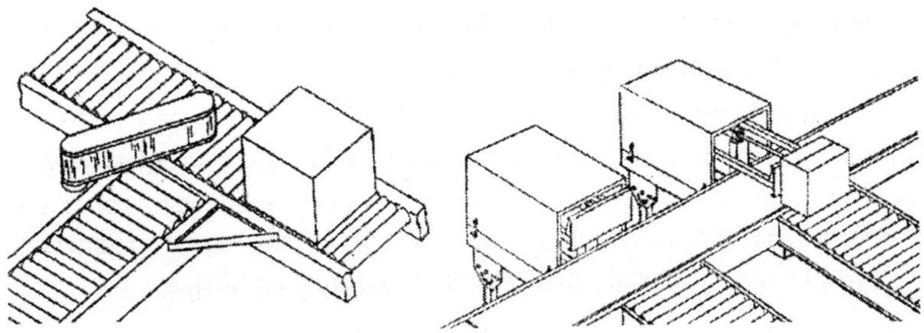

Figure 13.2

(b) Sortation Conveyor : Pop-up Device

One or more rows of powered rollers or wheels or chains that pop up above surface of conveyor to lift and guide it off conveyor at an angle; wheels are lowered when products not required to be diverted. Only capable of sorting flat-bottomed items. Pop-up rollers are generally faster than pop-up wheels.

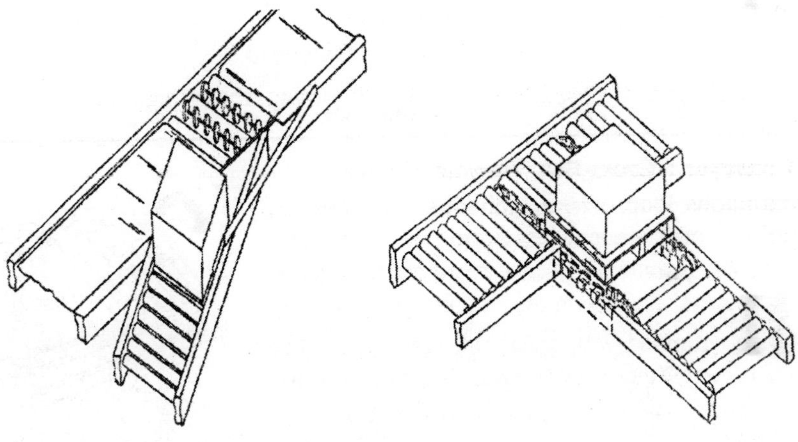

Figure 13.3

(c) Sortation Conveyor : Sliding Shoe Sorter

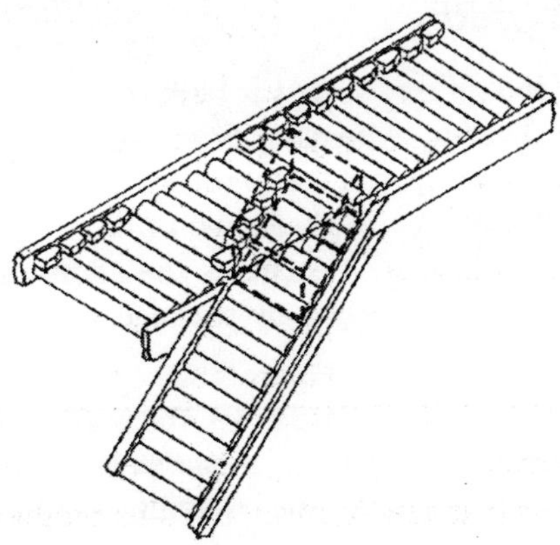

Figure 13.4

Sliding shoe sorter (moving slat sorter) uses series of diverter slats that slide across the horizontal surface to engage product and guide it off conveyor.

Slats move from side to side as product flows in order to divert the product to either side.

Gentle and gradual handling of products.

(*d*) Sortation Conveyor : Tilting Device

Trays or slats provide combined sorting mechanism and product transporter. Can accommodate elevation changes. Tilt tray sorters usually designed in continuous loops with a compact layout and recirculation of products not sorted the first time. Tilt slat sorters carry products on flat-surface slat conveyor and can handle wider variety of products compared to tilt tray.

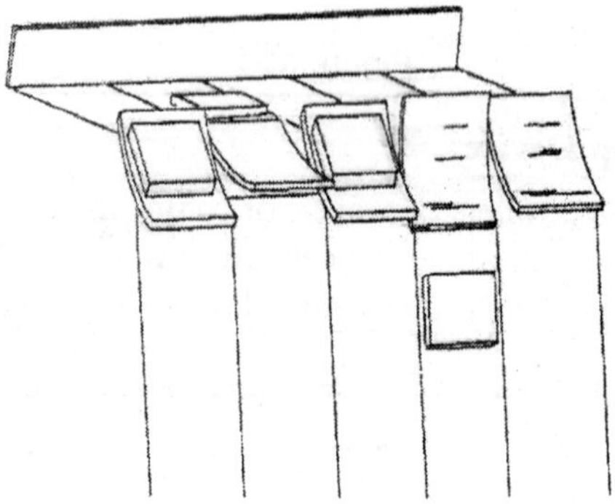

Figure 13.5

(*e*) Sortation Conveyor : Cross-Belt Transfer Device

Either continuous loop, where individual carriages are linked together to form an endless loop, or train style (asynchronous), where a small number of carriers tied together with potential for several trains running track simultaneously.

Each carriage equipped with small belt conveyor, called the cell, which is mounted perpendicular to direction of travel of loop and discharges product at appropriate destination.

Automatically separates single line of products into multiple in-line discharge lines.

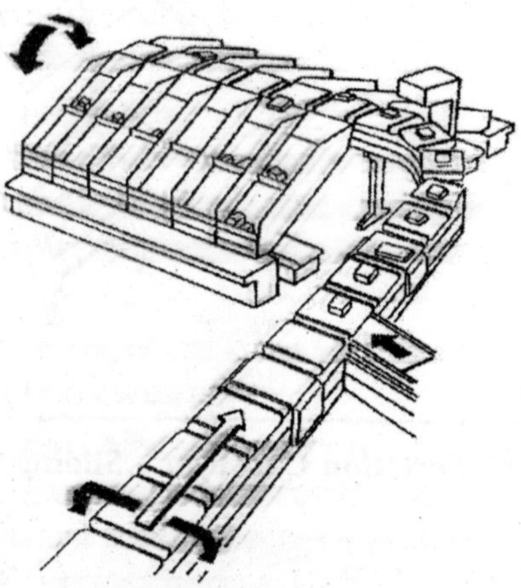

Figure 13.6

13.7 SORTATION SOFTWARE

The software used to drive the tilt tray sorters is built on a Windows 2000 system platform using Microsoft SQL for data storage. Aside from being user friendly, our sorter software in concert with the sorter controls permit a high degree of programming flexibility. Included with these systems is the additional functionality to allow you to keep track of product by chute and subsequently by package.

Safeguards

Software safeguards are in place the integrity of the sort as well as the maintenance of the equipment. We presently check that a solenoid was triggered to confirm the sort request and stop the sorter under extreme conditions that may affect the reliability of the sort or the equipment condition, such as a possible chain jam.

Tracking Software

- Every irregularity seen by the scanner.
- The time it was seen.
- Whether it got a valid read or not.
- The tray it traveled on.
- To which drop it was directed.
- Mode of operation (automatic or manual)
- On which cycle of the sorter.
- Where it was dispensed.
- Was the solenoid triggered to drop or cancelled.

This information is available for examination by network or modem to assist in diagnosing irregularities or for generating productivity or other analysis reports. Remote access software and modem are supplied with each sorter computer. An Internet connection is the preferred method for access and will be utilized if provided.

UNSOLVED QUESTION

1. Why there is a need of sorting system ?
2. Explain different types of sorting systems.
3. Explain different types of sorting conveyors with neat sketch.
4. Explain the procedure used for sorting.
5. Find out the advances which are going on about the sorting systems.

❑❑❑❑

14 Storage Systems

14.1 INTRODUCTION

Storage equipment are used for holding or buffering materials over a period of time. The general objective of a storage system is obvious enough to store by most manufacturing firms. Most production plants use manual methods for depositing and retrieving items into and from storage. The storage function is generally accomplished in ways that are inefficient and inadequate in terms of control. Automated methods are available to those firms willing to treat the problem of storage with the attention it deserves.

Different companies have different reasons for installing an automated system for storing material. A list of possible objectives that a company may want to achieve by installing an automated storage system. The most common reason for storing a product allows the other elements of production to operate more efficiently on a per-unit basis because the fixed costs associated with utilizing the element can be spread over more products; *e.g.,* storing up to a truckload of product in a facility reduces the per-unit costs of shipping; and buffering or storage of WIP enables production, which reduces the per-unit setup costs.

FUNCTION

The function of the storage system is to transfer the materials for a period of time and to permit access to those materials when require.

14.2 OBJECTIVES OF STORAGE SYSTEM

Possible objectives of installing an storage system in a Factory or Warehouse are as under :

(*a*) Increase storage capacity.

(*b*) Increase floor space utilization.

(*c*) Recover space for manufacturing facilities.

(*d*) Improve security and manufacturing facilities.

(*e*) Reduce labour cost in storage systems.

(*f*) Improve safety in storage function.

(*g*) Increase labour productivity in storage operations.

(*h*) Improve safety in storage function.

(*i*) Increase stock rotation.

(*j*) Improve customer service.

Other potential reasons for storage include : time bridging–allows product to be available when it is needed (*e.g.*, storing spare machine parts at the facility); processing–for some products (*e.g.*, wine), storage can be considered as a processing operation because the product undergoes a required change during storage; and securing–*e.g.*, nuclear waste storage.

Types of Material Typically Stored in Material Handling System

Types	Description
(*a*) Raw Materials	Raw stock to be processed-bar stock
(*b*) Purchased Parts	Parts from vendors to be processed or assembled.
(*c*) Work in Progress	Partially completed parts
(*d*) Finished Product	Completed product ready for shipment.
(*e*) Rework and Scrap	Parts that are out of specification
(*f*) Refuse	Other waste products left over processing
(*g*) Tooling	Cutting tools, jigs and fixtures, moulds
(*h*) Spare Parts	Parts needed for maintenance
(*i*) Plant Records	Records on products, equipment and personnel records.

Categories 1–5......... product

Categories 6–8......... process

14.3 PERFORMANCE OF STORAGE SYSTEM

The performance of the storage system in accomplishing its function must be sufficient to justify its investment and operating function.

Parameters

(*i*) Storage capacity

(*ii*) Density

(*iii*) Accessibility

(*iv*) Throughput.

Standard parameters for storage system for mechanized and automated system include : Utilization.

14.4 TYPES OF STORAGE SYSTEMS

(*i*) Block stacking (no equipment).

(*ii*) Selective pallet rack.

(*iii*) Drive-through rack.

(*iv*) Drive-in rack.

(*v*) Flow-through rack.

(*vi*) Pushback rack.

(*vii*) Sliding rack.

(*viii*) Cantilever rack.

(*ix*) Stacking frame.

(*x*) Shelves/bins/drawers.

(*xi*) Storage carousel.

(*xii*) Automatic Storage/retrieval systems (AS/RS)

(*a*) Unit load AS/RS

(*b*) Miniload AS/RS

(*c*) Mon-on-board AS/RS

(*d*) Deep-lane AS/RS

(*xiii*) Split case order picking system.

(*xiv*) Mezzanine.

(*i*) Block Stacking (No Equipment)

Bulk storage using block stacking can result in the minimum cost of storage since cube utilization is high and no storage medium is required, but material accessibility is low since only the top of the front stack is accessible and loads at bottom of a stack must not require support.

Storage racks are used when support and/or material accessibility is required.

(*ii*) Selective Pallet Rack

Most popular type of storage rack. Pallets are supported between load-supporting beams.

Special attachments and decking can be used to make the racks capable of supporting other types of unit loads besides pallets (*e.g.,* coils, drums, skids).

Selective racks can be used for the following types of storage :

Standard : Single-deep storage using a counterbalanced lift truck.

Narrow-Aisle : Storage using a narrow-aisle lift truck.

Deep-Reach : Greater than single-deep storage. (Typically double-deep storage)

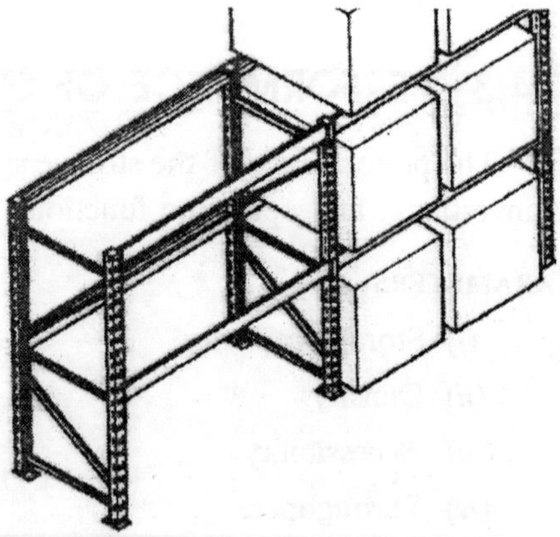

Figure 14.1 *Block stacking.*

(*iii*) Drive-Through Rack

Loads are supported by rails attached to the upright beams.

Lift trucks are driven between the uprights beams.

Requires similar-width loads.

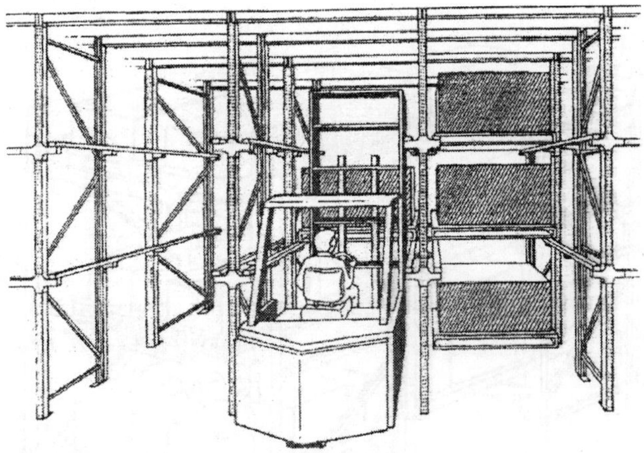

Figure 14.2 *Drive through rack.*

(iv) Drive-In Rack

Same as drive-through rack, except closed at one end, allowing entry from only one end (LIFO).

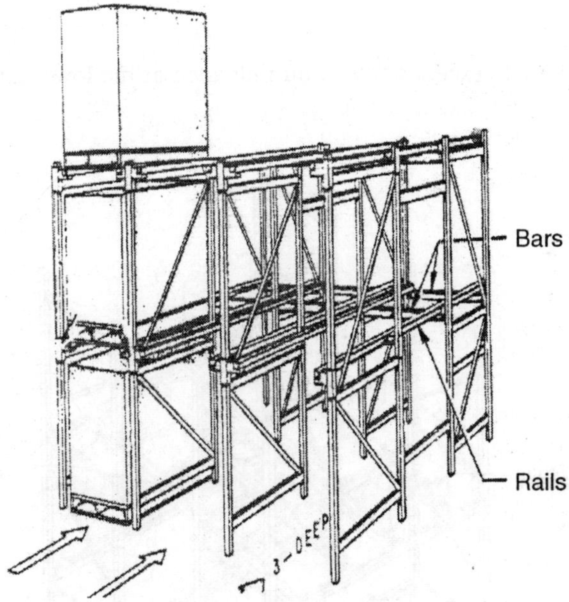

Figure 14.3

(v) Flow-Through Rack

Loads are supported on an incline to enable gravity-based movement of the loads within the rack (via, *e.g.,* a gravity roller conveyor).

Loaded at the higher end and unloaded at the lower end (FIFO).

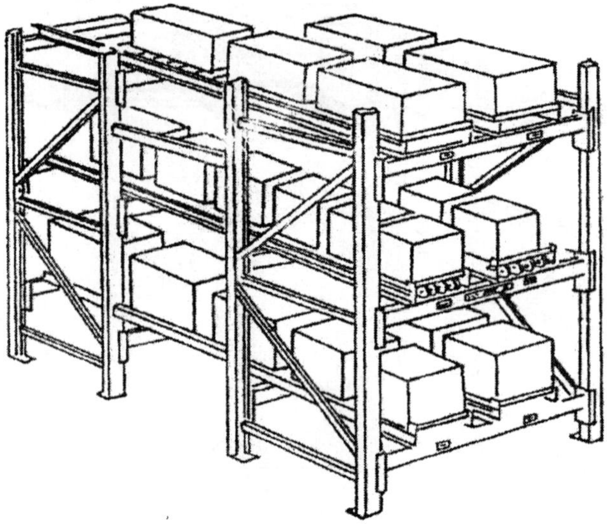

Figure 14.4

(vi) Push-Back Rack

Same as push-back rack, except loaded and unloaded at the lower and closed at the higher end. (LIFO).

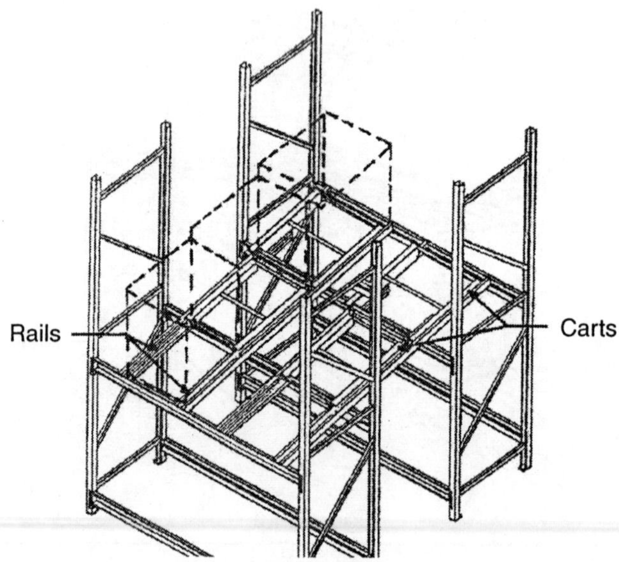

Figure 14.5

(vii) Sliding Rack

Only one mobile aisle is used to access several rows of racks. Location of the aisle is changed by sliding the rows of racks along guide rails in the floor. Typically found in library stacks.

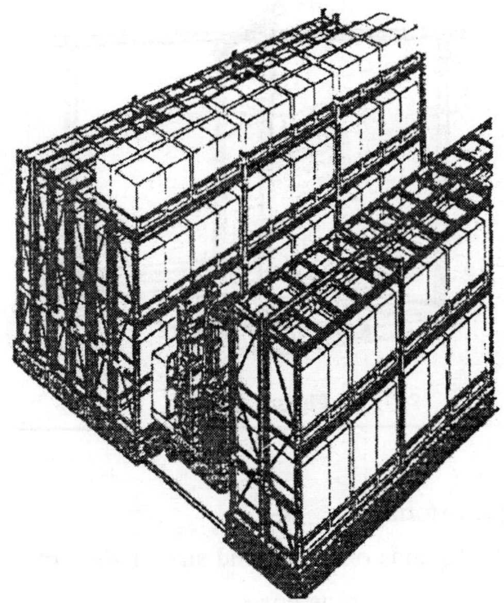

Figure 14.6

(viii) Cantilever Rack

Loads are supported by cantilever "arms". Used to store long loads (*e.g.,* bar stock, pipes, lumber). Similar to pallet racks, except the front upright beams and the front supporting beams are eliminated.

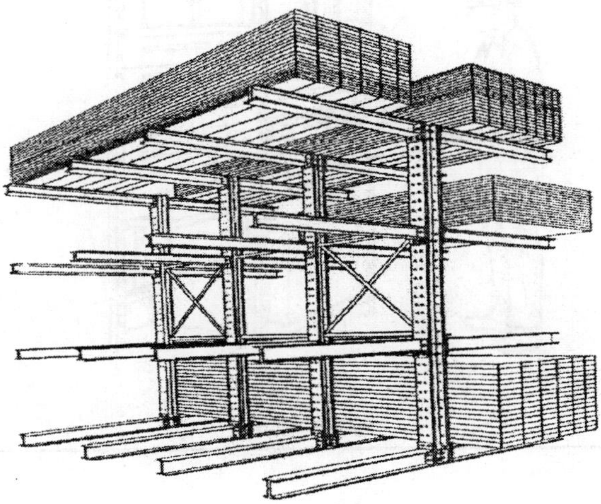

Figure 14.7

(ix) Stacking Frame

Interlocking units that enable stacking of a load so that crushing does not occur.

Can be disassembled and stored compactly when not in used.

Pallet frames can be used to enable multilevel block stacking.

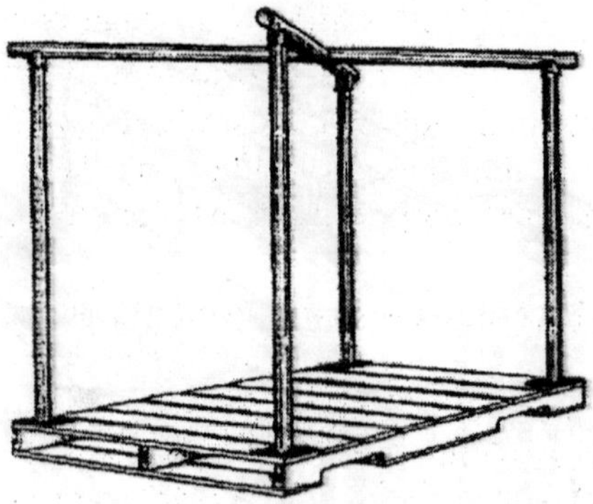

Figure 14.8

(x) Shelves/Bins/Drawers

These are used for compact storing.

Shape and size of shelves depends on shape and size of the product.

Shelves are used for unit storing such as boxes.

Drawers are used for bulk storage of small items like nuts, bolts, screws etc.

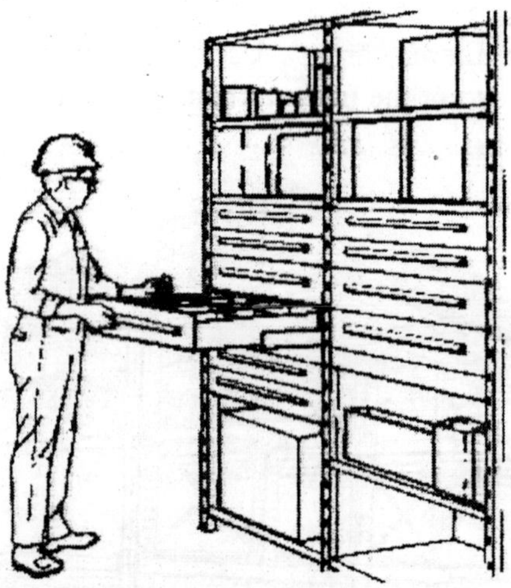

Figure 14.9

(xi) Storage Carousel

Carousel consists of a set of vertically or horizontally revolving storage baskets or bins. Materials (and the storage medium) move to the operator, "part-to-man". For end-of-aisle picking. Each level of the carousel can rotate independently in a clockwise or counter-clockwise direction. Control ranges from
manually activated push buttons to automated computer controlled systems. Provides an alternative to typical "man-to-part" AS/RS, where the S/R machine moves to the part. Similar to a trolley conveyor with storage baskets.

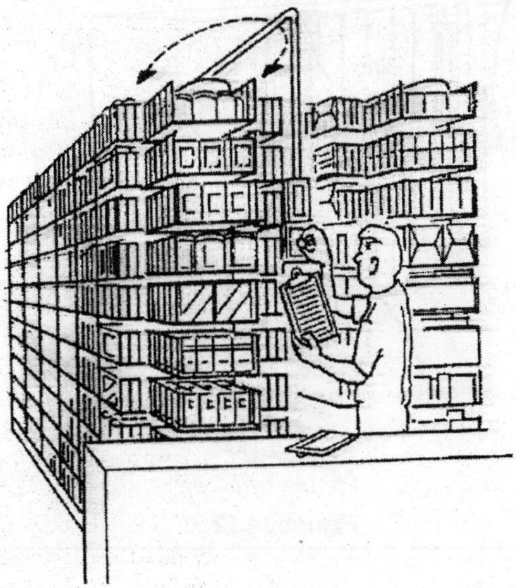

Figure 14.10

(*xii*) Automatic Storage/Retrieve Systems

Types of AS/RS Systems

(a) Unit Load AS/RS

Used to store/retrieve loads that are palletized or unitized and weigh over 250 kg.

Stacking heights up to 40 m high, with most ranging from 18 to 25 m high; 1.5 to 1.8 m wide aisles; single or double-deep storage racks.

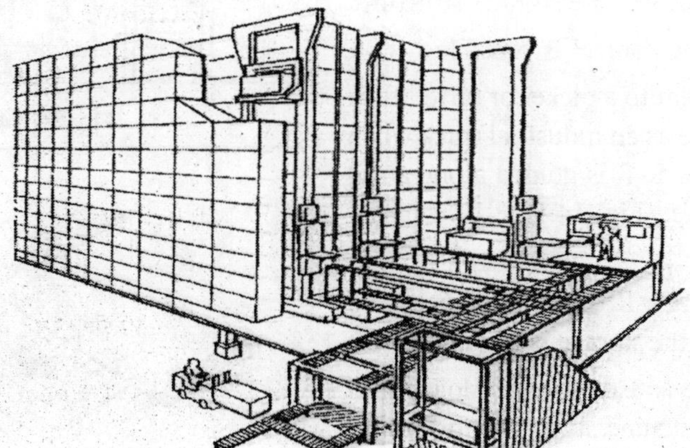

Figure 14.11

(b) Miniload AS/RS

Used to store/retrieve small parts and tools that can be stored in a storage bin or drawer. End-of-aisle order picking and replenishment. Stacking heights range from 3.5 to 6 m; bin capacities range from 100 to 350 kg.

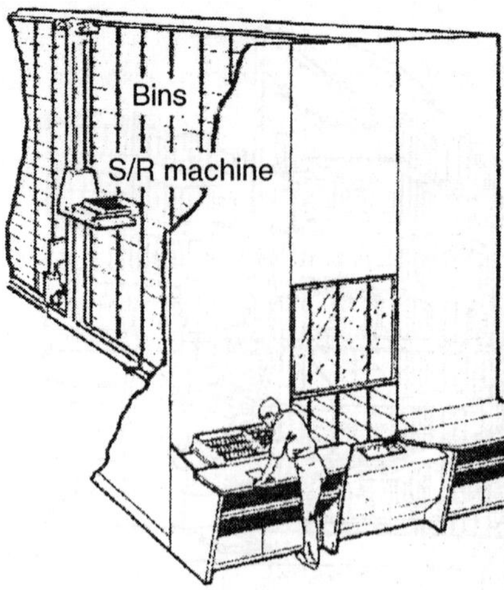

Figure 14.12

Termed a "microload AS/RS" when used in assembly, kitting and testing operations to deliver small containers of parts to individual workstations, where workstations are typically located on the sides of a pair of racks and the S/R machine operates between the racks to move containers to openings in the racks located next to each station.

(c) Man-On-Board AS/RS

Used for in-aisle picking; operator picks from shelves, bins or drawers within the storage structure.

Manual or automatic control is possible.

S/R machine is similar to a picker or turret truck and can sometimes operate as an industrial truck when outside an aisle, except the S/R is guided along a rail when operating in an aisle. (Fig. 14.13)

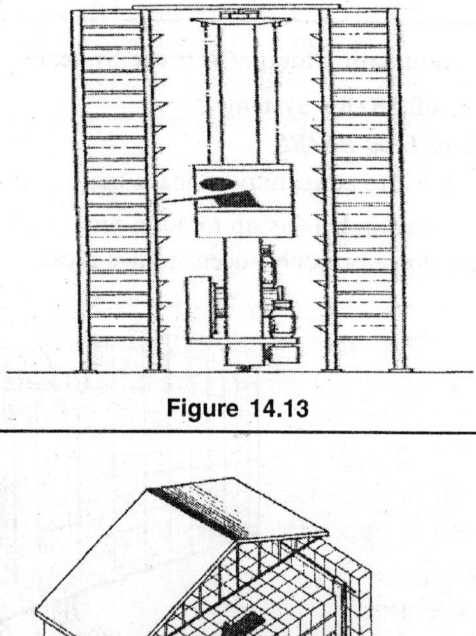

Figure 14.13

(d) Deep-Lane AS/RS

Similar to unit load AS/RS, except loads can be stored to greater depths in the storage rack. (Fig. 14.14)

A rack-entry vehicle is used to carry loads into the racks from the S/R machine, and is controlled by the S/R machine.

Termed an "automated item retrieval system" when used to automatically retrieve individual items or cases, with replenishment (storage) taking place manually from the rear of a flow-through storage lane and items are pushed forward with a rear-mounted pusher bar for automatic picking from the front of the storage lane.

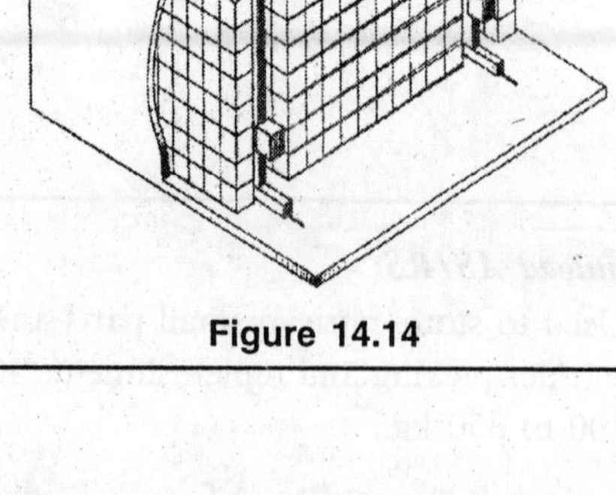

Figure 14.14

Applications of AS/RS

Most application of AS/RS technology has been associated with warehousing and similar distribution operation. A growing number of AS/RS systems are used for sorting materials between operations in manufacturing. We can distinguish three application areas for storage and retrieval systems.

1. **Unit load storage and handling :** Unit load storage and retrieval applications are represented by the unit load AS/RS and the deep lane storage systems. These types of application are commonly found in warehousing for finished goods in a distribution centre.
2. **Order picking :** Order picking is used to store and retrieve materials in less than full unit load quantities. Miniload, Man-on-board, and item retrieval system are used for this second application area, in which the individual items are stored and retrieved.
3. **Work-in-process storage systems :** Work-in-process is an important problem in manufacturing. It is the most recent area of application in manufacturing automation.

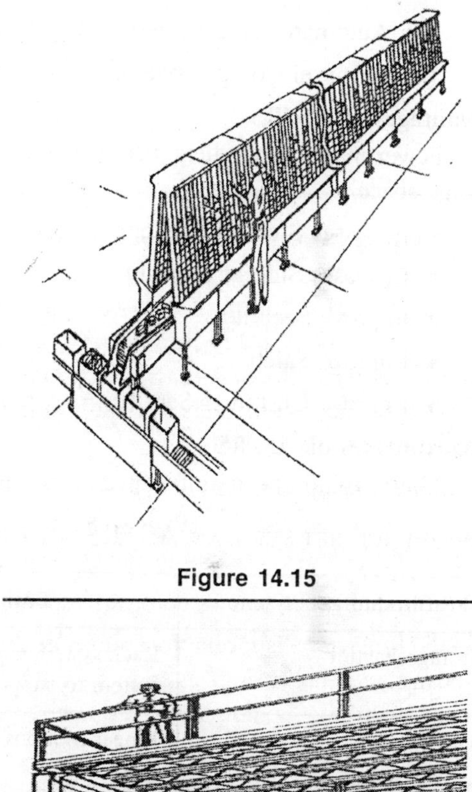

Figure 14.15

(*xiii*) Split-Case Order Picking System

Unlike an AS/RS, a split-case order picking system enables fully automated picking of individual items.

Two general categories of split-case order picking system are robotic based systems and magazine/dispenser based systems.

Robotic based systems are similar in construction to robotic pick and place palletizers. Magazine/dispenser based systems are similar to vending machines, but larger in scale.

"A-Frame" dispenser system (pictured) is popular within pharmaceutical distribution centers; items are dispensed onto a belt conveyor that carries them into a container.

(*xiv*) Mezzanine

Inexpensive means of providing additional storage or office space.

Makes use of clear space over activities not requiring much headroom (*e.g.,* restrooms, block storage, etc.)

At least 4 m of clear space is needed for a mezzanine.

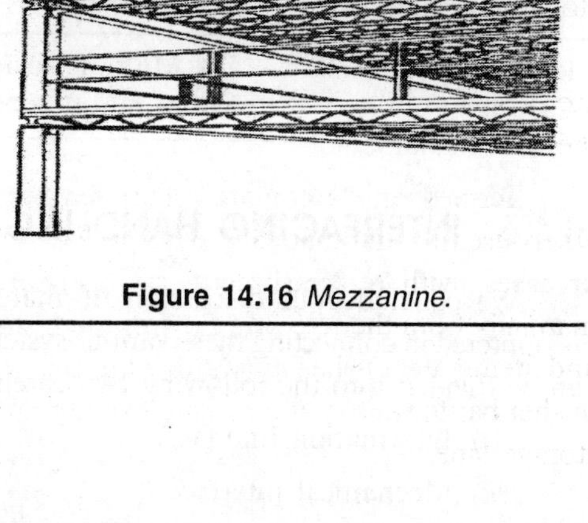

Figure 14.16 *Mezzanine.*

14.5 AUTOMATIC STORAGE/RETRIEVAL SYSTEMS (AS/RS)

Consists of an integrated computer-controlled system that combines the storage medium, transport mechanism and controls with various levels of automation for fast and accurate random storage of products and materials.

Storage/retrieval (S/R) machine in an AS/RS operates in narrow aisle, serving rack slots on both sides of aisle; can travel in horizontal (along the aisle) and vertical (up and down a rack) directions at same time.

Automated storage system are generally divided in two types :

(*i*) Automated storage/retrieval system.

(*ii*) Carousel storage system.

Advantages of AS/RS

Fewer material handlers, better material control (including security), and more efficient use of storage space :

- Increases Storage Capacity.
- Increases Storage Density
- Improve Security
- Improve Safety
- Improve Customer Services.

Disadvantages of AS/RS

High capital and maintenance costs, and difficult to modify.

DIFFERENCE BETWEEN AN AS/RS AND A CAROUSEL STORAGE SYSTEM

Feature Storage System	Basic AS/RS	Basic Carousel
Storage Feature	Rack system to support pallets shelf system to support tote bins.	Baskets suspended from overhead conveyor trolleys.
Motions	Linear motions of S/R machines.	Revolution of overhead conveyor trolleys around oval track.
Storage Retrieval Operation	S/R machines travels to compartments in rack structure.	Conveyors revolves to bring baskets to load/unload station.
Replication of Storage Capacity	Multiple aisles, each consisting of rack structure and S/R machine.	Multiple carousels, each consisting of oval tracks and suspended bins.

14.6 INTERFACING HANDLING AND STORAGE WITH MANUFACTURING

We conclude our discussion of material handling and storage by considering the problems encountered in connecting these various systems to the manufacturing function. These interface problems can be divided into the following two categories :

(*i*) Information Interface.

(*ii*) Mechanical Interface.

The information interface is concerned with the flow of information that must accompany the movement and storage of materials in the factory. It encompasses the problem of materials identification and tracking, inventory control, production scheduling and the data communications required to coordinate and control the various systems in the plant.

The mechanical interface deals with the problem of transferring parts and loads between storage systems, material handling systems and production systems. The load transfer function is an integral part of any material handling system. The design of the mechanical interface depends on the type of handling equipment used, the system that is to be interfaced with the handling system, and whether the load/unload procedure is to be manually or automatically. Our discussion of mechanical interfacing will include two problem areas :

1. Position accuracy of the handling system at the transfer station.
2. Methods of transferring parts or loads to and from the handling system.

14.7 OPEN STORAGE OF BULK MATERIALS

14.7.1 General

During the past years there has been a considerable increase in the number of open storage installations for important raw materials like ores, coal, limestone, fertilisers, etc. This is not only due the fact that the requirements of these raw materials by the concerned processing industries have increased tremendously but also due to the equal increase in their movement from country to country. In our own country we have recently built major ore exporting facilities in ports like paradeep, Vishakhapatnam, Madras, Mangalore, Goa; fertiliser import complex in Kandla; large coal fired power stations using huge quantities of coal; iron and steel weorks using huge quantities of ore, coal and stone; cement plants using limestone etc. In all these, handling of the large quantities of bulk materials requires to be planned and designed carefully.

The bulk materials handling installations of modern industries essentially take care of one or more of the following types of transfers of the bulk material:

(*i*) from source to carrier
(*ii*) from carrier to carrier
(*iii*) from carrier to process
(*iv*) from process to carrier
(*v*) from carrier to user
(*vi*) from process to process.

Such a transfer should satisfy the following basic requirements:

(*i*) Receive the material as and when it arrives.
(*ii*) Dispatch the material as and when required.
(*iii*) Avoid hold-ups in any of the operations preceding or following the transfer.

The natres and rates of material flow under (*i*) and (*ii*) above, *i.e.,* the incoming and outgoing flow rates could be quite different. In order to balance out this difference as well as to have a stock of material to tide over nforeseen circumstances, *i.e.,* (*iii*) above, *storage* becomes necessary part of the installation. Since, we are here dealing with such materials that are not easily degraded due to weather, *open storage in piles* is the most economical in view of the large quantities involved. The amount of storage in such installations depends not only the natures and rates of the incoming and outgoing material flows but also on uncertain factors such as (*i*) Breakdowns, bottlenecks in traffic lines; inclement weather; strikes etc. which affect the external carriers.

(*ii*) Breakdowns or difficulties in handling equipment which cause delay in clearing the external carriers.

(*iii*) Breakdowns in connected process equipment

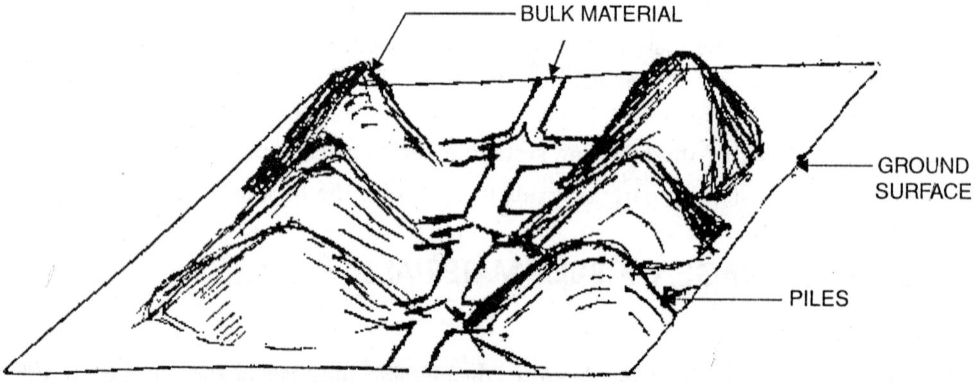

Figure 14.17 *Open storage.*

The amount of storage necessary to guard against such eventualities can be expressed as so many weeks (or months) requirement of the material based on the average hourly rate of handling or processing. In view of the fact that this average ghourly rate could be anywhere from a few hundred to a few thousand tonnes per hour, the amount of storage works out to quite large figure.

Extensive storage and high handling rates are possible only through the use of mechanical equipment. Stackers to build up the stockpiles of materials and reclaimers to withdraw the material from the stockpiles.

Exporting Harbour Operations :

(*i*) Unloading the ore from inland transport carrier (in this case railway),

(*ii*) Stacking and relaiming the ore, and

(*iii*) Loading the ore into seagoing vessels.

Importing Harbour has Main Operations :

(*i*) Unloading the ore from seagoing vessels,

(*ii*) Stocking and reclaiming the ore, and

(*iii*) Loading into inland transport carrier (in this case river barges).

Steel Mill has following main operations :

(*i*) Unloading from inland transport carrier (in this case river barges),

(*ii*) Stackling and reclaiming the ore,

(*iii*) Preparation of the ore to suit the process (crushing, grading, blending, sintering), and

(*iv*) Charging the ore into production equipment.

14.7.2 Stockpile

The material is commonly stockpile on levelled and compacted ground. The ground must be prepared to withstand the pressure due to the stacked material as well as the pressure of **crawler wheel**

mounted equipment which may be employed. The tracks for **rail mounted** equipment are specially prepared for the higher track loading. The rails are laid on ballast or concrete footings. Wind and rain are incovenient in open air stockpiling. Although good drainage of the ground can reduce the problems caused by rain for many materials, storage under a roof may be required for other materials. Storage under roof also enables good control fo the environmental pollution due to dust; it is; however; out of question for very large stockpiles. Other solutions are then to be sought for the dust problem.

Properties like the natural angle of repose, particle size, flowability, fragility, tendency to self-ignite, etc. influence the size and formation of stockpiles as well as the equipment selection.

Different materials or different qualities of the same material are often required to be stacked separately. whent he incoming material shows considerable variation in composition or quality a blending effect can be achieved by suitable choice of the stacking and reclaiming methods. The main idea is to stack in several layers and reclaim by cutting across as many different layers of the pile as possible.

In cone shell stacking the pile is built up to full height at successive points along the length. The stacker discharge point need move only along and above the centreline of the pile.

In **strata** system the pile is made up of different strata sloping in one direction. The stacker discharge point needs to move longitudinally as well as across the pile.

In the **chevron** system, inverted vee shaped double strata are formed. The stacker discharge point needs to move only longitudinally above the centreline of the pile.

In the **window** system, the main pile is built up of several smaller piles. This suitable for prevention of segregation of different sized lumps in the pile. The stacker discharge point needs to move longitudinally as well as across the pile.

The chevrosn and window systems are commonly employed in large scale automated stockpiling and bucket-wheel reclaiming. The cone-shell and strata systems are suitable for smaller stockpiles using scraper reclaiming.

Single pile is rarer than **multiple** piles except when storage and reclaiming capacities are small enough. Crawler-mounted reclaimers are more suitable for wide piles than rail mounted ones because the **bucket** wheel reach need not be excessive. When more than two piles are present, the number of stackers and reclaimers can be reduced by using special means to transfer the equipment from one track to another depending on the pile to be worked upon.

Circular stockpiles are convenient for limited storage capacity.

14.7.3 Equipment

The main equipment required are

The **conveyors** for carrying the material to and from storage,

the **stacker** for building up the stockpile, and

the **reclaimer** for withdrawing the material from the stockpile.

14.7.3.1. Belt Conveyors

Belt conveyors are used throughout for material transport because of their high capacities and low power requirements. The **stockpiling** and the **reclaiming** conveyors may be different from each other in which case the stacker and reclaimer can operate independently of each other. If the same conveyor is used for stockpiling as wel as reclaiming then only one operation can be carried out at a time. Special attention is paid to the design of transfers from one belt to the other; feeding points which may be stationary or moving; trippels; discharge points etc. At the main transfer points, *e.g.*, reclaiming to shiploading, buffer storage in the form of **surge hoppers** are provided to even out the flow.

14.7.3.2 Stacking

Stacking may be done either by discharging the material from stationary overhead belt with a variable tripper stackers discharge to make long piles or by **boom stackers.** These latter travel onj rails between stockpiles and possess a **boom belt** conveyor. The boom belt received material from a tripper. The tripper is hitched to the stacker so that it travels with the latter. Instead of a simple tripper, an **elevating belt** preceded by a tripper can be used to feed the boom belt. This enables a higher feed point as well as onward conveyance at a present rate from the in coming flow to shiploading or process while the excess material is diverted to stockpile.

Fixed boom stcakers simple but possess a constant discharge height which is unfavourable from the point of view of dust pollution. Also for fragile materials like pelletised ore the large drop in height is not advisable. The **luffing boom** stackers enable discharge of the material from a small height. When two pile are to be made, on either side of the stockpiling conveyor, then the stacker can be fitted either with two booms or one **slewing boom.** The slewing bom also enables "Windrow" stacking and making a wide trapezoidal pile.

Railmounted stackers are run automatically according to programme to get a pile with optimum layer characteristics such that the reclaimed material is as homogeneous as possible.

14.7.3.3 Re-Claiming Methods

The method with **gates and feeders** at different points below the pile is suitable for only smaller capacities. The material in the pile cannot be fully reclaimed without additional help from other equipment like bulldozer.

The travelling **rotary plough reclaimer** is suitable for large rates. Inj both the above methods, the underground tunnel conveyor for reclaiming is a disadvantage. Reclaimers which can feed a surface conveyor are preferable.

Drag and scraper type reclaimers are suitable for covered storage piles and smaller capacities. The bridge mounted drum reclaimer has a long drum with buckets on its periphery and spanning over the whole width of the pile. While the reclaimer moves forward into the pile, the buckets of the rotating drums scrape up the material and pass in through the.....

The bucket wheel reclaimer is the most widely used and effective device. It can be either rail-mounted or crawler mounted. The rail-mounted variety can be further divided as

(*i*) **Boom** type, **Bridge** type. The crawler mounted variety is always of the boom type and has a higher mobility since two-directional travel is possible. These have a loading conveyor in addition to the bucket boom conveyor. Independent slewing of the bucket boom and the loading conveyor enable the crawler mounted reclaimer to operate on very wide piles with a relatively short bucket boom.

The rotating **bucket wheel** which scrapes up the material from the heap and discharges it onto the boom belt is the most important element of this machine. Boom type reclaimers have both luffing and slewing motions. Starting from the top of the pile, the material is reclaimed by simultaneous slew and travel movements. The reclaimed material is finally discharged onto the main reclaiming conveyor which runs alongside the pile. Bridge type reclaimers operate always at the base of the pile. The bucket wheel is mounted such that it can traverse across the pilewidth while the bridge travels along the pile. Reclaiming is done while carrying out by traversing across the pile. The material rolls down the pile to the reclaiming level. A **rake** assists in loosening up the material particles lying above the level of the bucket wheel. The reclaimed material is discharged onto the bridge belt either directly or with the help of a short belt.

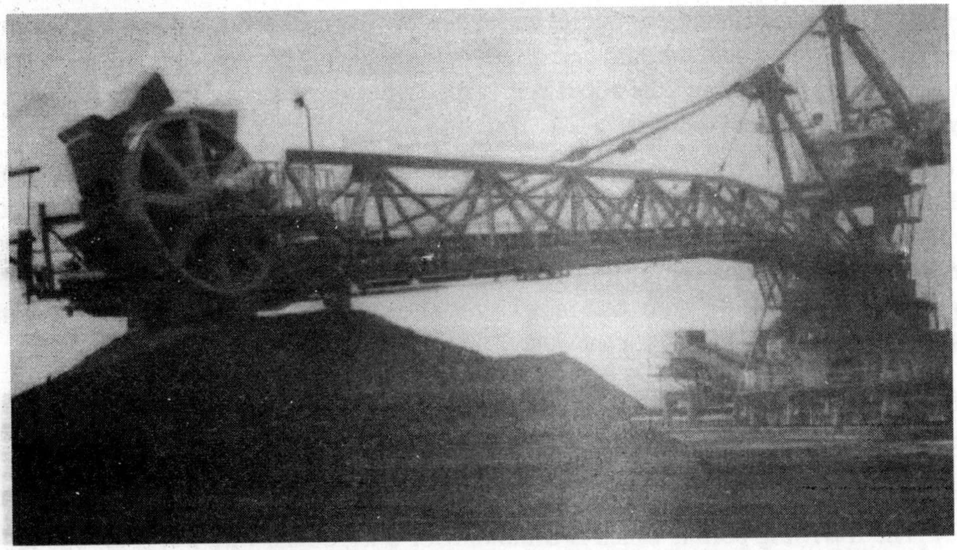

Figure 14.18 *Bucket Wheel Stacker and Reclaimer.*

14.7.3.4. Stacker/Reclaimers

The boom type bucket wheel reclaimer with a reversible boom belt can be used a stacker pile. Such combined **stacker/reclaimers** are quite popular. When two separate machines are operated along the same line, difficulties arise due to the fact that one machine cannot pass the other. The combined machines do not have this difficulty but, however, they can obviously operate only as a stacker or as a reclaimer at one time. Hence, careful planning of the stacking and reclaiming operations

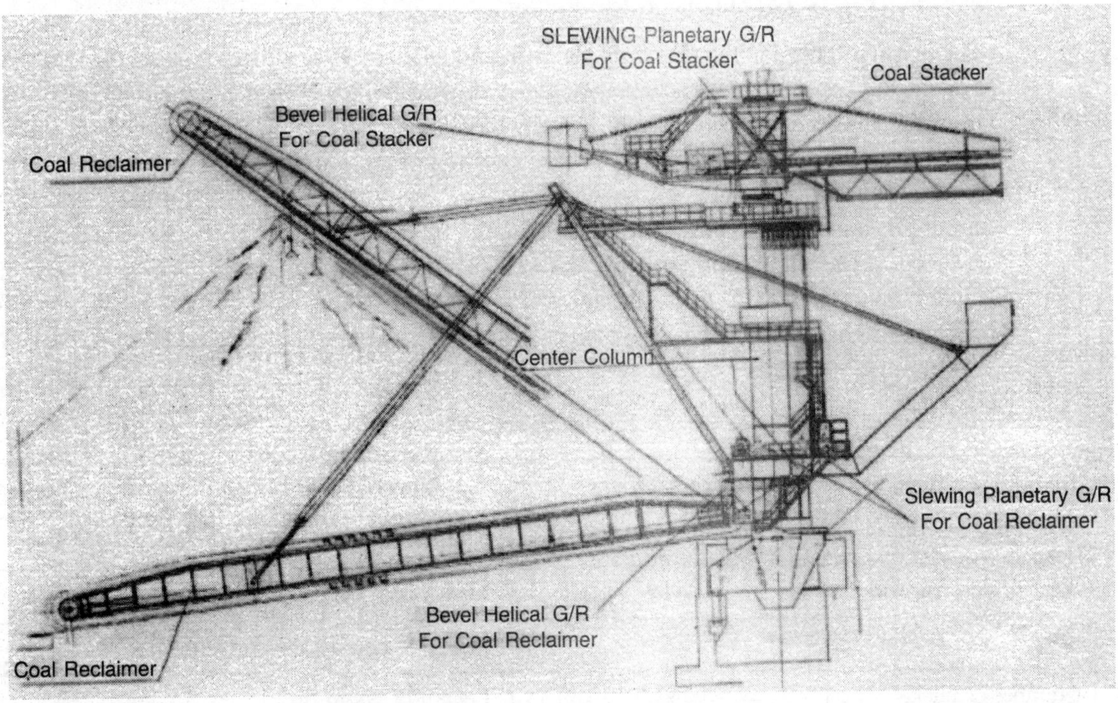

Figure 14.19 *Coal Stacker and Reclaimer.*

is necessary. During reclaiming, the bucket wheel digs the material and the boom belt is run inwards, the material discharging onto the yard belt conveyor via a feeding hopper/device. During stacking the bucket wheel is shut down and the boom belt is run outwards, receiving the material from the yard conveyor via a tripper/inclined belt.

Where the yard conveyor (which is used for both stack piling and reclaiming) runs only in one direction, the transfer of the material to the boom belt for stacking is possible. When, however, the yard belt is reversible, running in opposite directions while reclaiming and stacking then a tripper with a counter directional elevating conveyor or with a height adjustable discharge pulley is to be used. The former solution is suitable for short stockpiles as it enables the stockpile toe extend over the maximum possible length of the yard conveyor.

14.7.4. Rail-Mounted Bucket Wheel Reclaimer

Some constructional features of rail-mounted boom type equipment :

Essentially the boom type stacker or reclaimer consists of the following main parts :

(*a*) **The travelling portal,** which is commonly of 3 leg design with 2 legs on one rail and the other on the second rail. Even though less stable than the 4 leg design, this is preferred because it is statically determinate and accommodates level differences of the rail tops. The portal straddles the yard belt conveyor. The travelling wheels and mechanism are designed and provided in an essentially similar manner as in heavy cranes. In the case of stacker, the tripper is linked with the portal and is towed by it.

(*b*) **The revolving platform** is supported on the top of the portal through a large diameter double row rolling bearing. The platform as well as the portal top should be sufficiently rigidly designed as otherwise their deformations are transmitted to the bearing races, thus giving rise to undesirable loads on the rolling elements.

(*c*) The **boom** is hinged in the revolving platform and luffed by **rope winch** or **hydraulic cylinders.** For stacking operations, the slewing angle of the revolving platform is limited by the transfer from the tripper to the boom belt. Usually this angle is not more than ± 90° to ± 100°. In reclaimers, the bucket wheel with its drive arrangement is fitted at the extreme end of the boom. The boom weight and part of the live loads (viz., belt load, cutting force) are counterbalanced by a **couterweight** so that the moment loading on the slewing bearing is reduced at the same time ensuring the stability of the machine. It is usually located at a higher level so that it does not interfere with the pile profile during operations. The counterweight may be **fixed,** in which case its position (radius) is constant irrespective of the boom angle **movable.**

Technical Specifications

Manufacturer	Babcock-Moxey	Bucket wheel design	Cell-type
Customer	B.H.P. Australia	Max. boom inclination (degrees)	0
Year of construction	1969	Max. boom declination (degrees)	16
Wheel diameter (m)	5.85	Width of boom conveyor (mm)	914
Number of buckets	9	Length of boom (m)	36.5

Bucket capacity (liter)	280	Service weight (tons)	330
Discharges per minute	40.5	Max. wheel load (tons)	30
Cutting speed (m/sec)	1.38	Bucket wheel drive power (HP)	75
Material handled	Iron ore, lumps and pellets	Total installed power (kVA)	300
Stacking (tons/h)	—	Cost	Approx.
Reclaiming tons/h)	1000		5 crores

With a movable counterweight the following advantages are obtained :

(*i*) In case of the boom accidently coming to rest on gthe ground gthen its ground reaction is reduced.

(*ii*) The lifting work for luffing is reduced considerably. Addtional structural members suitably connect the boom and the counterweight.

In a reclaimer the operators cabin is located at the very end of the boom such that he has a good view of the bucket digging into the material.

14.7.5 Design Specifications

Apart from deadweight and wind loading, the load due to the presence of the conveyed material on the belts, transfer chutes and hoppers etc. can also be considerable. For reclaimers, the digging force at the bucket lips is an additional load. Being mainly outdoor equipment, both service wind and storm wind need to be taken into account. Accidental/intentional resting of the boom on the ground causes different load conditions which need to be investigated, as also the action loads—Stability requirements are gone into detail with great care for both operational as well as for non-operational conditions.

UNSOLVED QUESTIONS

1. What are the objectives of storage system ?
2. Explain different types of materials stored in material handling system.
3. Explain the parameters influencing performance of storage system.
4. Classify or note down types of storage systems.
5. Write short note on-storage carousel and mezzanine.
6. Explain with neat sketches Types of AS/RS systems and their applications.
7. What is meant by Automatic Storage/Retrieval Systems (AS/RS) ?
8. State difference between an AS/RS and a carousel storage system.
9. Explain how will you interface handling and storage with manufacturing.
10. Explain what is meant by stacker and its types.
11. Explain what is meant by reclaimer and its types.
12. Explain different methods of stockpiles.

□□□□

15 Miscellaneous Equipments

In this chapter we will discuss the equipments such as

15.1 CUPOLA/FURNACE CHARGER

Every furnace need to be charged by providing the materials continuously. These materials generally include scrap material, ores of metal and some additives. Since in foundry requirement of molten metal is continuous its charging must be continuous. Thus there is need of specialized cupola charger. Cupola chargers or furnace chargers are the equipments which are used for continuous feeding of the furnace to increased productivity.

These scrap feeders and charging systems gives precise automated batch control, optional scrap drying and other charge enhancement capabilities. Also automated systems help to improve worker safety. There are various designs and capacity according to the requirement and capacity generally ranges from 1 tonne to 25 tonnes capacity. Now days these type of chargers are equipped with many accessories to improve productivity such as :

1. Furnace Chargers With On-Board Weigh-Cell Systems
2. Charge Make Up Feeders
3. Preheat Systems
4. Two-Way Charge Make Up Feeders
5. Additions Feeders
6. Noise Reducing Designs
7. Abrasive Resistant Liners
8. Close Capture Hood
9. Low Energy Consumption Variable Speed Material Discharge.

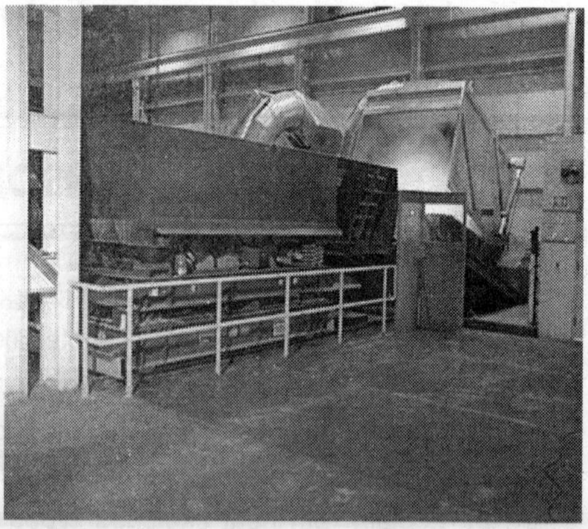

Figure 15.1

These chargers generally include bucket and cage elevators discharging onto the conveyor. These conveyors discharge the scrap into the top of the cupola. Speed of elevator and conveyors is synchronized to meet the production and capacity of furnace. Control of this is made either on-board or remote panel PLC control for the furnace charging operation for medium-frequency electric melting furnaces and cupola melting.

15.2 MOBILE LIFTING FRAMES

These surrounds their loads horizontally *i.e.,* load is put on to the frames and these frames moved from one place to another. Material is put on the frames using some kind of materials handling equipment (as shown in the figure the crane is putting pipes on the frame). Sometimes these frames carry its lifting equipments. These frames are designed for specific products. These are towed (pulled) by vehicles such as trucks or towing tractors.

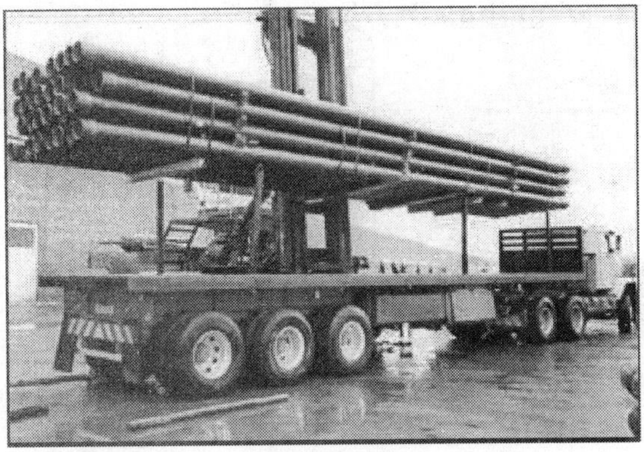

Figure 15.2

15.3 HOPPERS

Bulk loads are loaded into hoppers from motor lorries or directly by machines.

Hoppers are generally discharged through gates in their lower part into conveying machines of various types (conveyors, overhead track and railless transport devices) or to processing machines, to mixers, grinders and sorting machines, briquetting presses, calcinating furnaces, chemical plants, boiler furnaces, etc. Sometimes, hoppers are unloaded at the top end by means of special devices, for example, grab cranes. (See Fig. 15.3)

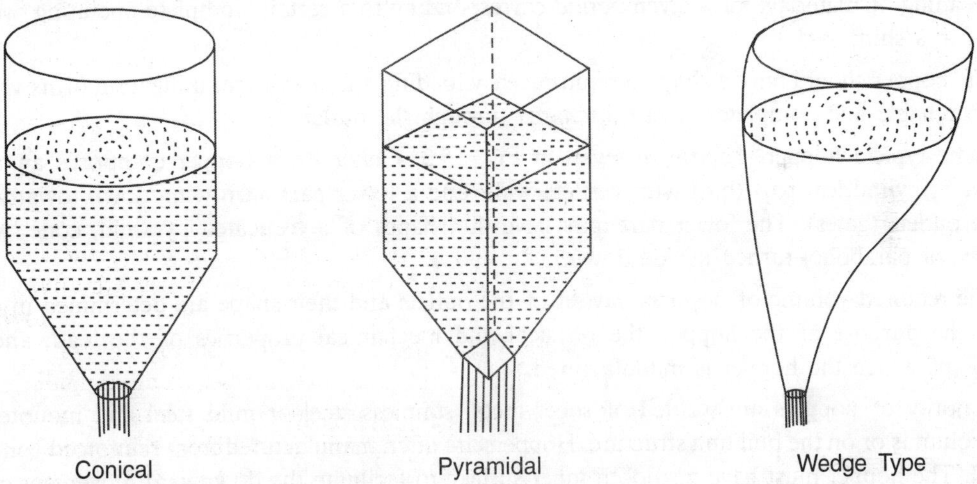

Figure 15.3 *Types of Hoppers.*

Interim storage hoppers in production and handling cycle make the operation of various individual transport and processing equipment less independent. Interim storage becomes essential when various mutually connected machines work at a different pace *i.e.,* work at different hours within a single shift or for a different number of shifts, or when certain devices work continuously (for example, conveyors, pneumatic or hydraulic transport installations) and others intermittently, delivering or receiving large or railless transport equipment. The pace at which interdependent machines operate and their output determine the volume and layout of hoppers in the system.

The greater the discrepancy in the operating cycles of machines and larger the hopper capacity per unit time, the larger becomes the required volume of hoppers.

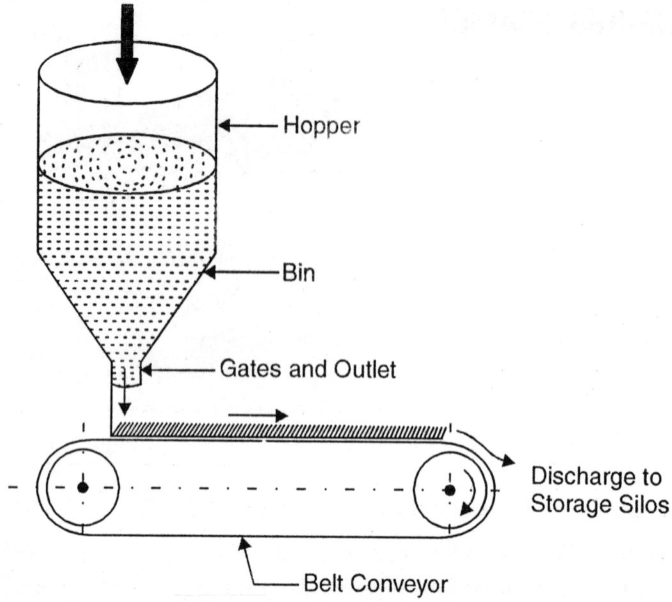

Figure 15.4 *Use of Hoppers.*

The required volume of a hopper is best determined by drawing a graph representing the loading and unloading of materials for a given period corresponding to a certain complete operating cycle in a day or in a shift.

The hopper should have a shape permitting easy loading and maximum utilization of its volume and also reliable and complete gravity discharge through the outlet.

Some typical hopper shapes are given in Fig. 15.3. Generally, a hopper comprises an upper prismatic or cylindrical part (bin) with vertical walls and a lower part narrowing down towards one or more outlets (gates). The lower part may have the shapes of a truncated pyramid, cone, wedge (prismatic or parabolic) turned upside down.

The required volume of hoppers, layout of the outlets and their shape are determined in every case by the purpose of the hopper, the physical and mechanical properties of the load, and the materials of which the hopper is manufactured.

Majority of hoppers are welded of sheet metal stainless steel or mild steel and mounted on special columns or on the building structure. Hoppers are often manufactured from reinforced concrete, or wood. The hopper must have a smooth inner surface to facilitate the downward movement of the bulk material contained. For abrasive material, vertical and especially inclined hopper walls subject to intensive wear are equipped with a removable lining. The lining is generally made of removable steel plates or Tiscal or Clinker brick. At present extra-strong glass lining is also used.

The inner faces of the hopper bottom and walls are subjected to pressure by the material contained. The flat walls are subjected to bending stress and should be reinforced with stiffners.

(*i*) The conical, cylindrical and parabolic walls are subjected mainly to tension are of lighter design.

(*ii*) A parabolic hopper is most practical in this respect as it repeats the shape that a flexible hopper wall would naturally assume under the action of a hydrostatic pressure. Parabolic hoppers are generally suspended on supporting structures by their upper edges.

The value of the bulk load pressure 'p' on the vertical walls is proportional to the vertical distance from the surface of the load 'h' m etc., the bulk weight of the load 'y' tonnes per m^3 and its flow factor $k_1 (k_1 < 1)$:

$$p = k_1 hy \text{ tonnes/m}^2$$

Where,

$$k_1 = \frac{1 - \sin\phi}{1 + \sin\phi}$$

Where ϕ = angle of repose of the bulk material (for liquids $\phi = 0$ and $K_1 = 1$).

A bulk load flows through the outlet at the hopper bottom.

(*i*) In a hopper of symmetric shape, the vertical column of the load above the outlet, when set in motion, forms a funnel on the upper surface of the material, along the periphery of which the load particles slide as the material is discharged. Last to enter the funnel are the particles near the inclined hopper walls.

(*ii*) In an asymmetric hopper, the outlet is near the vertical side wall, the bulk load slides along that wall and the funnel formed is asymmetrical.

For sized coal 45° to 50°, for fine coal 60° or at least 50° for moulding sand at least 65° will be angle of incline. Measure angle of repose of material and use hopper angle equal or greater than the angle of repose.

Economy of space (vertical or horizontal), type of loading (space required at top), capacity as per type of material (angle of repose) are criteria for bin design.

Arching in the hopper is a factor that often posses a serious problem. Handling up, freezing and plastic packing of the material and its sticking to the hopper walls are also problems. Arching is off set by vibrators mounted on the hopper walls, by jetting, if compressed air is available, or by motor-driven mechanical mixers; for powdered materials porous pneumatic aerating plates may be mounted on the in side hopper walls near the outlet. In certain cases, special hatches for a poker are provided. Hoppers for freezing materials are thermally insulated or heated. Hoppers of special shape with stepped walls are designed for materials that tend to arch or hang up.

15.3.1 Types of Flow Bins

1. Mass flow bins.
2. Funnel-flow bins.

1. Mass Flow Bins :

(*a*) Particles segregate but remit on discharge.

(*b*) Powders deaerate and do not flood when the system discharges.

(*c*) Flow is uniform.

(*d*) Density of flow is constant.

(*e*) Level indicators work reliably.

(*f*) Product does not remain in dead zones, where degradation can occur.

(*g*) Bin can be designed to yield non-segregating storage or to function as a blender.

2. Funnel Flow Bins :

(*a*) Particles segregate and remain segregated.

(*b*) First portion in is last one out.

(*c*) Product can remain in dead zones until complete cleanout of the system.

(*d*) Product tends to bridge or arch, and then to rat-hole when discharging.

(*e*) Flow is erratic.

(*f*) Density can vary.

(*g*) Level indicators must be placed in critical positions so they will work properly.

(*h*) Bins perform satisfactorily with free-flowing, large-particle solids.

15.3.2 Specifying Bulk Materials for Best Flow

Many flow problems can be eliminated at the source by rigid, accurate arid sensible specifications of the physical characteristics of the material.

Particle size is one of the most common and controllable factors which affect the flowability of a given material. The larger the particle size and the more free the material from fines, the more easily the material will flow. Stage grinding in the plant can reduce waste and improve flowability by producing a ground material with a minimum of fines, but this involves extra operations which may not be economically feasible.

Handling ease is often enhanced by pelleting the raw materials. The large particle size, uniformity of particle size and hard smooth surface of pellets all contribute to good flow.

Most materials can safely absorb moisture up to a certain point; further addition of moisture can cause significant flow problems. Specifications can control the amount of moisture content present in purchased raw materials. Moisture content can be lowered in the plant by including a drying operation in the process line. The costs incurred in drying may be offset by more efficient flow, lower shipping cost and control of deterioration losses.

15.3.3 Concepts of Vibratory Hoppers

15.3.3.1 Theory of Bin Discharge

The primary problems in discharging materials from storage bins are (*i*) classification, (*ii*) selectivity and (*iii*) bridging.

(*i*) Classification is a separation of aggregate and fines which occurs when material is placed into storage bins. Since the larger particles have more mass, they tend to fall to the outside periphery of the bin (See Fig. 15.5).

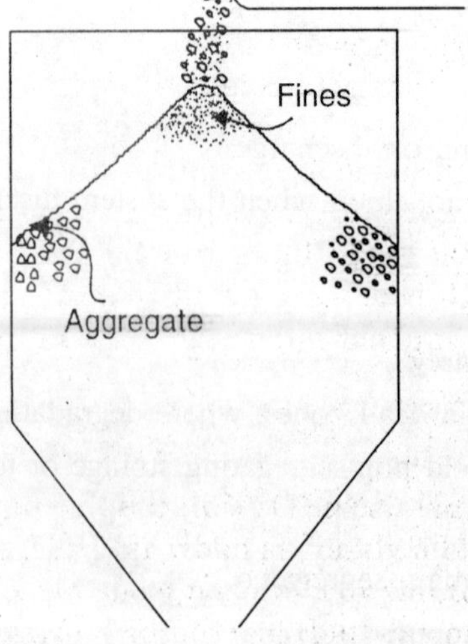

Figure 15.5 *Classification.*

(*ii*) Selectivity is the tendency of a bin design to discharge the center portion of the material first and the outside walls last. This is commonly referred to as coring or plug flow, (See Fig. 15.6).

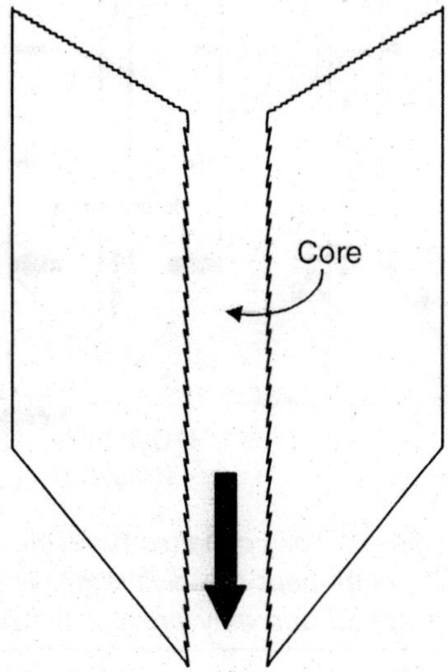

Figure 15.6 *Selectivity*

(*iii*) Bridging renders a flowable material to an unflowable state. In processes, this is costly because product flow is shut off. Generally, bridging comes about for the following reasons :

(*a*) **Adhesion to walls and cohesion to particles.** If a material has an adhesive quality to the bin wall, it will tend to cling to that wall. Then, if the particles will cohere to one another, a successful bridge will be formed. Most pressure-sensitive materials, such as puttylike polyvinyl resin, fall into this category.

(*b*) **Irregular particle structure-interlocking.** When odd-shaped particles are placed in a confined space, they tend to interlock. Coal or wood chips are examples.

(*c*) **Set factor.** "Set" takes place because of an interchemical reaction of the particles in the bin with time of storage. If a very hot material is placed in the storgage bin and allowed to cool, set will take place. Moisture would result and cause a greater cohesive bond. Materials so affected include spent tea leaves, foundry, sand and gypsum.

(*d*) **Static pressures.** When materials are placed in a confined space, they have a tendency to exert pressures in all directions, much like water. These pressures contribute to the formation of a structural bond.

(*e*) **Dynamic pressures.** These are the additional pressures that come about when the bin is discharging. All the material rushes to get into the flow stream, and the forces thus exerted have their equal and opposite force reflected to the bin side wall, (See Fig. 15.7). In grain silos, these dynamic pressures have been known to be as much as three times the pressures that existed statically—enough to crack large concrete silos. Furthermore, these same dynamic pressures continue to exist after the flow has been stopped. This is the reason why some materials will flow out of a bin initially; but then when flow is interrupted and stopped, it will not start of its own accord again.

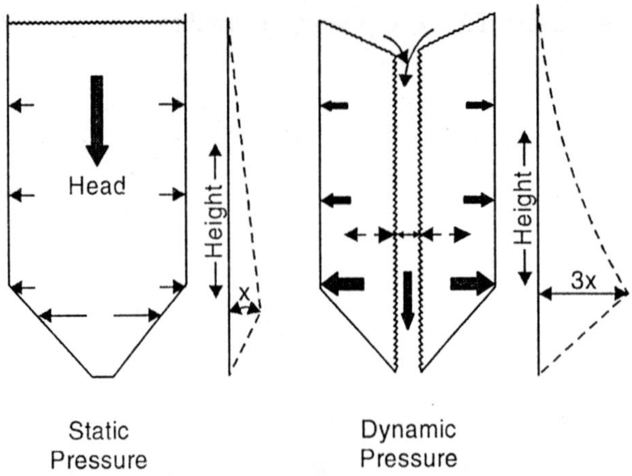

Figure 15.7 *When material begins to flow, dynamic pressures build up against the sides of a storage bin, especially at points near the discharge. The pressures continue even after the flow has stopped and may prevent the flow from resuming.*

To overcome the problem of storage bin discharge, the process design engineer has essentially four alternatives.

(*i*) He can change the design of his process to avoid the bin;

(*ii*) change the chemical properties of his material;

(*iii*) customize or change the geometry of the bin, (Fig. 15.8);

(*iv*) or apply some external force. Namely vibrations.

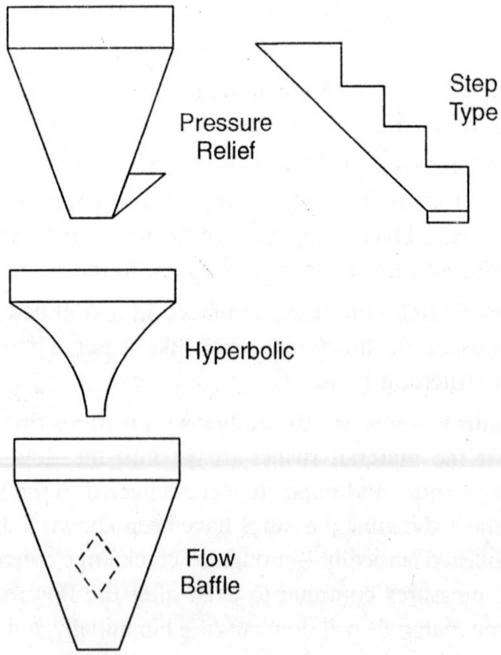

Typical Bin Designs to Relieve Pressure

Figure 15.8 *Many types of bin design.*

These vibratory actions should destroy any structural bridge formation, maintain a uniform flow from a full of material, and do this with a minimum amount of energy. To minimize the energy requirement, the vibratory action should "induce" material flow rather than "force" it. This is best achieved by meeting the following conditions :

(*a*) **Fluidization.** The materials should take on the flow characteristics of water. Any form of vibratory action tends to do this inherently.

(*b*) **Maintain constant shear.** All solid materials are weakest in shear as compared to their compressive or impact strengths. Therefore, the vibrations should be applied in such a manner as to keep the material in constant shear.

(*c*) **Relative motion must be minimized.** All materials resist work. Relative motion reflects work or energy consumption, so it follows that relative motion must be minimized to keep the material's resistance to flow at a minimum.

(*d*) The mass inertia of the material must be directed toward the outlet. The external force should be applied so that it takes advantage of the inertia of the material and directs it toward the outlet of the hopper.

Vibration offers the inherent characteristic of fluidization. However, these vibrations must be applied in a specific direction. If you arbitrarily vibrate a hopper you are obtaining fluidization, but you are not necessarily doing anything more. These vibratory actions must be applied in such directions as to maintain the material in constant shear, eliminate relative motion, and take advantage of the mass inertia characteristic of the material in the direction of the outlet.

The stroke of the vibratory action is not arbitrary, either. Experiments show that short strokes (3 mm or less) have a tendency to activate only the surface of the material. The material near the wall of the hopper will be fluidized, but the material in the center will not. Deep penetration requires larger strokes—as a rule, 6 to 10 mm.

The frequency of vibration needs to be relatively low, to permit the wave formation to penetrate deep into the mass. Low frequencies are anything less than 1800 cpm. A frequency of 860 cpm seems to be practical.

Because a vibratory hopper applied to a storage silo can be subjected to a tremendous change from no-load to full-load condition, it necessitates a departure from the "positive" type vibratory system. If the system were positive, the connecting vibratory arm would be subjected to ferocious stress conditions because of these changing heads and shock loadings. So, the "free mass" vibratory system is mandatory. Here, the complete system floats atop isolators and is free to give with any shock or heavy head load, regardless of directions. There are two disadvantages to the free mass system. It will move as a separate entity, so that there is no control over the system once it is energized. This requires precise design so that it will do exactly what you want it to do. In addition, high structural stresses can be generated by the excitor system, and this requires special care in the hopper design and fabrication.

15.3.3.2 Two Types of Vibrating Hopper

The whirlpool hopper meets all the conditions listed previously—fluidization, constant shear, elimination of relative motion, and taking advantage of the material's mass inertia effect in the direction of the outlet. As a result of meeting all these conditions, the material appears to spin around in a vortex-like action, like water going down a drain, (See Fig. 15.9).

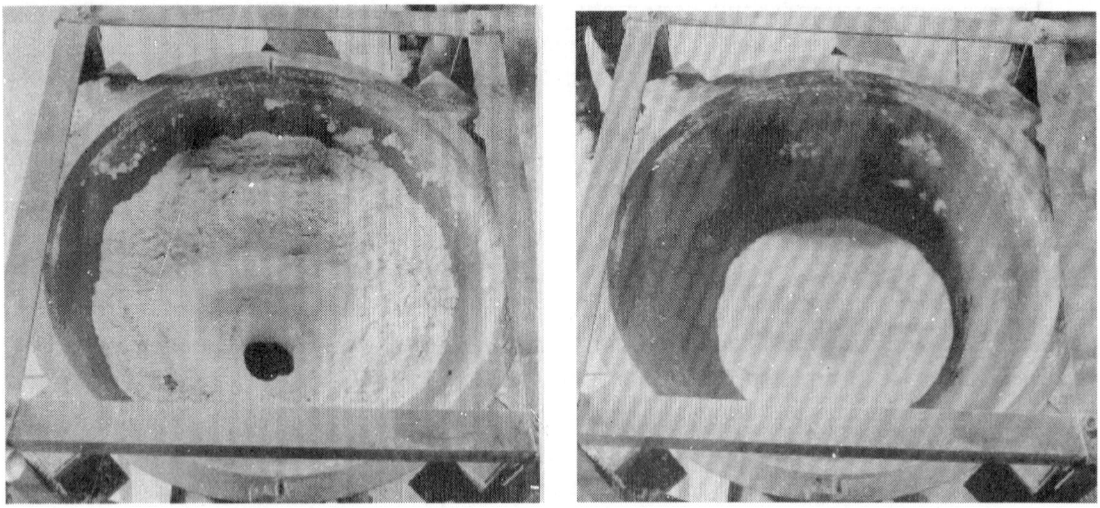

(a) (b)

Figure 15.9 *Contrast between static discharge (a) and vibratory-hopper discharge (b) which meets conditions of fluidization, constant shear, minimal relative motion, and inertia of material directed toward outlet.*

To obtain a constant shear relationship, an unbalanced couple is introduced in the hopper's horizontal plane. This causes the hopper to vibrate in a horizontal twisting motion, Fig. 15.10.

At the same time, a vertical component is introduced to cause the hopper to move up and down. When the hopper is on its up stroke, the material is subjected to dynamic forces caused by inertia effect. These dynamic forces cause the material to "think" it weighs many times more than it actually does. As a result, it causes the material's bridge to act like a heavily and uniformly loaded structural beam. The structural bond cannot negotiate these additional high stresses, and the bridge is broken. The material is accelerated downward toward the outlet.

In the whirlpool hopper, two motors are mounted 180 deg apart in such a way that all unwanted forces are balanced. Any relative motion is eliminated and the hopper moves perfectly about the vertical axis as a complete entity, Fig. 15.11. If both the motors are rotated in exactly the same direction, the motors will balance when their weights are pointed inward. Because they follow the path of least energy, the always try to balance themselves. So any impacting forces to the material are eliminated. Rotating the motors 90 deg causes one eccentric to point 180 deg opposite, and the twist or shear action is obtained. As the eccentric weight rotates the next 90 deg, the impacting forces are again balanced throughout the complete arc displacement. In the following 90 deg, the eccentrics find themselves unbalanced and in opposite directions, and thus the twist type of action is again obtained. Note that all unwanted forces are balanced completely throughout the rotary cycle. Only the twist action remains unbalanced, and this is exactly what is desired for the constant shear effect.

By mounting the motors at a "cocked" angle, the vertical component is introduced. Significantly, the vertical component is at its maximum value at exactly the same instant the shear or twist force reaches a maximum. Thus a resultant couple is generated within the hopper, as is shown in the lower center of Fig. 15.10. The angle is the resultant stroke angle and reflects the quantitative values of shear and inertia forces. Because the hopper is a free mass vibratory system, and because the phasing of the motors comes about by the motors seeking the path of least horsepower.

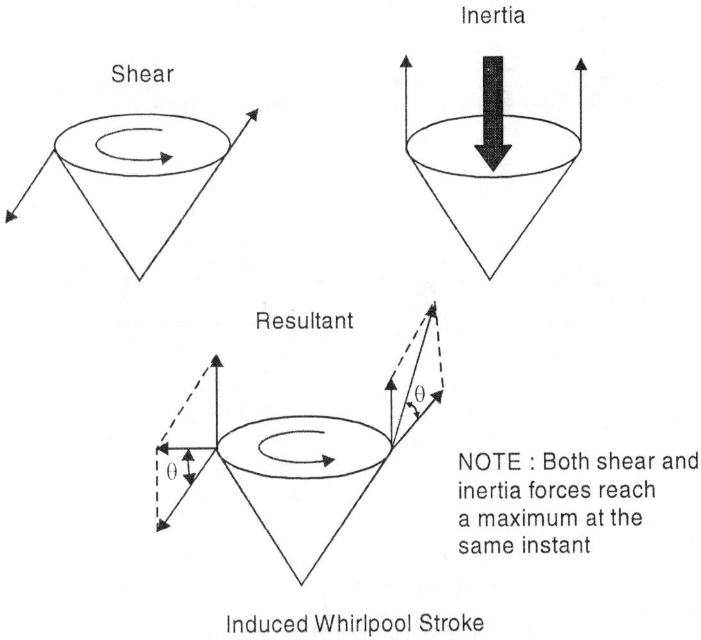

Figure 15.10. *In the induced stroke of whirlpool hopper, both shear (twisting) and inertia (downward) forces reach a maximum at same instant.*

The disadvantages of the type of hopper described here are two—the tremendous structural relationships existing within the hopper structure itself, and the need to design the vibratory system so that the motors have full command of the system throughout their accelerating characteristic speed-torque relationships.

Isolators provide the whirlpool hopper's only connection with the "outside world", leaving the entire system free to move in any direction to react to head loadings or shock. A sturdy excitor ring is also needed because of the strong structural relationships that exist. On most whirlpool hoppers there are three inverted angular rings. Essentially, this type construction forms an excitor ring which would remain intact if the hopper shell itself were removed. Thus, stresses on the hopper shell are little more than what they would be under a static operating bin. The quality of welding is also vital.

The high lateral pressures within the hopper will cause a tendency for the material to arch. These pressure points can be reduced greatly by placing pressure cones within the hopper. Best results have been obtained by placing the cone at the location of the hopper center of gravity. The diameter of the cone should be 1/3 of the diameter of the hopper at the center of gravity.

The heart of the whirlpool hopper is the motor. Special care in the construction is necessary to obtain a motor that will withstand the severe vibratory stresses. Also, the rotor must have its lamination configuration altered to compensate for the extremely large shaft. The bearings are designed for an average life of 85,000 hr under maximum rated load and are of the spherical roller extremely heavy-duty type.

Because they are of high-torque design, the motors are rated in bearing load capacity rather than horsepower. A VM-6000, for example, means a vibratory motor with a bearing load of 30000 N. Or, this motor could develop a total of 60000 N of dynamic forces. The spherical roller bearing found in a VM-6000, which is only a nominal 1 or 1 1/2-hp motor, has the same bearing normally found in a 500-hp motor.

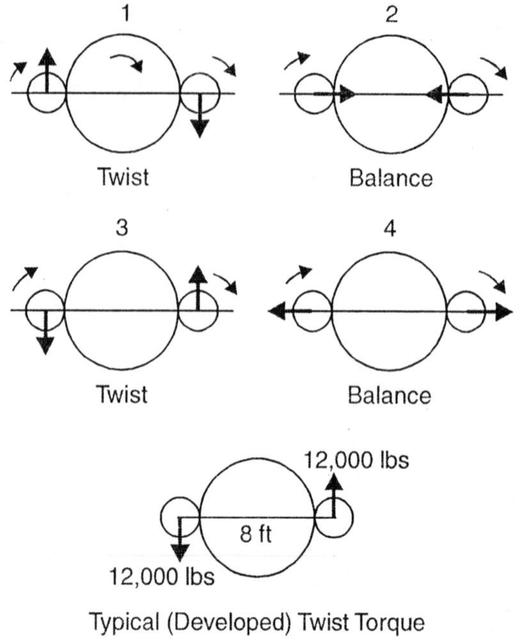

Figure 15.11. *Diagrams shows how eccentric weights of motors on whirlpool hopper are offset to eliminate any relative motion, leaving only the constant shear effect throughout the cycle.*

The second type of vibratory hopper design is represented by the gyrated hopper. This is not as versatile a design as that utilized in the whirlpool, but provides adequate and economical results with many materials.

A single motor excites the gyrated hopper, driving the hopper essentially in the horizontal plane. Thus, it does not necessarily have an inertia effect in the direction of the outlet, and it does introduce some relative motion, (See Fig. 15.12). When the eccentric weight is outboard from the hopper, it tends to yank the hopper in that direction. The material is thus slammed against the hopper wall. As the eccentric weight rotates 90 deg, it then is tangent to the hopper and causes a twisting or shearing action which enhances the flow. Another 90 deg of rotation finds the eccentric weight driving through the horizontal centerline of the hopper and shoves the hopper in that direction. Again the material is impacted against the wall. In the last 90 deg rotation, the hopper again obtains a shear action. Note that the elliptical stroke pattern is obtained, and the greater stroke occurs along the line of shear, with the least stroke along the line through the center of the hopper. Since, the same force exists throughout the cycle, this shows that the least resistance lies in the shear direction, with the greatest resistance in impact.

To ensure that the entire hopper moves an an entity, the excitor drive is mounted so it drives through the hopper center of gravity. This keeps relative motion to a minimum and makes the entire assembly move truly in the horizontal plane.

Because the hopper is a free mass system, with no positive connection between the eccentric drive and the hopper, the assembly ranks as a low energy arrangement. However, it should be noted that the unwanted forces of the excitor motor are not cancelled, as they are in the whirlpool hopper type design.

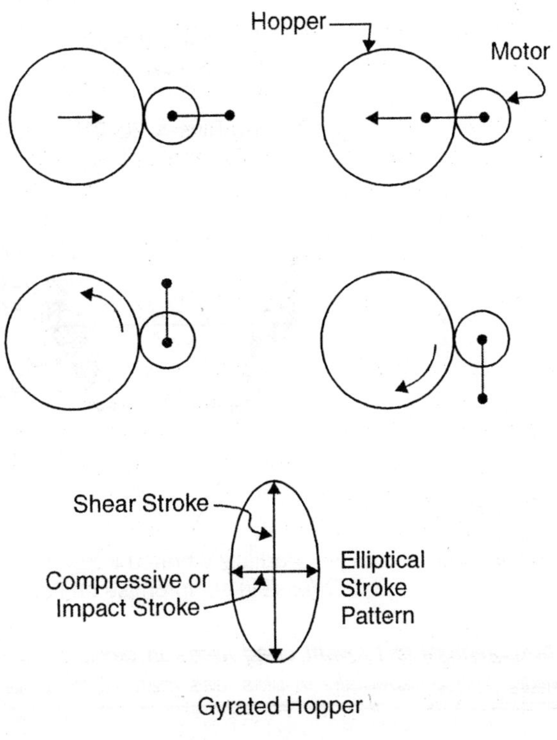

Figure 15.12. *Single motor produces elliptical stroke pattern on gyrated hopper, which vibrates in horizontal plane only.*

In the Gyrated Hopper design, two excitor rings are utilized, although the structural relationships are relatively small. The hoppers are built on liquid-filled isolators, and pressure cones are utilized to reduce pressure points and arching.

15.3.3.3 Application of the Two Hopper Types

In general, however, materials which will tolerate a certain amount of relative motion, and materials which have enough mass to draw them towards the outlet, can be handled successfully with a hopper of the gyrated hopper design. For pressure-sensitive materials, however, which have a strong tendency to resist flow, which have a high "set", or which are so light that the mass is not strong enough to draw them to the outlet, a hopper of the whirlpool hopper type is required.

15.3.3.4 Design of Vibratory Hoppers

Various factors must be considered in determining the proper size and design of vibratory hopper for the particular application.

(*a*) Diameter

The best approach is to make the bin cylindrical, with the vibrating hopper representing the entire conical bottom portion. This is particularly recommended for bins 300 cm in diameter or less. If smaller hoppers than this must be used for reasons of cost, the hopper diameter should be 25 percent larger than the known bridging length of the material. For bins of 3 m to 6 m in diameter, the hopper diameter should be at least 1/2 of the upper bin diameter. For bins of more than 6 m in diameter, the vibratory hopper diameter should be at least 1/3 that of the bin, Fig. 5.13.

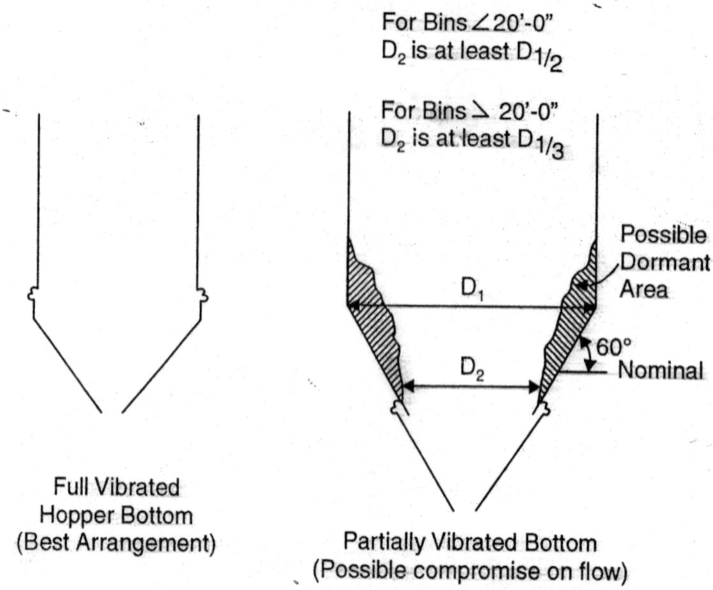

Figure 15.13 *Ideal arrangement to prevent dead areas in bin is to have hopper diameter equal to bin, especially in bins less than 10 ft in dia.*

(b) Effect of Head

The head of material in a silo affects the operation of a vibratory hopper, and a portion of the head is reflected into the isolation mounts supporting the hoppers. A hopper of the gyrating hopper design, which vibrates only in the horizontal plane, "picks up" a head equal only to about 150 percent of the hopper diameter, showing that the vibrations fluidize little material above the hopper. With a hopper of the whirlpool hopper design, the horizontal or shear stroke is little affected by head, while the vertical component will diminish slightly under full load. However, the dynamic forces are still present. The actual effect of head on the hopper stroke depnds on the material—the more flexible the material, the smaller the effect. A hopper of the whirlpool type will pick up a head qual to at least 4 times its own diameter, indicating it fluidizes a great deal of the material above it.

(c) Slope

A hopper with steeply sloping sides tends to draw from across the entire cross section of the bin. If the slope is shallow, it tends to draw from directly above the outlet, and coring could result. If shallow hoppers are necessary, baffle or pressure cones can be utilized to offset this, (See Fig. 15.14).

(d) Method of Support

The general philosophy of supporting a free mass system is to make the supports either very rigid or very flexible. Anything in between will cause difficulties. Various types of hanger supports or under supports can be utilized, depending on the installation.

(e) Isolated Bin

If the structural supports are inadequate or marginal, an isolated bin design should be considered. Here, the upper storage bin is set atop its own set of isolating springs while the vibrating hopper is hung from the isolated upper storage bin, Fig. 15.15. The equal and opposite reaction of the springs of the vibratory hopper is passed into the isolated bin above. Since, it is free to move, it absorbs this reaction

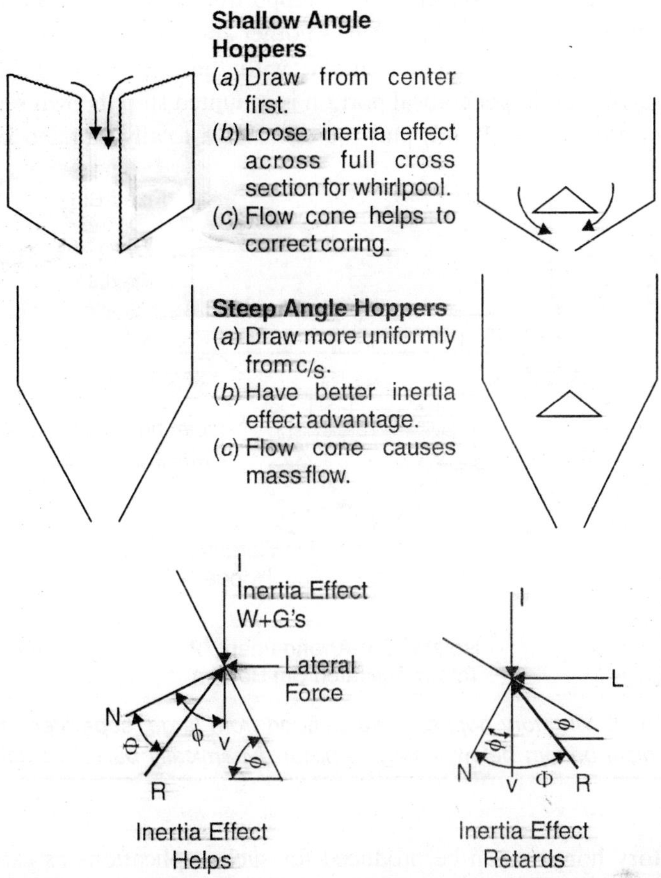

Figure 15.14 *Steep angle hoppers encourage better flow than shallow-angle hoppers; flow cones may promote maximum results with either.*

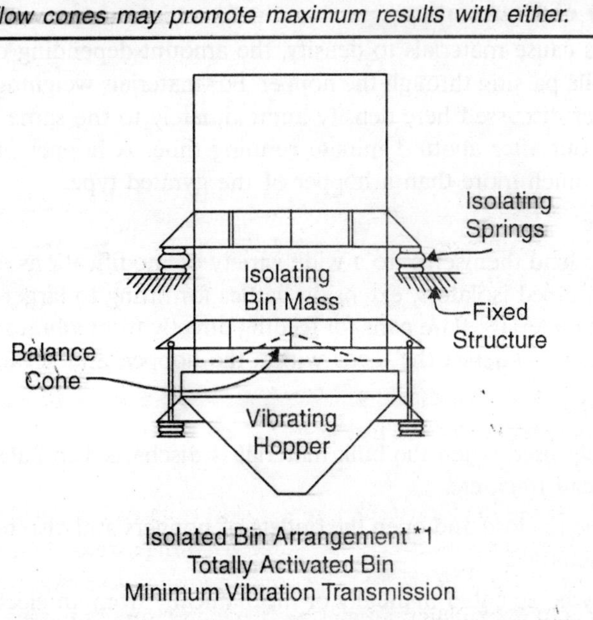

Figure 15.15 *Upper storage bin rests on own isolators, while vibrating hopper is suspended from bin, so virtually no dynamic forces reach supporting structure.*

and little or no dynamic forces are transmitted to the supporting structure. As a result, this arrangement has the advantage of giving a totally fluidized bin. Another approach is to use a large conical bottom. A smaller vibratory hopper bottom could be built, with the upper conical portion acting as its counterbalance, Fig. 15.16. The upper conical portion is mounted stop its own set of isolation springs. The result is a dynamically balanced unit, plus the effect of a totally vibrated large hopper bottom.

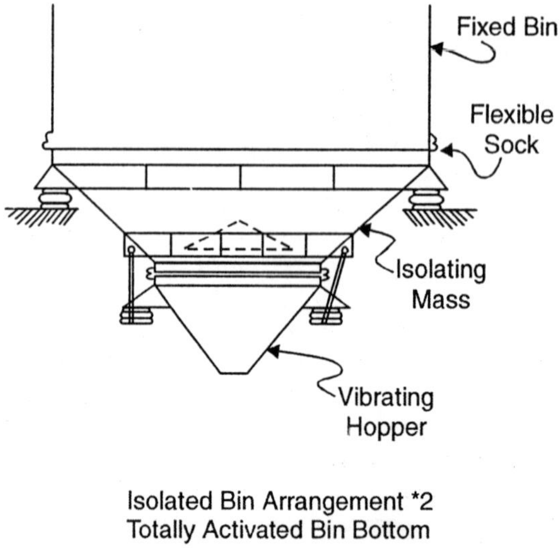

Figure 15.16 *Vibratory hopper is suspended from large, separately mounted conical bottom below storage bin, for dynamically balanced unit.*

(*f*) Miniature Units

Miniature vibratory hoppers can be produced for such applications as extruders. These are of high frequency and short stroke design, because deep penetration is not required.

15.3.3.5 Densification of De-Aeration

Vibrating hoppers cause materials to densify, the amount depending on the length of time the material is vibrated while passing through the hopper. For materials weighing more than about 50 pcf, the two types of hopper discussed here densify approximately to the same extent. The densification curve appears to level out after abotu 3 minute running time. A hopper of the whirlpool type will densify light materials much more than a hopper of the gyrated type.

15.3.3.6 Modifications

Vibratory hoppers lend themselves to a wide variety of modifications. These include slide gates, adjustable outlet cones, piped isolators, extension baffles for fitting to larger diameter bins, belt skirts for feeding non-dusting materials, flare pans for feeding directly from vibratory hoppers and regulating the flow, pressure switches to detect the loads within the hopper, and a variety of feeding chutes.

15.3.4 Hopper Gates

Gates are generally used when the bulk material is discharged in batches into truck card on a ground level or over head track etc.

Hopper gates serve to close and open the outlets of hoppers and also to adjust the stream of the bulk material being discharged.

Hopper gates may be actuated manually or mechanically from an electric motor or air cylinder, the latter being of simpler design and easier to control. Both types may be remote-controlled. Remote controi is especially desirable for hard and large lumped materials when the gates are of large dimensions and their shifting by hand requires great effort.

The dimensions of the gate outlet depend on the lump size and degree of flowability of the bulk material and on the required transfer capacity. The empirical formula given below may be used to determine the minimum dimensions of the side of a square outlet.

$$B = k\,(80 + a'_{max})\tan\phi \text{ mm}$$

Where k = empirical factor (k = 2.6 for sorted and k = 2.4 for unsorted bulk materials) :

a'_{max} = size of the largest lump, mm;

ϕ = static angle of repose.

If the outlet is round, the diameter should not be smaller than B determined as above; if rectangular—the sides should not be smaller than 0.75 B and 1.5 B.

Experience has shown that for large-lumped materials, the side of a square outlet should not be less than four to five times the size of a'_{max}. At the same time, the outlet should not be too large to avoid a sudden discharge of a large quantity of the material and also to keep gate size small. Heavy gates are difficult to handle.

Horizontal type is better for table feeders while vertical type is better for electromagnetic feeders.

15.3.5 Types of Hopper Gates

According to their principle of action, hopper gates may be divided into three main groups : (1) simple slide gates (2) trough gates and (3) pivoted gates (4) finger gate.

1. Simple Slide Gate

A simple slide gate is a flat valve sliding in grooves. It is actuated by a hand-operated rack or a leverage and is simple and compact. The gates of this type have a number of serious disadvantages-resistance in the grooves makes it difficult to shift the gate, lumps of material are likely to wedge in when the gate is closed. Simple slide gates are used mainly for small-lumped free-flowing materials and when repositioning of the gate is not frequent.

Crawler or Cut-belt Gate

The crawler or cut-belt gate is an alternative to the slide gate. It consists of an endless rubberized textile conveyor belt one side of which is secured at the edge of the outside or the other. The belt runs over the roller and open or closes the outlet. There is no sliding friction of the belt on the material.

2. Trough Gate

The trough gate consists of a trough hinged to the outlet. When the trough is lowered, it directs the stream of material from the outlet and when raised keeps it from flowing. Trough gates permit adjustment of material flow and exclude jamming but their great height is a marked disadvantage.

3. Pivoted Gate

The pivoted gate has a cylindrical surface and pivots around its horizontal axis when opened or closed. As compared with flat gates, they have the advantage of excluding friction in the grooves, while friction in the pivot is negligible. They are, therefore, opened and closed with less effort than slide gates. For horizontal outlets, they may be made of one sector or two sectors opening in the middle. When mounted on inclined outlets, the gate may open upwards or downwards. In the former type the material flow is regulated by the extent to which the gate is raised. A disadvantage of this design is that lump of material may wedge in when the gate is closed. The gate of the second type does not possess the above disadvantages but material flow cannot be regulated because it forms an obstacle when not fully open.

Double Pivoted Gate

A double pivoted gate may be used for large-dumped materials. It eliminates jamming and enables to adjust the stream of material. When the gate is opened, the lower sector opens fully while the stream is regulated by the extent to which the upper sector is opened. To close the gate, the upper sector is partly lowered, while the lower sector is raised to form an obstacle and to prevent further flow of material through the outlet.

4. Finger Gate

The finger gate is an alternative to the pivoted gate. It consists of individual heavy bent and pointed levers (fingers) each suspended on a chain. When lowered, the levers easily cut into the material flowing down a chute, obstruct its movement and close the outlet. They are raised by means of an air cylinder, first those that have penetrated deepest, then all at once.

15.3.6 Applications of Hoppers

For following materials : Oil seeds, Rice, Sugar, Herbs, Miscellaneous powders, Animal feeds, Pet food, Cereals, Wheat, Corn, Meals, Hakes, Flour, Tea, Milk powder, Spices, Sand, Flyash, Ore, Salt, Gypsum, Coal, Starch, Cement, Resins.

Hoppers are used in almost all industries handling bulk material as listed above.

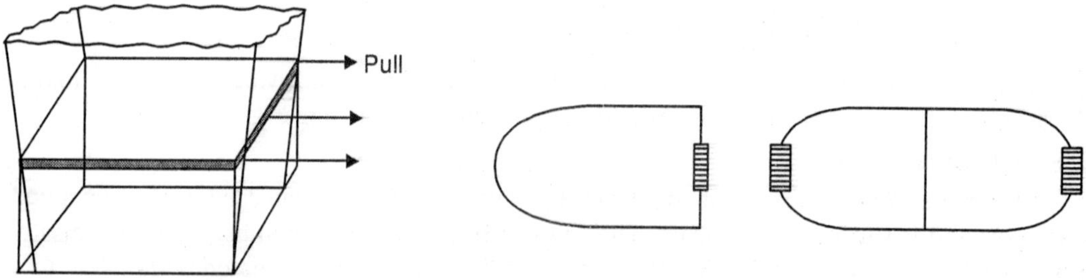

Slide Gate Pivoted Gates

Figure 15.17

15.4 CRUSHERS

Crushers are used to crush the material on a small size so as to make it easily flowable. Crushers can be classified as stationery and mobile. Stationery one is used in industrial applications while mobile one is used in construction, mining, coal, ore, concrete industries.

The material is crushed by impact, shear or attrition.

There are three main types of crushers used by industries for material handling. They are :

(*i*) Roll type

(*ii*) Ring type

(*iii*) Hammer type or

Jaw type.

(*i*) Two roll type is used for sizing coal down to 40 or 30 mm for use on travelling, chaingate or under feed stokers.

The single-roll type with fixed breaker plate is used for similar purpose for size less than 25 mm.

(*ii*) The **ring or hammer type** depends on centrifugal action to do its crushing.

Wear plates are provided on the bolts of the crusher which can be easily replaced when they wear out. These are made of hard material such as manganese steel or high carbon steel.

The rotating elements, which carry the rings and hammers, need regular maintenance in case of ring and hammer type crushers. They need a balanced replacement of worn parts so that high vibrations are not caused due to any unbalance while replacing worn out parts.

(*iii*) **Jaw type crushers.** These crushes the material between two opposing forces, the jaws. The movable (swivel) jaw compresses the material against stationery jaw.

15.5 FEEDERS

When it becomes necessary to have the material flow continuously operating processing equipment, powered feeders are used.

Feeders are mechanical devices mounted near the outlet of a hopper and serve to unload the hopper. Discharge is partly by gravity, partly induced. Feeders regulate the stream of material much more accurately than gates. The stream is adjusted either by changing gate positions, *i.e.,* by narrowing or widening the cross-section of the outlet, or by changing the rate at which the loading element operates (speed of rotation or translatory, etc.) This is effected with steeped or infinitely variable speed changers, adjustable—speed electric motors, etc. When the supply of the motor is cut *i.e.,* when the motor is stopped, gravity flow of the material ceases, and the feeder acts as a gate.

If the hopper has several outlets discharging alternately, the feeders are sometimes carriage—mounted and moved from one outlet to the other. Special feeders for slotted hoppers (so—called dischargers) are always mobile and service certain area of the hopper length.

The type of the feeder selected depends mainly on the properties of the material, its lump-size, flowability, specific weight, and also on the required capacity, hopper shape, etc.

(*a*) Belts, (*b*) apron, (*c*) screw and (*d*) oscillating feeders, (*e*) Vibrating feeder, (*f*) Disk feeder are alternatives of conveyors of the same name. They are comparatively short, of increased rigidity, because they are subjected to higher load pressures than conveyors (the load column above the outlet bears on that feeder) and also to higher resistances coming into play when the material flows out of the hopper outlet.

Conveyor type feeders may be extended if the receiver is situated at some distance from the hopper. They have no prototypes among materials handling machines and can be utilized only for the discharge of materials into a receiver directly below the hopper outlet.

(*a*) Belt Feeder

It may be horizontal or inclined. The features distinguishing belt feeders from ordinary belt conveyors are : the working strand runs over a fixed runway or bed of closely spaced rollers (with a pitch of 0.25 to 0.3 m), there are no idlers on the lower strand, the feeder has stationary skirt boards, belt speed is low (from 0.1 to 0.3 m/sec), as the material leaves the outlet slowly and forms a comparatively heavy layer on the belt. Belt feeders are used for granular and small-lumped, less often for medium-lumped materials. A flow-control valve regulates capacity. To eliminate rapid belt wear, the feed chute is shaped to diminish the active pressure of the material on the belt.

(*b*) Apron Feeder

Apron Feeders, may be horizontal or inclined, the angle of inclination of an apron feeder is larger than that of a belt feeder. Apron feeders may have travelling or stationary rollers, the later are spaced unevenly, at a smaller pitch in area under the hopper outlet where the apron is subjected to active load pressure, and at a larger pitch in the other sections.

Apron feeders are predominantly used for heavy and large lumped and also for medium-lumped materials. Apron feeders of specific design with a runway of increased strength are used to handle materials consisting of very large lumps, for example for stone and ore charged to jaw crushers.

Ordinary apron feeders are used for run-of-mine and sized coal, limestone, ore, etc.

The type of apron employed depends on the nature of the material and may be flat or corrugated, with moving or stationary skirt boards (heavy type).

Apron speed generally ranges from 0.05 to 0.25 m/sec. Low speeds are specified for low capacities and also for heavy and large lumped materials is large and the column of the material bearing on the apron is heavy.

Advantages. Apron feeders are strong. They permit high load pressures on the apron (important for large- sized outlets) are reliable with difficult loads and unload the material in a uniform stream. Their disadvantage is that they are comparatively complicated in design and have large weight resulting in high initial cost.

(*c*) Screw Feeder

Screw feeder consists of a continuous screw rotating in a pipe. The shaft of the screw turns in two terminal bearings mounted outside the pipe.

Screw feeders are used for powered, granular and small lumped bulk materials which do not have a threat of further crushing. A flow control valve regulates capacity. The pitch of the screw is generally smaller than in-screw conveyors; the screw is sometimes double-thread. When the screw diameter is determined, the trough loading efficiency ψ equals 0.8 to 0.9; high loading efficiency is explained by the absence of intermediate bearings.

(*d*) Oscillating Feeder

An oscillating feeder has a horizontal or slightly declined table with stationary side-boards. The table is generally supported by stationary rollers or may be suspended on rods. The table is set in reciprocating motion by an eccentric gear.

The operating principle of an oscillating feeder differs somewhat from that of an oscillating conveyor. When the table moves forward, the material is carried along by friction and a certain amount of material from the helicopter fills the free space formed below the outlet. When the table moves backwards, the material is prevented from following the table by the rear wall of the discharge spout and, therefore, spill over the front edge of the table.

The capacity of an oscillating feeder is :

$$Q = 60\ Bhs\ n\ \psi\ y \text{ tonnes per hour.}$$

Where,

B = distance between the side-boards (the efficient width of the feeder table), m.

H = distance from the lower edge of the flow-control valve to the table surface (along the normal to the surface), m.

N = number of strokes per minute (generally ranging from 20 to 60).

ψ = correction factor similar to the loading efficiency in oscillating conveyors (generally ψ = 0.65 to 0.70).

Oscillating feeders are used for materials of medium and small lump-size, extra-strong designs for large-lumped materials; the handling capacity of the feeder is adjusted by a flow-control valve or by regulating the speed of the table.

Advantages. Simple design and low initial cost have made the oscillating feeder a frequently used feeding devices.

(e) Vibratory Feeder

The vibratory feeder is an alternative of the vibratory conveyor and its operating principle reminds that of the conveyor. A low amplitude of vibrations (generally 1 to 3 mm) and high vibration (table stroke) frequency (up to 3000 per minute) are typical of the vibratory feeder. The trough is suspended on rods and supported by spring bars; it is driven by an electric motor or an electromagnet. Vibratory feeders are preferable used for small-lumped, less often for medium-lumped materials.

(f) Disk Feeder

The disk feeder consists of a motor-driven rotating disk-shaped table and a telescopic spout suspended above it. The spout is suspended under the round hopper outlet and does not reach the disk surface. A scraper clears the disk of the material falling on it in a manner similar to the operation of a discharge plough in belt conveyors.

By adjusting the spout and scraper positions, it is easy to regulate the capacity of the feeder with great accuracy within a wide range. Besides, these feeders do not pack the material as do screw feeders. Disk feeders are most often used with powered, small- and medium-lumped materials.

15.6 DUMPERS

As the name implies these are used for dumping of the material. These are essentially trucks with tilting arrangement on its back side. Dumpers are very common in construction, mining and powder industries. Important point of consideration in the dumpers is their capacity, movement and shape of the platform, speed of travel and dumping. These are ideal for dumping bulk parts, foods, liquids or waste materials, floor level loading. Dumping can be stopped at any point in the cycle ensuring a controlled flow of materials to prevent damage. Increases productivity, reduce plant handling costs.

15.6.1 Important Parts of Dumper

(*i*) **Engine :** Engines are the main driving or power source. These are generally multi cylinder, water cooled, turbocharged engines.

(*ii*) **Hydraulic cylinders :** These are installed below the platform. They are used to control the lifting and lowering of the platform during loading and unloading. These hydraulic cylinder ram arrangement is used for high capacity. These are controlled by operator or sometimes automatic raising and lowering is done.

(*iii*) **Transmission :** power is transmitted from engine to rear wheels using synchromesh reverse gear box.

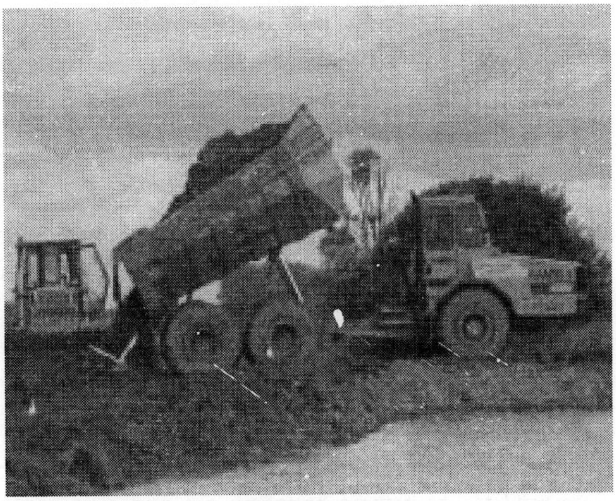

Figure 15.18

(*iv*) **Wheels and tyres :** are important parts of any dumper because when working it has to take all the loads. Generally, they are designed in Unified lug pattern, single in front and dual fitment at rear because they have to work in the soft terrains.

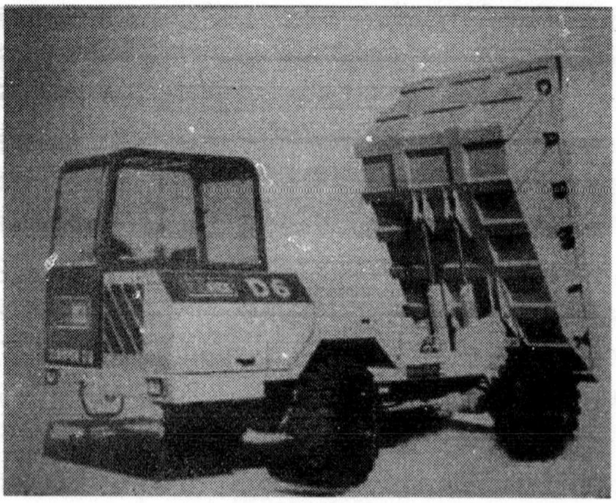

Figure 15.19

(*v*) **Brakes :** Pneumatic brakes are used.

(*vi*) **Suspension system :** These are important for shock absorption. In Front telescopic shock absorbers are used and in Rear leaf springs (semi-elliptical) are used.

(*vii*) **Cabin :** These are used for the driver. Position of it is important to control the vehicle.

(*viii*) **Frame and bumper :** Heavy duty welded frame and bumpers made of high strength structural steel channel are used for safety of vehicle.

Maximum speed 60 to 70 km/h.

Tilt angle : It varies from dumper to dumper, maximum achieved up till now is 75 degrees.

Movement of the platform : This is essential for loading and unloading. As shown in the figure in almost all the conventional dumpers it is done longitudinally. Nowadays dumpers with movement in transverse directions also been developed.

Capacity : It is having wide range of capacity from kg to tonnes.

15.7 SHIP LOADER

It is a device which is specially designed for loading and unloading of ships at the port or quayside. This is designed for flexibility and for rapid operation. Design varies from manufacturer to manufacturer with basic aim of rapid loading and unloading.

As figure indicates every ship loader have :

(*a*) Hoppers or Feeding Spouts

These are used to feed the conveyors. Feeding is done according to the speed of the conveyors.

(*b*) Chute or Conveyor

Conveyors carries the material from feeder to the spiral chute. There may be one or two intermediate conveyors between last and first conveyors which depends on distance to be covered. Conveyor discharges the material into the chute. Chute here is essentially used to transfer the material from conveyors to the ship. Normally spiral chute is used in ship loader.

(*c*) Hoist for Luffing Motion

Almost all types of ship loaders are equipped with the hoist for raising and lowering the conveyor boom (platform). Due to hoist luffing motion is given to the boom up to the angle of 60 degree with respect to the ground.

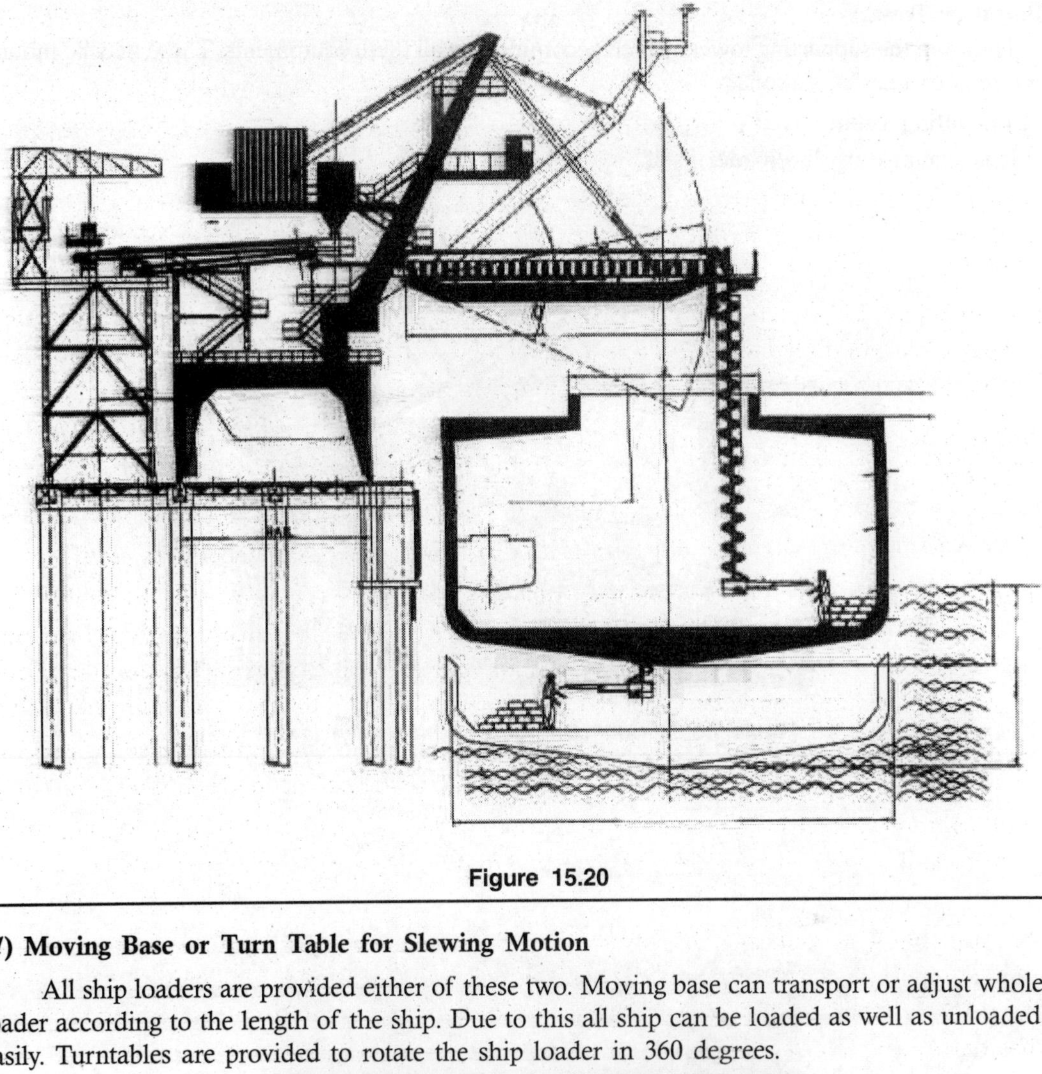

Figure 15.20

(*d*) Moving Base or Turn Table for Slewing Motion

All ship loaders are provided either of these two. Moving base can transport or adjust whole ship loader according to the length of the ship. Due to this all ship can be loaded as well as unloaded very easily. Turntables are provided to rotate the ship loader in 360 degrees.

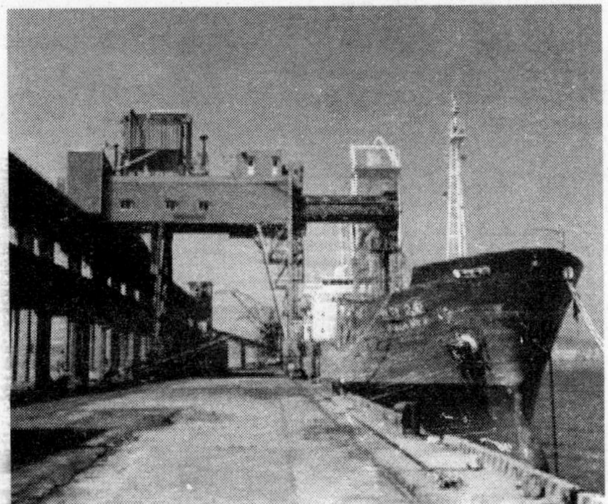

Figure 15.21 *Horizontal motion of ship loader.*

(e) Portal or Towers

These are the supporting towers which accommodate all these equipments. These may be mounted on the rails or may be stationary.

(f) Controlling cabin

It is available in ship loader itself.

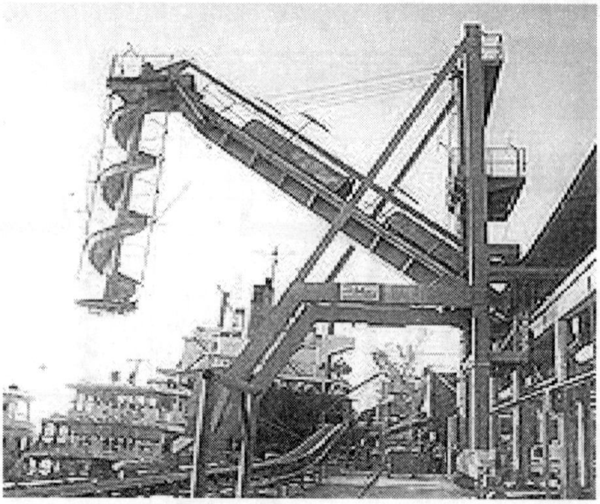

Figure 15.22 *Controlling cabin mounted on towers.*

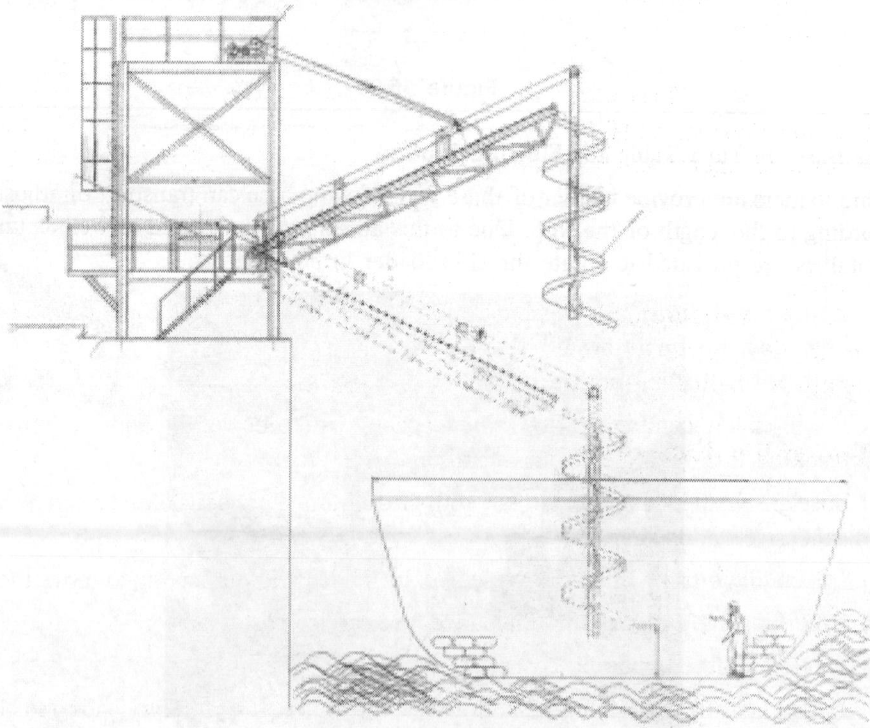

Figure 15.23 *Ship loader with slewing motion.*

(g) Capacity

Loading and unloading systems for bulk material (up to 1000 tonnes/hr) and bags (up to 2400 bags/hr) which cater for ships of all sizes can be done.

(h) Applications

Loading and unloading of bags or bulk material and the combined loading system for bags and bulk material can also be done. The wide range of systems designed and constructed allows loading and unloading of all types of material onto barges or sea vessels.

15.7.1 Cambered Boom Ship Loader

It is used to achieve continuous flow, high capacity and zero down time from loading operation. This consists of a conveyor boom with mobile vehicle and rotary turntable. It is specially used to handle bulk load only. The boom is inclined at the end towards the ship and it can turn at 360 degrees from one place this gives flexibility to this kind of ship loader.

Application of Ship Loader

Loading ships and/or stockpiling material at the quayside with it.

(*i*) High speed of loading and unloading.

(*ii*) Loading can take place into all areas of a vessel, regardless of tidal conditions.

(*iii*) Ships or barges remain stationary whilst its mobility offers flexible loading via remote control.

(*iv*) Depending on the models it can be designed to have a telescopic outreach from 26 to 42 metres, luffing to an angle of 17 degrees and slewing through 270 degrees.

(*v*) Machine operation can be directed via radio remote control, from on board the ship, meaning only one person is required to oversee the loading operation.

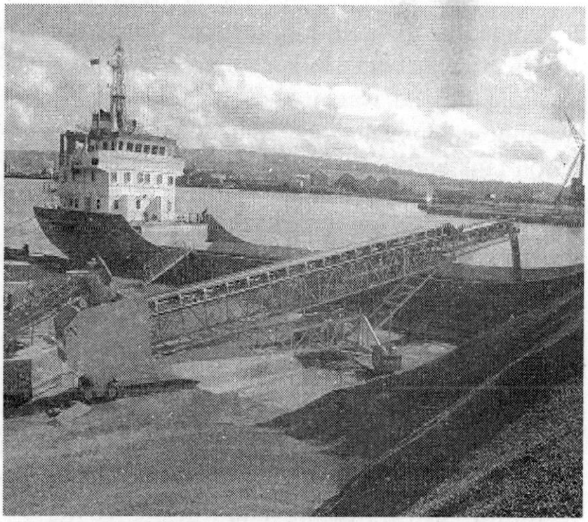

Figure 15.24

(*vi*) It will enable continuous flow, high capacities, zero downtime and a minimal workforce, resulting in a highly efficient operation.

(*vii*) Specialist ship loading equipment with greater tonne per hour outputs can be a much more expensive option than it.

(*viii*) Depending on size, an entire vessel can be loaded without having to move the ship loader.

(*ix*) It offers greatly reduced maintenance and service costs.

(*x*) It offers reduced running costs.

(*xi*) It conforms to environmental requirements with a low noise ship loader/stacker complete with extensive safety features.

15.8 WAGON TIPPLERS

These are used to unload the wagons carrying bulk material such as coal in power plants, steel plants that's why it is known as wagon tippler. The tippler consists of two circular rings, a platform with travel rails, support rollers, two girders as well as clamping device, which retains the wagon from the top as well as from the side during tippling.

A loaded wagon is placed on the tippler platform by an in-haul beetle or other means. After correct placement is achieved, the in-haul beetle returns and clears the tippler table. The hydraulic system is now actuated, and the upper and lateral beams rest on the wagon coping and side walls. The hydraulic system incorporates a pressure relief which balances the pressure of the wagon springs, thus avoiding any damage to the wagon.

Important parts of the wagon tippler are :

(*a*) Circular Ring

This a backbone of any wagon tippler because it carries the platform which can be turned through some angle to unload the wagon. It is made up of cast iron and is of box C section. The cross-section of this arc is trapezoidal. Each wagon tippler has at least two circular rings.

(*b*) Clamping Device

Once the wagon comes on the platform it is clamped to the platform so that it does not slip from platform. For clamping space is provided on the wagon itself. The complete hydraulic equipment, including the motor and the oil tank are located on the tippler itself and rotate along with the tippler. Top clamps and side beam are hydraulically operated. Wagon is clamped before tippling operation. The top clamps and side beam are lined with rubber pads throughout the length of the tippler which reduces the pressure on wagon walls.

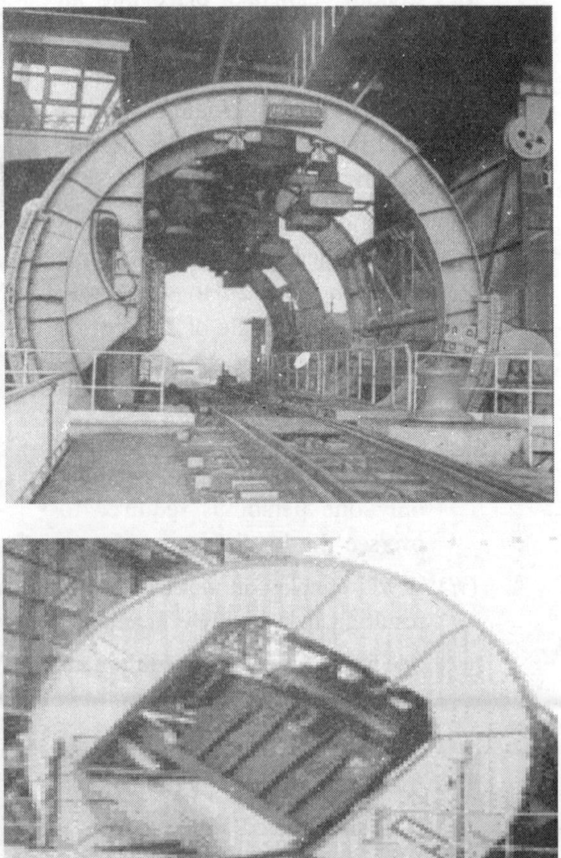

Figure 15.25

(*c*) Power Supply and Transmission

The tippler is driven by a drive unit located on one side of the tippler. The drive unit consists of motor—with flexible coupling, thruster operated brake, helical gearbox, pinion and toothed rings. The type of drive is decided on the number of tips/hour required. These are run by electrical supply. Important parts in transmission are heavy duty motor, huge segmental gear and gear box. Electric motor is run by electric supply which drives the gear box through the belt. This reduction gear box drives the segmental gears through spur and bevel gearing. The speed of the segmental gear is very slow. This gear do not rotate through 360 degrees but it oscillates from initial to final position like wiper used in the cars. Two motors and two segmental gears are used with two reduction gear box.

All these devices are mounted in the pit. When wagon is tippled it discharges into storage hopper below. Belt conveyors carry coal from hopper to the boiler. The tipping rate can be varied to suit particular plant requirements, up to a maximum of about 60 wagons per hour. To achieve a throughput of more than 12 wagons per hour, it is necessary to incorporate integrated wagon handling equipment in the form of a beetle charger.

Maximum Load capacity, Maximum Rotation angle and Time of rotation is 132 tonnes, 175° and about 52s respectively. These are used in UP POWER CORPORATION PLANT in PARICHA near JHANSI, in all NTPC PLANTS and in BOKARO STEEL PLANT for unloading of the coal.

15.9 ESCALATORS

HISTORY

Around 1900, Charles Seeberger, who designed the forerunner of the modern escalator, came up with the name that finally stuck. His term "escalator" is a combination of "elevator" and "scala", the Latin word for steps.

Escalators are one of the largest, most expensive machines people use on a regular basis, but they're also one of the simplest. At its most basic level, an escalator is just a simple variation on the conveyor belt. A pair of rotating chain loops pull a series of stairs in a constant cycle, moving a lot of people a short distance at a good speed.

15.9.1 WORKING AND CONSTRUCTION

The core of an escalator is a pair of chains, looped around two pairs of gears. An electric motor turns the drive gears at the top, which rotate the chain loops. A typical escalator uses a 100 horsepower motor to rotate the gears. The motor and chain system are housed inside the truss, a metal structure extending between two floors.

The chain loops move a series of steps. The interesting thing about an escalator is the way these steps move. As the chains move, the steps always stay level. At the top and bottom of the escalator, the steps collapse on each other, creating a flat platform. This makes it easier to get on and off the escalator.

Each step in the escalator has two sets of wheels, which roll along two separate tracks. The upper set (the wheels near the top of the step) are connected to the rotating chains, and so are pulled by the drive gear at the top of the escalator. The other set of wheels simply glides along its track, following behind the first set.

The tracks are spaced apart in such a way that each step will always remain level. At the top and bottom of the escalator, the tracks level off to a horizontal position, flattening the stairway. Each step has a series of grooves in it, so it will fit together with the steps behind it and in front of it during this flattening.

In addition to rotating the main chain loops, the electric motor in an escalator also moves the handrails. A handrail is simply a rubber conveyor belt that is looped around a series of wheels. This belt is precisely configured so that it moves at exactly the same speed as the steps, to give riders some stability.

The escalator system is not nearly as good as an elevator at lifting people dozens of storyes, but it is much better at moving people a short distance. This is because of the escalator's high loading rate. Once an elevator is filled up, you have to wait for it to reach its floor and return before anybody else can get on. On an escalator, as soon as you load one person on, there's space for another.

15.9.2 Step Speed

Escalator speeds vary from about 27 m per minute to 54 m per minute. An escalator moving 44 m per minute can carry more than 10,000 people an hour+many more people than a standard elevator.

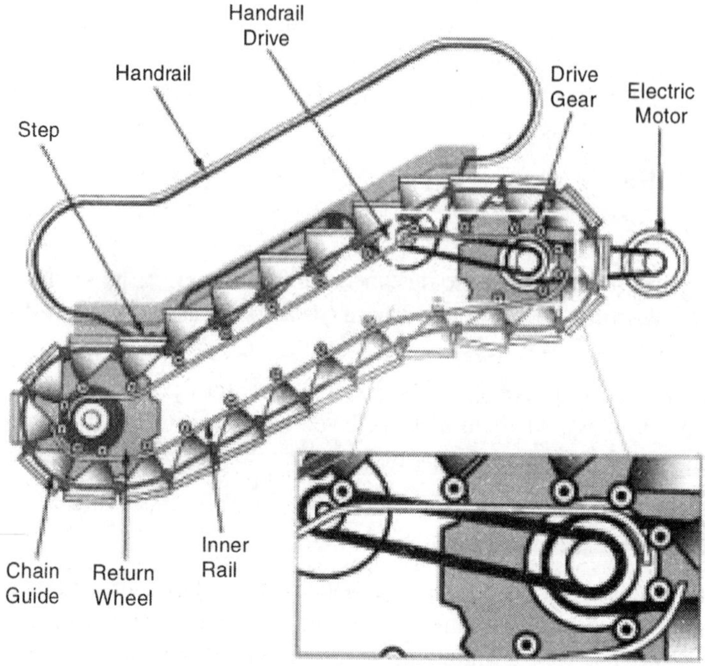

Figure 15.26 *Escalator details.*

15.9.3 Calculation of an Escalator

The throughput capacity of an escalator, men per hour, can be found by the formula :

$$ZX = 3600\ \varphi\ Av/a_s = 9000\ A\ \varphi\ v$$

Where φ = 0.6 to 0.96 is the coefficient of filling of steps (the highest value to a speed of 0.4 m/s and the lowest to 1 m/s); A = 1 to 2 is the number of passengers on a step; v = 0.4 to 1 m/s is the escalator speed; and a_s is the pitch of escalator steps, a_s = 0.4 to 0.405 m.

The calculation of the pull of an escalator is carried out in the same manner as for an apron conveyor; the coefficient of resistance to the motion of floor is taken of 0.025 to 0.028.

15.9.4 Table of Main Characteristics of Escalators

Characteristics	Floor to floor type		Tunnel type			
	A11–6k	A11–7k	AT–2	AT–3	AT–4	AT–5
Throughput capacity men/hour	9720	4860	16200	16200	16200	16030
Maximum height of lift, m	8	12	65	45	25	12
Stair step width (mm)	1000	666	1000	1000	1000	1000
Stair step speed (m/s)	0.54	0.54	0.9	0.9	0.9	0.9

Application Example

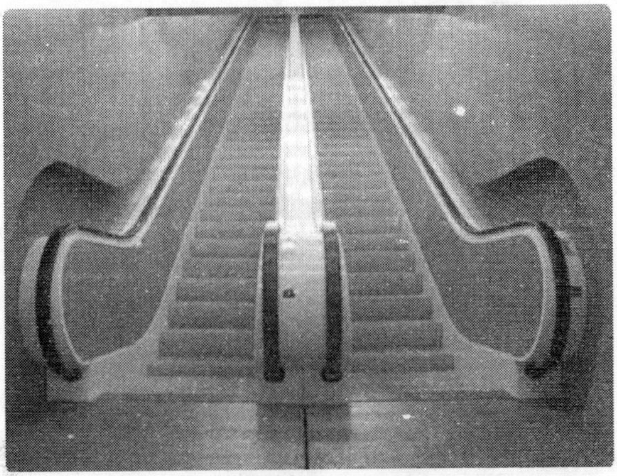

Figure 15.27 *Escalator.*

15.10 APRON AND SCRAPPER

These are used for the farm work. These are always coupled with the some kind of tuggers such as tractors or truck.

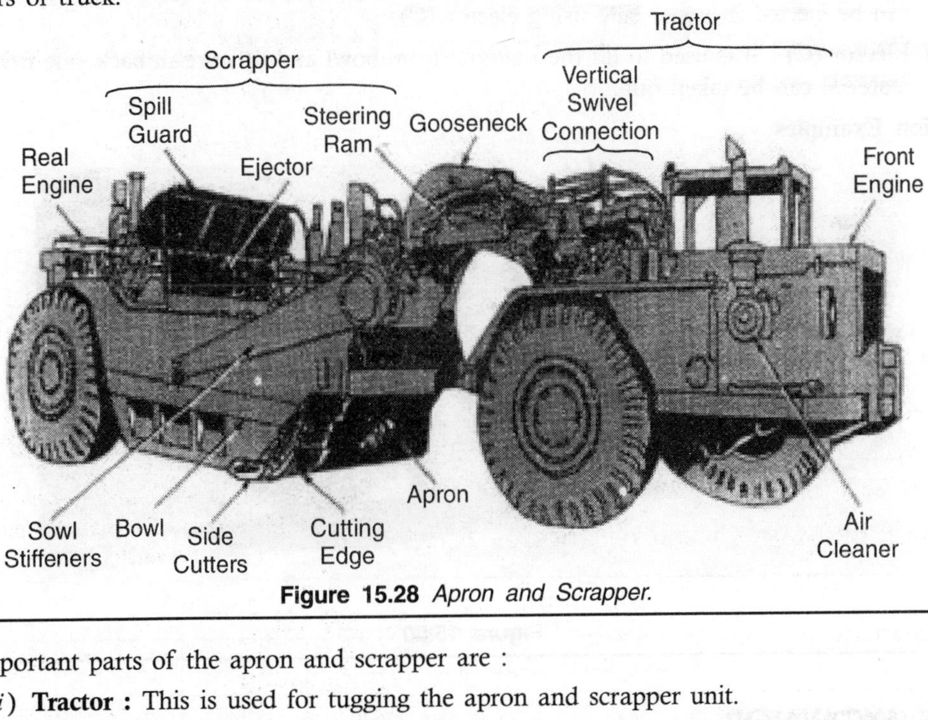

Figure 15.28 *Apron and Scrapper.*

Important parts of the apron and scrapper are :

(*i*) **Tractor :** This is used for tugging the apron and scrapper unit.

(*ii*) **Goose neck coupling :** This coupling is used to connect the tractor and apron and scrapper. This is made up of curved beam of cast iron or steel. It contains the steering arm in it which is connected to the steering wheel.

(*iii*) **Rear engine :** This is a auxiliary engine situated at the rear side of scrapper. This engine operates apron.

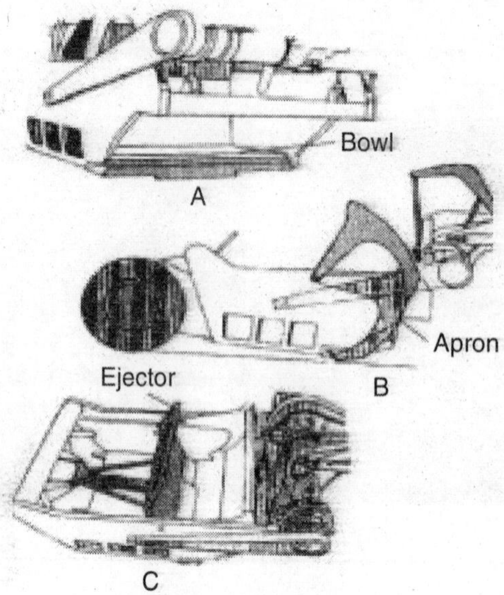

Figure 15.29 *Details.*

(*iv*) **Apron (B) :** It contains cutting edges (tool), side cutter and bowl. Cutting tool cuts the crop and puts it into the bowl (A). The bowl is used for temporary storage of cut crop. This bowl can be ejected in upper side using ejector (C).

(*v*) **Ejector (C) :** It is used to lift the material from bowl and to store in back side from where material can be taken out.

Application Examples

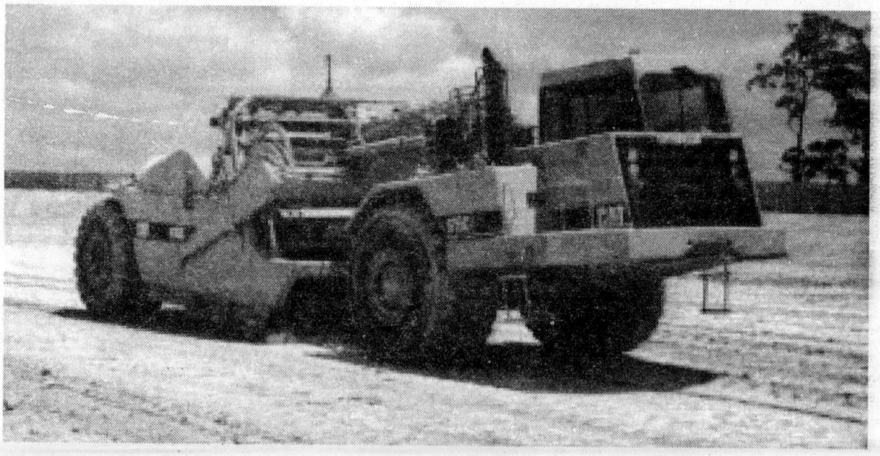

Figure 15.30

15.11 EXCAVATOR

This is a device which is used for digging and carrying purpose. This is a earth moving equipment which can be used for various purposes. Its working is same as that of hydraulic crane except that crane is not used for digging purpose. It is generally mounted on the caterpillar. It can travel in any direction. It is provided with turntable so that it can move or rotate in 360 degrees.

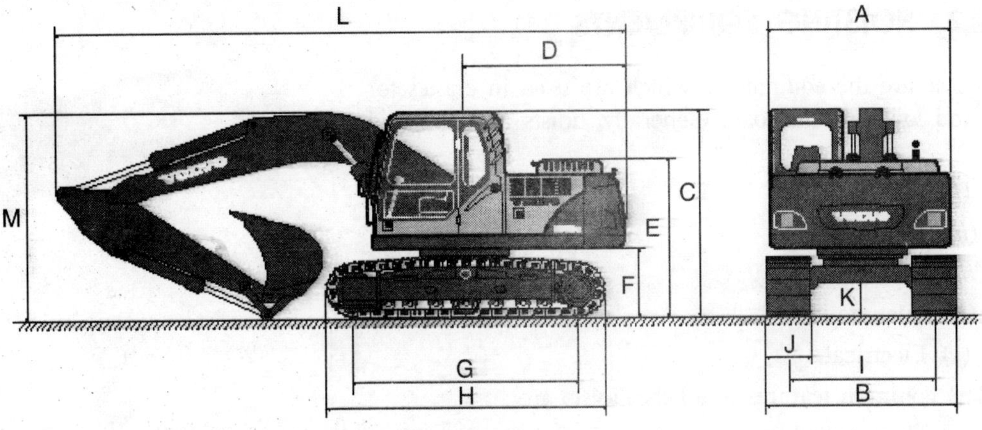

Figure 15.31

It is propelled due to caterpillar at the bottom. The bucket is actuated through the series of hydraulic ram and cylinder arrangement. This equipment can be attached with various attachments like cutter having shearer and crusher attachment, breaker and grappler. In earthquake hit areas, flood hit areas this is very useful equipment. Road construction, construction projects, tank desiliting, general irrigation, canal works public works, etc.

This figure shows the various movements and path of the bucket, ram and boom.

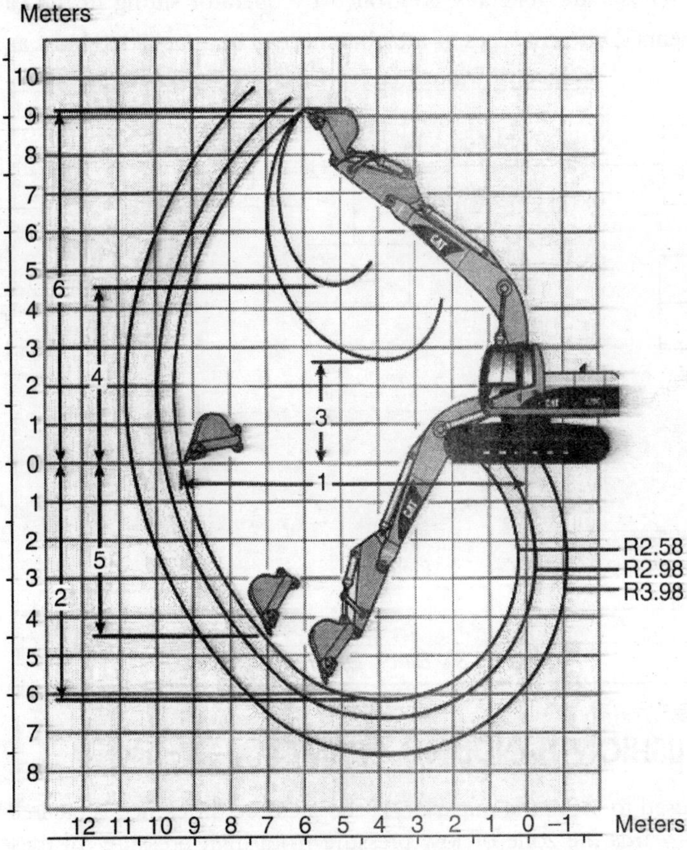

Figure 15.32

15.12 HOISTING EQUIPMENTS

These are the equipments which are used in cranes for lifting and lowering the load. Generally, hoists are classified as :

(*i*) Manual Hoist

(*ii*) Hydraulic

(*iii*) Mechanical

(*iv*) Pneumatic

(*v*) Electrical.

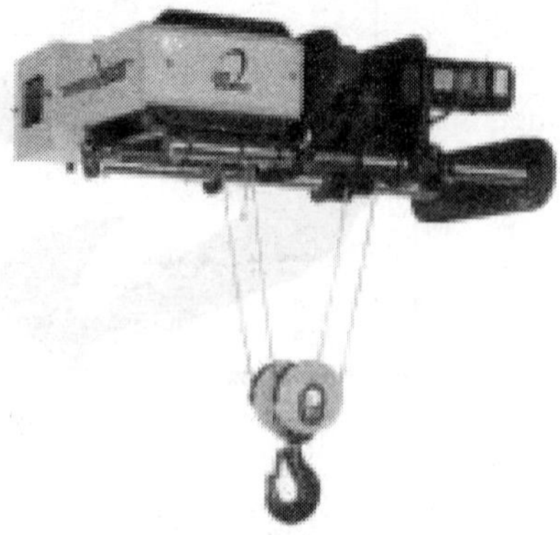

Figure 15.33 *Electric Hoist.*

The common features in all the hoists are :

(*i*) Drum which carries wire rope.

(*ii*) Means of moving this drum generally, hand wheel or motor is used.

(*iii*) Wheels to move on rails/ground as applicable.

(*iv*) Crane hook for lifting or attaching load.

(*v*) Drive to transmit motion. Gear drive with different speed reduction is suitably designed to suit the application.

(*vi*) **Control.** Hoists are generally controlled by operator sitting in the cabin.

(*vii*) **Attachments.** Different types of attachments can be attached to hoist as per the requirement.

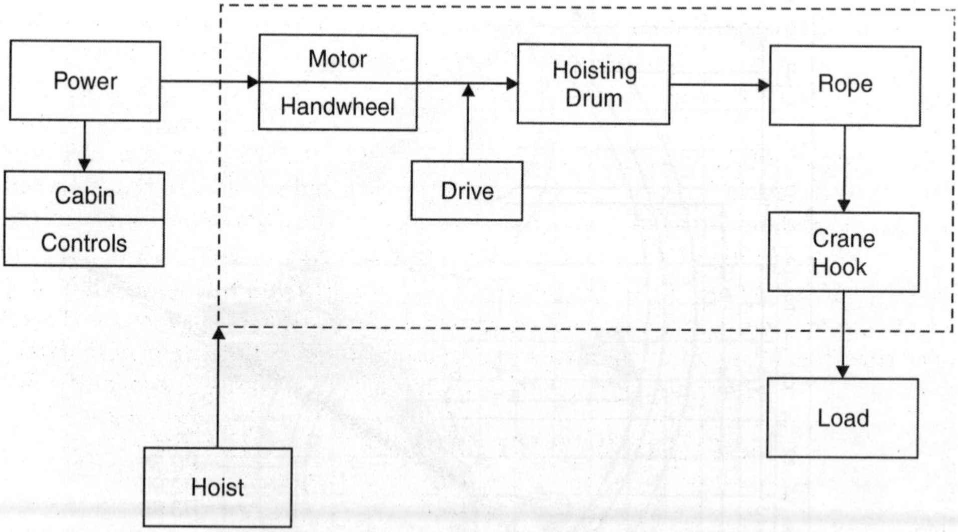

Figure 15.34 *Block Diagram of Hoist Parts.*

15.13 AIR CUSHION HANDLING FRAMES

In these air is used to move the material. If the pressure difference is created between two points the load is carried towards the zone of low pressure from high pressure. In these load is put on the simple frames.

It consists of rubber diaphragm and the structure. The diaphragm is bolted or fixed on the structure or load carrying surface. The structure has a specified geometry to allow the air to pass through the holes made in it. The air may be supplied from top or from side of the structure. The air passing through the structure passes into the diaphragm. This diaphragm has tiny holes at the bottom. When air with pressure fills diaphragm it comes out through these holes with pressure. The air coming out of tiny holes exerts tremendous pressure on the ground. Due to this action whole frame gets lifted and separates from ground. The frame is propelled due to pressure difference between front and end points.

The lift is not much bigger but it is sufficient to reduce the friction between ground and frame to considerable extent.

Advantages

(*a*) This can be moved in any direction.

(*b*) Low is fractive force is required.

(*c*) Very low operating and maintenance cost.

Disadvantages

(*a*) Continuous air supply is needed.

(*b*) Problem of air leakage affects performance.

(*c*) Good quality ground surface is required.

(*d*) Speed of transportation is very slow like snail.

(*e*) No vertical movement of the load is possible only horizontal movement is possible.

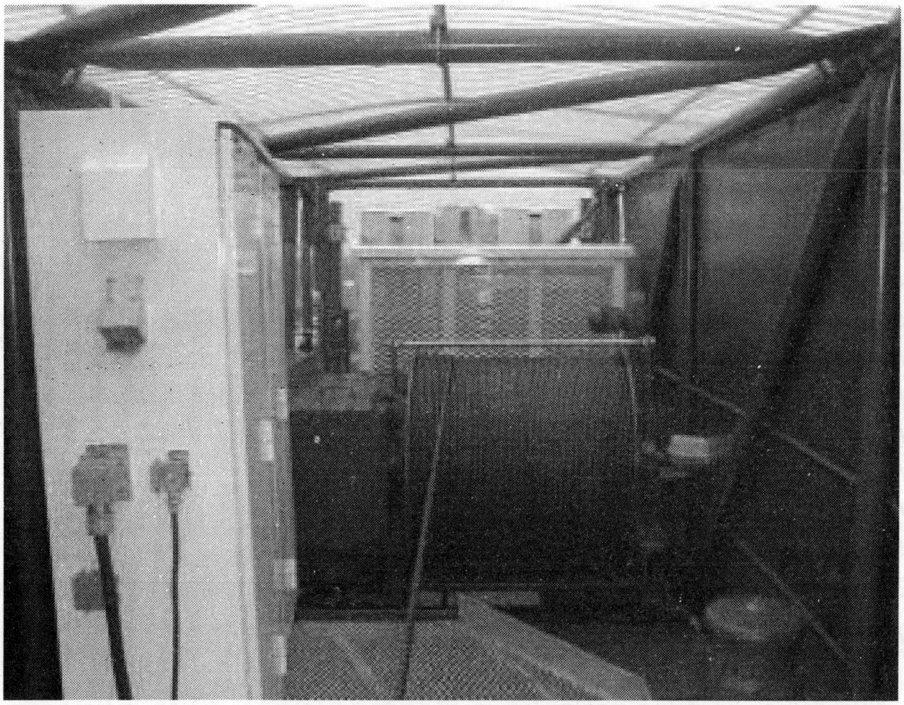

Figure 15.35 *Hoist Drum and Wire Rope.*

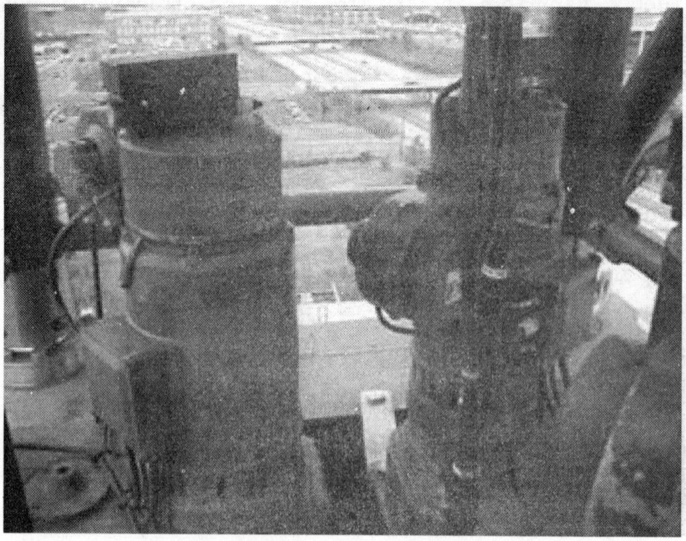

Figure 15.36 *Hoisting Motor.*

UNSOLVED QUESTIONS

1. Write about cupola furnace charger.
2. Write about mobile lifting frames.
3. Write about air cushion handling frames.
4. Write about hoppers.
5. Write about crushers.
6. Write about feeders.
7. Write about flow bins.
8. Write about hopper gats.
9. Write about apron feeder, screw feeder, oscillating feeder and vibrating feeder.
10. Write about dumpers.
11. Write about ship loader.
12. Write about wagon tripplers.
13. Write about escalators.
14. Write about apron and scraper.
15. Write about excavators.
16. Explain meaning of hoist.

❑❑❑❑

16 Maintenance of Material Handling Equipments

INTRODUCTION

Every living and non-living has a definite period of life. Point of stress is that no one can remain as it is for the longer duration because change is the rule of nature. Considering the machine or human, the status or condition is bound to change. It is now essential to become to restore the condition and is done by maintenance. As human being needs doctor machines also needs doctors, commonly known as maintenance engineer.

16.1 DEFINITION

Maintenance can be defined as the art of restoring the condition of the machine or equipment by applying standard procedures done in stated conditions.

16.2 OBJECT OF MAINTENANCE

Object of maintenance is not only to do the maintenance but also to remove the causes which affect the condition of the machine. Although all the causes cannot be eliminated but efforts can be made to reduce to the considerable extent. If efforts are made to identify the cause, maintenance tasks will generally consist of a large number of minor jobs and a relatively small number of major jobs. Generally, it concentrates on the cause finding and collection of large amount of data.

16.3 TYPES OF MAINTENANCE

Maintenance can be broadly classified in two categories :

Planned maintenance

Unplanned maintenance.

16.4 PLANNED AND UNPLANNED MAINTENANCE

Consider an example of a patient of low blood pressure and a normal person. A patient regularly or periodically visits the doctor that is his visits to the doctors were already planned irrespective of the condition. This is a planned maintenance. Now consider example of normal man, he doesn't visit the doctor unless he sees the symptoms of some kind of disease, this is known as unplanned maintenance. Planned maintenance costs more but keeps the machine fit and reduces breakdown. Unplanned maintenance costs less but breakdown costs are high.

16.5 PLANNED MAINTENANCE

This includes :

(*a*) Planned preventive maintenance.

(*b*) Planned breakdown maintenance.

(*a*) Planned Preventive Maintenance

Preventive maintenance means acts of maintenance used to prevent the malfunctions or breakdown of machines. To prevent means to protect. We protect ourselves from disease by cleaning, washing, or by some form of exercise. Similarly machine needs to be protected from failures or malfunctions. The aim of this is to prevent the machine from damage. It is done by preventing the causes of failure by applying different policies.

While this seems a sound idea, it may be found that when such planned preventive—maintenance programs are developed and followed rigiously, the frequency and extent of overhauling or re-conditioning equipment may not be justified particularly if the downtime required to perform the work, results in loss of production. These include :

- ➢ Regular inspection or checkups.
- ➢ Proper lubrication of parts.
- ➢ Minor adjustments.

There are broad two categories of the planned preventive maintenance :

- **Corrective maintenance**

  Corrective maintenance means is to correct or adjust. Consider a worker does regular inspection and check up of machine and finds something which is not in expected state he does the corrective works. It generally includes minor adjustments. It just like servicing of the vehicle, *e.g.*, tightening of the nuts and bolts, brakes, adjustment of pressure, adjustments of levers and fits, adjusting level of lubricant etc.

- **Predictive maintenance**

  Predict means reading the future or knowing something before it happens. It is a type in which possible failures or malfunctions were predicted and preventive actions were taken to avoid breakdown. This is done either by past experience or continuous monitoring. In big industries or mass production industries more stress is given on continuous monitoring. In industrial world it is known as condition monitoring.

  Condition Based Monitoring (CBM) recognize a change in physical conditions of the machine by the use of sensing, measuring or monitoring devices to determine significant changes. This technique can minimize the need for disassembly and inspection of internal parts. Just as important, through periodic measurement or monitoring it is possible to identify to conditions that require correction before a major problem develops.

Condition Based Monitoring or Signature Analysis

Condition based monitoring is concerned with extracting information from the machines to indicate their condition and to enable them to be operated and maintained with safety and economy. The main function of condition monitoring is to provide the knowledge of machine condition and of its rate of change. This helps in carrying out preventive maintenance wastes, the productive time of the machine and increases the risk of troubles arising from human errors is reassembly. Too long intervals in maintenance result in an unacceptable number of machine failures during operation. So a correct time for the condition of the machine. This type of maintenance is called preventive-predictive maintenance or condition based monitoring.

Every machine gives some form of signals if it doesn't function properly. Condition monitoring technique is used to detect these signals using some sensing and portable device. After analysis of these signals condition of machine can be predicted, hence sometimes it is known as signature analysis.

Generally machine gives following signatures and analysis of these gives the condition of machines. It is not necessary to measure every signature because generally this requires costly equipments.

A. Vibration Monitoring

Large number of machine failures take place because of slow deterioration during which the machine gets progressively worse over a period of time. It has been observed that about 70–80% of machine failures in the case of rotating machinery lead to a significant change in vibrations.

Vibration monitoring is based on the concept that provided the operating conditions not been changed, an increase in vibration is an indication of an impending failure. The greater the increase in vibration level greater is deterioration.

Knowing the machine details (such as, shaft speed, number of gear teeth, number of impeller blades, etc.) it is possible to calculate the frequency at which vibration will be produced by a particular component.

Thus, a vibration monitoring component.

(*a*) A measure of the increase in vibration level to indicate the urgency of the need of attention.

(*b*) A measure of the frequency at which the increase occurs to permit diagnosis of the problem.

(*c*) Thus, monitoring and analysis of the following vibration parameters can be used as an effective indicator of the machine health and the need for maintenance :

(*d*) Overall vibration levels

(*e*) Critical frequencies.

(*f*) Vibration amplitude at the critical frequencies.

Failure of material handling equipments bearing may be detected well in advance by suitably installing vibration transducers on the bearings, studies need to be done before hand for compiling data about the normal vibration levels and the change in the same as fault proceeds. So that recommendations can be made to the maintenance person.

B. Temperature Monitoring

Temperature changes are usually the result of some changes in the condition of component. In material handling equipments bearing, wear or incorrect lubrication always results in a change in temperature. Changes in surface temperature can also be the result of damage to insulation or removal of material as a result of wear. Rise in temperature beyond a specific value would mean break-up of oil film in the bearing and may lead to seizure.

Different types of contact thermometers and thermocouples or resistance transducers available and most of them are simple to use and give quite reliable results.

Non-contact infrared detectors and the thermometers work by sensing for the infrared radiation emitted by different materials at different temperatures.

For monitoring of exhaust temperature a proximity type I.R. temperature measuring instrument is recommended. This is very convenient to handle and is accurate. Monitoring of exhaust temperature can indicate inadequate combustion or, improper valve seating etc.

Thermopoint – 80 is a precision manual scanning radiometer manufactured by M/S AGEMA infrared systems A.B. Sweden is used. It is based on microprocessor technique combined with a refined optical range of – 30°C to above 38°C is ± 1°C and for reading 38°C ± 1.4°C. It has digital display of temperature. The wide range of temperature is continuously and conveniently available without changing filters, apparatuses or electronics. It can simply be pointed towards the object and temperature can be measured. It has programmed memory and the software permits five important temperature values simultaneously *i.e.,* point temperature, average temperature, maximum temperature, minimum temperature and differences of highest and lowest temperature. It has facility to hold about 10,000 in the memory.

C. Oil Analysis

Wear debris in lubricating oil is being analyzed using spectrometer on a periodic basis. This has helped as some depots in diagnosing impending failures of the material handling equipments bearings, main engine bearings, connection rod bearing. In addition to monitoring of piston ring wear. There appears to be same what reduced faith in the techniques at a few places. It is suggested that a detailed study may be carried-out the technique used for collection of oil samples and interpretation of results and whether adequate training has been imparted to the concerned personnel.

In order to identify the source of wear debris adequate tracer elements will have to be used on the bearing elements. The use of tracer elements is critical engine bearing may also be tried. Oil viscosity tests are being done in addition to the wear debris test.

Oil Conditioning—Degradation

In addition to the presence of foreign particles in fluids, the lubrication oil may loose it's original properties through chemical changes. The chemical stability of a fluid is measured mainly in terms of its resistance to oxidation. When oxidation occurs oxygen combines with oil molecules to set off a chain of chemical reactions, that creates soluble and insoluble products of degradation. As a result the colour of fluid becomes dark, the viscosity and acidity of the oil increase and gums and sludges are deposited in the system. The tendency for the oil to oxidize is greatly increased by the presence of air, wear and certain fine metal particles produced by wear also when oil is subjected to a high temperature.

The characteristics most likely to indicate the condition of the oil are :

- Acidity
- Appearance
- Odour
- Viscosity
- Water content

Acidity is usually measured in mgh.—KOH/gm., that is called the Neutralization number and Total Acid Number. Viscosity may be measured by means of a number of proprietary viscometer.

D. Crack Monitoring

Detection Techniques

Techniques like Eddy current, Ultrasonic testing and Dye penetration test is used for crack detection in material handling equipment's parts.

Ultrasonic Measurement

By this method, it is possible to measure material thickness from one side only, and can be used for measurement of material thickness and investigation of crack formation. It is suitable for use on metals, plastics, concrete and other materials.

Measurement is effected by a high frequency sound signal, applied to the wall of the pipe or vessel to be measured, and reflected from the far wall when it has passed through the material and be reflected can be read off directly on the instrument.

Ultrasonic measurements can be used for investigation of erosion, pitting, corrosion and cracks in pipes, parts of machinery, pressure vessels and other equipments.

Dye Penetrant Test

Penetrating liquid is often used to indicate the presence of cracks. Parts are first carefully cleaned before being sprayed by the penetrating liquid over the area to be investigated. The penetrating liquid, which is coloured, has a very high capillary attraction and penetrate all cracks, pores and similar defects in the material, is then sprayed on the part. A cleaning liquid is sprayed over the surface once the surface is dry. It mixes with the penetrate dye, enabling it to wriped-off. The cleaning liquid cannot penetrate into the cracks and pores, and so the penetrate remains intact in them. Finally, a white developing liquid in the cracks and pores, the presence of which is revealed by the dye showing up clearly against the white developer.

Florescent penetrate is also available. It can be seen under ultraviolet light. When it appears as bright yellow-green lines or spots where fine cracks, crevices, process, etc. are present in non-porous materials.

By this method, cracks down to 0.000025 mm can be detected.

E. Sound Monitoring

Sound is measured by using microphones, signal amplifier and analyzer. The sound is characterized according to the cause such as loosening, sliding, rolling etc. condition is assessed by knowing increase in frequency and amplitude.

F. Smell and Leakage Monitoring

Smells and leakage monitoring equipments are used for monitoring of pneumatic and hydraulic equipments. Also they are used in boilers, refrigeration of plants.

(*b*) Planned Breakdown Maintenance

It is generally done in big industries such as power plants (NTPC, NHPC, NPCIL), heavy electrical (BHEL) etc. In this type all the machines of a unit were shutdown on a planned date and their maintenance is done. The period of two successive breakdown maintenance depends on the equipment and its history. Generally the time period is 3, 6, 12 months or may be 24 in some cases.

16.6 UNPLANNED MAINTENANCE

It consists of :

Unplanned Corrective Maintenance

Includes repair work which is simply doing maintenance work as the need develops. This elementary approach has sometimes been replaced by periodic overhauls and other preventive maintenance practices. But, however elementary, it can be the most logical approach to maintaining non-critical equipment-individual units or parts of a production system.

Breakdown Maintenance

Breakdown means complete stoppage. Machine completely stops working. Doctors can't do anything if human stops working (dies) but maintenance engineer can restore the whole machine to its operating condition after breakdown. This type of maintenance may be planned and unplanned.

Unplanned Breakdown Maintenance

It is not planned but is to be done to make the machine to operate. This type takes more time than planned one. It may cause damage to the life and property so it is not desirable and should be avoided.

Breakdown maintenance in general involves repairs and replacement. Parts, which are in good condition, are repaired and more damaged parts are replaced. Bearing, gears are replaced.

16.7 MAINTENANCE OF MATERIAL HANDLING EQUIPMENTS

Due to proper maintenance following can be achieved :

1. Ensures a longer life of equipment.
2. Prevent downtime due to equipment failure.
3. Make operations more reliable.
4. Get better information on the status of equipment.
5. Utilize people on other tasks while getting more specialized, upgraded maintenance on material handling equipment.
6. Prevent failures by heading off problems before they can occur.
7. Improvements in your operations.

Equipments

The proper maintenance of material handling equipments is extremely essential for preventing the occurrence of bottlenecks or points of congestion. Production line flow can be maintained only if the material handling equipment is in the proper working order. Preventive maintenance is by far ~e of the best maintenance techniques suggested for material handling equipments.

interruptio. preventive maintenance, the equipments can be kept running there by minimizing costly enough to preve in the production schedule. A little periodic inspection and minor adjustments may be and examination of equipment breakdown. Preventive maintenance consists of frequent inspections requiring it. The aim is to un. material handling equipments, with special attention to the components maintenance also includes fabric ver conditions leading to breakdown or harmful depreciation. Preventive stage of defect. on, adjustment, or repair while the equipment is still in a minor

Three stages of preventive maintenance are :

(*a*) Inspection

(*b*) Repair, and

(*c*) Overhaul

16.7.1 Maintenance of Hoists and Cranes

Maintenance on cranes should be performed only by formally qualified personnel, unless authorized by an authorizing officer on a single case basis for unusual repairs. You must do all the maintenance requirements according to the instructions provided on the applicable MRCs. If the MRCs do not exist for a particular piece of equipment, your supervisor or you should institute interim maintenance according to the manufacturer's recommendations. The following maintenance items should be included in the crane.

Preventive maintenance package : Lubrication, Safety inspection, Lube oil maintenance, Wire rope, Brakes, Instrumentation, Electrical.

(*a*) **Inspection :** All parts, open or covered are inspected for wear and tear. Worn out or unworkable components like wire ropes, wheels, bearings, bolts, etc. are removed. Brakes are adjusted and lubrication is provided wherever necessary.

(*b*) **Repair :** The repairable parts of the system, after inspection is corrected for small repairs and minor defects are rectified. Systems like open gear transmission, small repairs and minor defects are rectified. Systems like open gear transmission, couplings, riveted and bolted joints, trolley, brakes, guards, etc. may be repaired.

(*c*) **Overhaul :** Overhauling involves dismantling the complete mechanism and replacing all damaged components. Crane structure buffers, rails, open gear transmission, pulley blocks, etc. may be replaced and various sub-mechanisms may be aligned and adjusted.

16.7.2 Maintenance of Conveyors

(*a*) **Inspection : Belts** or **rollers** are inspected for tensions and wear and tear. Gear box is properly lubricated, various fasteners are tightened and safety guards are checked.

(*b*) **Repair : Rollers** and **belts** are checked, adjusted or repaired. Couplings, packing, safety guards are checked.

(*c*) **Overhaul :** The conveyor system is completely dismantled. Components, worn out and beyond repair like belts, bearings, packings, oil seals, rollers, drums, fasteners and couplings are replaced. Structures, safety guards, etc. may be repaired as per their conditions.

Typical scheduled conveyor maintenance plan

1. Check/lubricate all bearings, universal joints, and pulleys.
2. Check chain tension, wear and lubricate.
3. Check sprocket alignment, wear, and screw set.
4. Check flat belt tension, wear and lacing.
5. Check V-belt tension, wear and sheave alignment.
6. Check electrical connections at conveyor.
7. Check gearbox and fill oil to proper level.
8. Check general condition of system.
9. Operate entire system after service.
10. List any items requiring replacement or repair.

The time interval depends on the management policy.

16.7.3 Maintenance of Trucks

System of Checks

1. Safety equipment check

   (Guards, backrest extension, horn, warning beacon, fire extinguisher, rating plate).

2. Pre-operation check
   (Fluid levels; condition of horn, lights, tyres; gauges; brakes check; hydraulic controls)
3. Periodic inspection
   (Tension/condition of fan belt; level of forks; tightness of components; lift chains; hydraulic system under load; refueling equipment).

16.7.4 Maintenance of Lift

1. Check all safety devices on unit, such as up/down limits, door switches, pressure relief valves, etc.
2. Clean all debris from the pit or from the vicinity of floor mounted units in order to avoid interference with the lift mechanism or rollers.
3. Check for presence and proper setting of all snap rings and clips on axles, cylinders and rollers.
4. Check rollers, pins and bushings for any signs of wear such as flat spots, missing fasteners, or dislodged bearing material.
5. Inspect all welds under and around the lift for fatigue or failure. Particularly, inspect the structural welds.
6. Check the hydraulic fittings for cracks or leaks and clean up any seepage on or beneath the cylinders.
7. Check hoses and electrical lines for abrasions or other abuse and check for snug connectors.
8. Semi-annually, change hydraulic fluid in cylinders.
9. Check general condition of the lift. Operate the lift after service. Check for any abnormal noise or vibrations.
10. List any items requiring replacement or repair.

16.8 PREVENTIVE MAINTENANCE OF HIGH-SPEED BELT CONVEYORS

16.8.1 A Case Study

Introduction

'Prevention is better than cure', is a universal traditional maxim, both in the medical and the engineering fields. In the engineering field experts in management science like Fayol speak of 'prevoyance' as the essence of management. This prevoyance, looking ahead, includes both assessing the future and making provision for it. This is the basic principle underlying the concept of maintenance which has found widespread application on belt conveyor installations.

Belt Conveyor—Its Importance in the Mining Industry

In the highly advanced open cast mining industry, belt conveyors form the main artery of the transportation system mainly due to its reliability, simplicity, safety and economy besides its low maintenance cost and low power consumption.

The Neyveli Scene

The Neyveli complex comprising thermal power station, fertilizer briquetting and carbonizing plants is pivoted on the continuous supply of lignite—the basic raw material. The removal of overburden at a collosal rate of 18 million tonnes and lignite at 4 million tonnes annually is by means of conveyor transport system. The belt conveyor is therefore an indispensable part of this integrated project.

For the transportation of overburden and lignite, 20 km of belt conveyors costing nearly Rs. 7.5 crores, have been installed. Fig. 16.1 shows the high percentage of total cost of belts on conveyor installations. Considering that a sizable investment has been made on these conveyor systems, it is of paramount importance that these are maintained at a high degree of operating efficiency by adopting a sound preventive maintenance programme.

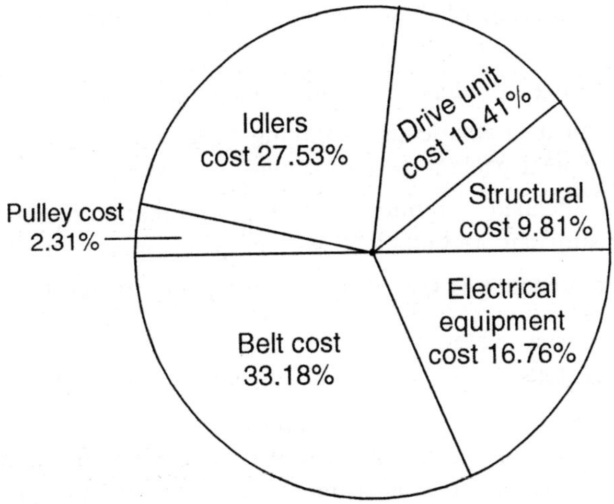

150 cm. conveyor 500 m. long
Speed 4 m. per sec. capacity 4,800 time per hr.
5 motors drive

Sl.No.		Unit cost, lakhs	%, total cost
	Overall/total cost of the conveyor system including electrical and mechanical equipment	19.87	100%
1.	Structural cost	1.95	9.81%
2.	Drive unit cost	2.07	10.41%
3.	Idlers cost	5.47	27.53%
4.	Pulley cost	0.45	2.31%
5.	Belt cost	6.60	33.18%
6.	Electrical equipment cost	3.33	16.76%

Figure 16.1 *Break up particulars of capital cost for a typical conveyor installation.*

Characteristic Problems Encountered on Belt Conveyors at Neyveli

(*a*) The conveyors are exposed to the vagaries of nature which adversely affect the operation and maintenance activities.

(*b*) The widespread nature of belt conveyors leads to the problem of transportation of men, materials and equipment.

(*c*) The belt conveyors run in sequence and hence any weak link in the chain can bring the entire operation to a grinding halt, causing appreciable production loss. For example, a stoppage of one hour on the lignite conveyor system costs Rs. 24,000 besides leading to other consequential ill effects.

(*d*) The materials handled being heavy, abrasive and sticky, difficulties are experienced in operation and maintenance due to frequent choking, cleaning, over loading, belt wandering, etc.

(*e*) Since thousands of rotating parts like rollers, drums, gear boxes, fluid couplings, are in operation over a wide area regular inspection and patrolling of conveyors is a necessity.

(*f*) Frequent track shifting of these conveyors on unconsolidated ground leads to misalignment of conveyors and heavy spillage of materials.

(*g*) Handling of huge boulders results in damage to rollers, conveyor belts and structures.

(*h*) Working at great depths leads to sinking of ground due to artisan and rain water. This necessitates equent packing and alignment of conveyors.

(*i*) Provision of adequate communication and illumination facilities throughout the conveyor length to ensure safe and efficient working poses special problems.

Hence, unless maintenance is geared up to the highest standard, it is not possible to meet the targets of production.

Organizational Pattern

Taking into consideration the peculiar, technical geographical and personnel situations and the problems enumerated above, a suitable organizational pattern embracing the Ten Principles, has been evolved for the conveyor maintenance department. An adequate number of well trained supervisory and labour personnel have been employed to ensure efficient and effective preventive maintenance. A well organized and fully equipped repair shop for reconditioning of conveyor components and belt vulcanizing, has contributed much towards reduction in the cost of belt maintenance.

Maintenance Practices

The developmental evaluation of maintenance practices can be classified under the following categories from which choice of suitable maintenance practice can be made.

(*a*) Capital replacement

(*b*) Provision of stand-by equipment

(*c*) Break-down maintenance

(*d*) Scheduled maintenance

(*e*) Planned maintenance

(*f*) Preventive/corrective maintenance

(*g*) Productive/selective maintenance

(*h*) Maintenance prevention

Items 4 to 8 constitute 'systematic maintenance'.

The investment involved in capital replacement and stand-by provision being very high, these practices are not economical for conveyors. However, it is necessary to keep certain critical components such as fluid couplings, gear reducers, pulleys, motors as stand-by for purposes of unit replacement. This is one of the maintenance techniques adopted on conveyors to minimise downtime.

Though scheduled and planned maintenance are almost alike, the degree of planning is of a higher order in the latter. To achieve this and to ensure trouble free service, the maintenance and production staff hold joint discussions twice a week, so that the maintenance needs could be dovetailed to production convenience. This also helps in planning the proper utilization of men, materials, machinery, money and time.

While evolving a sound preventive maintenance programme, the following elements of good maintenance have been kept in mind :

(*i*) Inspection

Inspection being an essential function of the preventive maintenance programme, there is a well trained crew for this purpose. This crew is responsible for both external and internal inspection of conveyors to watch for the detect conditions like abnormal sound, vibration, looseness, heat, smoke, etc., while the conveyors are in operation. Internal inspection of gears, bearing tolerances, etc., are carried out during pre-planned shut downs. Check lists are effectively used so that not only the causes of failure could be analyzed, but also the responsibility could be pin-pointed. Such meticulous care is essential so that no critical point is missed leading to failure of costly parts or equipment without warning. A typical inspection card is given in Appendix 1.

(*ii*) Lubrication

Lubrication in its broadest and simplest sense, is the application of the right type of lubricant at the right place, at the right time, and in the right quantity. A suitable lubrication schedule has been formulated to suit the conveyors working at Neyveli. Taking into consideration the arduous and dusty conditions of working, pneumatic devices are used for oil changing, in gear reducers and fluid couplings, thereby avoiding contamination by dust and moisture.

The inspection and lubrication check lists are combined for convenience and economy. A specimen form is shown in Appendix 2.

(*iii*) Planning and Scheduling

Since planning is the keynote of any preventive maintenance scheme every work to be carried out on conveyors is analyzed and pre-planned in detail so that there may not be any delay in commissioning the conveyors. For this, a pending works register or a back-log register is maintained up-to-date and a detailed job instruction sheet (Appendix 2) is prepared, clearly indicating the step-by-step procedure to be followed.

(*iv*) Engineering Records

A good preventive maintenance programme should have with it a very sound record keeping. Basic engineering records are essential for forecasting maintenance and allocation of man-power. A few essential records maintained for this purpose are (*i*) Instruction Manuals (Operational and Maintenance Manuals); (*ii*) Equipment Records/Equipment Data Cards; (*iii*) Equipment History Comprising History Cards History Registers, Delay Analysis Charts and Spares Procurement Register; (*iv*) Drawings; (*v*) Inspection and Lubrication Check-lists and Inspection Register; (*vi*) Log Books and Defects Register; (*vii*) Wire Rope and Oil Changing Registers; (*viii*) Job Instruction and Inspection Cards.

(*v*) Engineering Analysis

Inspite of the best inspection and attention it has not been possible to completely eliminate failures. A careful study and analysis of plant failures reveals many interesting and valuable facts and aid for better maintenance programme. With this in view all stoppages and failures on conveyors are recorded and reviewed through a 'Delay Analysis Chart' (Appendix 3).

Possible causes for major repetitive failures are examined critically and without prejudice, during joint discussions between the maintenance and the production staff and suitable corrective measures taken to improve the performance.

Case Study

An interesting case study is presented below :

Problem

Frequent failure of a particular fluid coupling duc to cxcessive oil leakage on lignite conveyor (13 times in 5 years) was quite alarming.

Analysis

Of the 20 failures during the past 5 years, the western side coupling alone had failed 13 times, eastern side 4 times and tail end side 3 times. At first the reasons for the failure were generally attributed to: (*i*) bad workmanship, (*ii*) defective seal, (*iii*) misalignment, (*iv*) improper overhaul, etc.

Remedy

Finally a Vibrograph reading indicated excessive vibration to the tune of 250 microns. Subsequently by suitable design changes, the cantilever portion of the structure was supported on four additional columns and vibrations reduced to about 50 microns. This eliminated the frequent failures of the fluid coupling and saved appreciable down-time on conveyors.

(*vi*) Training and Maintenance Personnel

Sound training is an essential pre-requisite for the success of preventive maintenance. Hence, the technicians and supervisors, who were associated since the erection and commissioning of conveyors, have been selected and indoctrained with the philosophy and importance of preventive maintenance and are trained to carry out maintenance, inspection and repair works in an orderly manner.

(*vii*) Application of Work Study Techniques on Belt Conveyors

However, carefully a conveyor installation may have been designed, it has to be modified to suit site conditions for maximization of efficiency and reliability. By the application of method study and work measurement techniques, a number of modifications and improvements have been carried out on belt conveyors not only to reduce downtime, but also to increase the percentage availability of conveyors. A few of the modifications carried out are detailed below.

Methods Improvements

(*a*) Effective and improved belt centralising devices such as (*i*) fine adjustment brackets for return rollers; (*ii*) improved barrel type return rollers; (*iii*) widening of take-up carriage wheel base, centralised pulling arrangement and use of rails in place of guide channels, etc., to prevent tilting of carriage; (*iv*) vertical guide rollers, 15° inverted troughing idlers and self-aligning roller frames on top and return strands of belts; have appreciably reduced belt-wandering spillage, and edge-damages, to conveyor belts, resulting in higher operating efficiency.

(*b*) Use of improved lubrication arrangement on conveyor idler shafts and transfer-feeder wheel bushes has considerably reduced the downtime on conveyors.

(*c*) Effective and modified belt cleaning devices such as (*i*) installation of modified, spring loaded, heavy duty wipers on all head-pulleys of conveyors, and use of multi-wipers at drive heads; (*ii*) special type of finger-scrapers to ensure efficient and quick cleaning of pulleys; (*iii*) use of bare idle/snub pulleys, in place of lagged pulleys; and (*iv*) use of improved wiper rubbers, with optimum thickness and shore hardness and use of Linatex have minimized the frequent choking and cleaning problems, thereby increasing the output and working hours on conveyors.

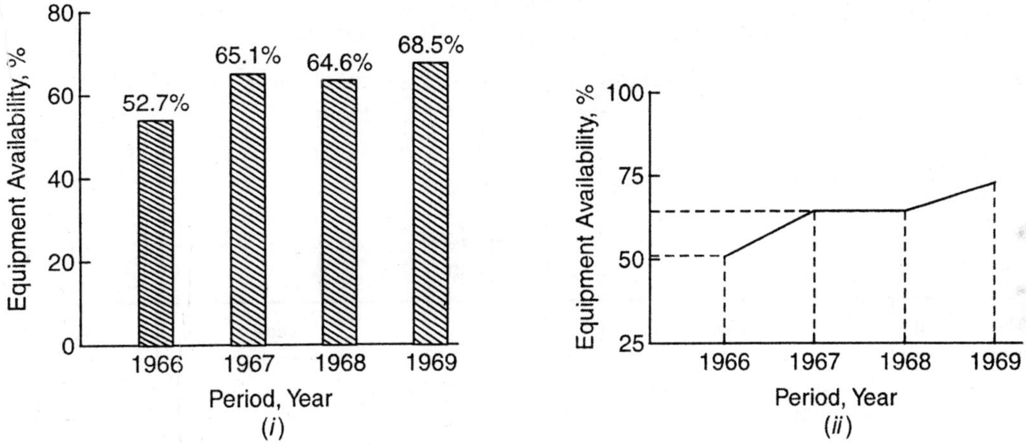

Figure 16.2 *Equipment availability of conveyors.*

(*viii*) Inventory Control of Maintenance Stores and Spares.

(*ix*) Standardization and Variety Reduction.

(*x*) Maintenance Standards.

(*xi*) Machine Replacement Policy.

(*xii*) Budgeting and Cost Control.

(*xiii*) Shops, Tools and Maintenance Facilities.

Time Saving Devices

Time saving devices such as (*i*) pre-vulcanizing method of joining long conveyor belts; (*ii*) use of 'clip-on' type of wiper rubbers and 'Linatex' instead of the conventional type; (*iii*) installation of finger type scrapers in place of metallic wipers; (*iv*) use of mobile transfer conveyor to avoid double handling; (*v*) installation of reversible conveyor on lignite conveyors instead of the uni-directional feeder, etc. have appreciably reduced the downtime on conveyors.

To take care of the emergency breakdown work on conveyors a "Breakdown crew" equipped with modern facilities has been established to reduce the downtime to the minimum.

The use of bearings 'sealed for life' on some conveyor idlers is a good example of the latest technique of maintenance prevention.

Design Changes

Improvements on conveyors by suitable design changes like (*i*) use of worm drive winches in place of the spur gear winches on counterweight towers; (*ii*) double sheave pulley arrangements on take-up sections instead of single sheave arrangements; (*iii*) installation of a slewable feeder conveyor, instead of a two-way chute at transfer points; (*iv*) use of synthetic rubberised gasket which has considerable flexibility and is oil resistant in place of ordinary gaskets on conveyor idlers.

PERT/CPM Techniques

Modern techniques like PERT has been effectively used in the erection, overhaul, modification works and trackshifting of conveyors to ensure optimum utilization of the available resources thereby reduce downtime and increase equipment availability ratio.

The remarkable increase in the equipment availability of conveyors, achieved as a result of the implementation of the above techniques is illustrated in Fig. 16.2.

APPENDIX 1 (A)
INSPECTION CARD/INSPECTION REPORT

Plant	Assembly	Inspection Cycle	
Conveyor No. T/5	Jones Gear Box	1250 Hours	
Equipment Group	High-speed and	(Quarterly)	Scheduled Inspection
Primary Drive Gearbox	Ist Reduction		No. 2525

Drive Head
Section

Work Item	Description of Work	Report	Signature
1.	Check when operating for noise, heat and vibration under load.		
2.	Check for oil leakage from seals and labyrinths also; record condition of oil and amount of sediment.		
3.	Check for tightness of bolts, including motor holding down bolts.		
4.	Remove inspection cover and check wear and mesh of gear teeth. Record backlash and extent of wear.		
5.	Check conditions of and clearance in bearings. Record clearance.		
6.	Check condition of gear box housing plate for cracks, etc.		
7.	Where possible, check the lift and end float in first motion shaft, with dial indicator. Record clearance.		
8.	Check alignment of gear box and coupling and correct where necessary.		

APPENDIX 1 (B)
PREVENTIVE MAINTENANCE CHECKLIST FOR C-60 CONVEYORS—MECHANICAL 'BI-WEEKLY'

Conveyors	1	2	3	4	5	6	7	Remarks
Zone A (Drive station)								
Z I. Grease the following points with Shell Multipurpose Grease No. 3 till old grease is extruded on the sides.								
1. Primary drive drum bearing LHS 1								
2. Secondary drive drum bearing RHS 1								
3. Bend pulley I LHS 1								
4. Bend pulley II (take up wagon pulley) RHS 1								
5. Take up pulley LHS 1								

Conveyors		1	2	3	4	5	6	7	Remarks
6. Return idlers, 4 nos.	LHS 1								
	RHS 1								
7. Counterweight rope sheaves	8								
8. Take up wagon shell bushings	4								
9. Grease the gear reducer labyrinths at the outer slow speed shaft till fresh grease is forced out									
(*a*) Primary drive drum	LHS 1								
	RHS 1								
(*b*) Secondary drive drum	LHS 1								
	RHS 1								
10. Grease winch on take up tower	2								
11. Apply grease on the winch teeth	4								
12. Wiper support bearings	LHS 1								
	RHS 1								
II. Inspect oil levels in gears and top up if necessary. Use mecoma oil 72									
1. Primary drive	LHS 1								
	RHS 1								
2. Secondary	LHS 1								
	RHS 1								
III. Inspect wiper at discharge pulley (the rubber should project at least 5 mm beyond the metallic portion); change if necessary									
IV. Inspect take up wagon wire rope for cut in strands and check for free movement of counterweight. (the guides should be free and should not touch the vertical structure)									
V. Replace damaged idler supports troughing idlers and return idlers at D.H. portion									
VI. Check rope clamps for tightness	4								
VII. Check oil samples from gear reducers for moisture and acidity; if beyond safety limit change oil.									
VIII. Check rope for cut in strand. If damaged replace with new one	1								
IX. Apply cardium compound to wire rope									
X. Inspect the coupling oil and adjust the level for correct load sharing									
1. Primary drive drum	LHS 1								
	RHS 1								

Conveyors	1	2	3	4	5	6	7	Remarks
2. Secondary drive drum LHS 1 RHS 1								
XI. Clean the breathers and check for leaks 4								
Zone B (Intermediate section)								
I. Replace damaged 1. Return idlers 2. Troughing idlers 3. Idler supports								
II. Check for free movement of self aligning roller brackets. If restricted remove restriction.								
III. Grease the following points in self aligning return rollers (4 Nos.) 4 (*a*) Centre bushing 4 (*b*) Vertical guide rollers LHS 4 RHS 4								
IV. Grease the self aligning return idlers with Shell Multipurpose Grease No. 1–3 4								
Zone C (Tail station)								
I. Grease the following points with Shell Multipurpose Grease No. 3 1. Return pulley LHS 1 RHS 1 2. Return idlers (2 Nos.) LHS 2 RHS 2								
II. Adjust plough wiper. Check the mounting belts								
III. Adjust the side wipers at hopper to get a clearance of 2 mm between belts and wiper LHS 1 RHS 1								
IV. Replace the damaged 1. Impact idlers 2. Idler supports								
V. Grease the following points using Shell Multipurpose Grease No. 3 Impact idler (24 Nos.) LHS 8×2 RHS 8×2 Middle 8×2								
Check and tighten bolts where necessary								
I. Drive station 1. Primary drive drum pillow blocks 8 2. Secondary drive drum pillow blocks 8 3. Bend pulley I 8								

Conveyors	1	2	3	4	5	6	7	Remarks
4. Bend pulley II 8								
5. Take up wagon drum 8								
6. Mounting belts for screw jacks LHS 8								
RHS 8								
7. Mounting bolts for rope sheave brackets 8								
8. Mounting bolts for winch in take up tower 8								
II. Tail end								
1. Return pulley pillow block mounting bolts 8								
2. Mounting bolts for impact idler supports								
LHS 8								
RHS 8								
III. Inspect brakes for proper action and adjust if necessary								
1. Primary drum LHS 1								
RHS 1								
2. Secondary drum LHS 1								
RHS 1								
In addition to the Daily P.M.C. & S. attend to the following :								
Zone A								
I. Grease the following points in addition to the daily points with Shell Multipurpose Grease No. 3								
1. Counterweight rope sheaves 8								
2. Take-up wagon wheel bushings 4								
3. Grease the gear reducer labyrinths at the outer slow speed shaft till fresh grease is forced out								
(*a*) Primary drive drum LHS 1								
RHS 1								
(*b*) Secondary drive drum LHS 1								
RHS 1								
4. Grease the winch on the take-up tower 2								
5. Apply grease on the winch teeth 4								
6. Wiper support bearings LHS 1								
RHS 1								
II. Check rope clamps for tightness 4								
III. Check rope for cuts in strands. If found damaged replace with a new one 1								
IV. Check oil samples from gear reducers for moisture and acidity. If found beyond safety limit, change oil								

Conveyors	1	2	3	4	5	6	7	Remarks
V. Apply cardium compound to wire rope								
VI. Inspect the Coupling oil and adjust the level for correct load sharing								
1. Primary drive drum LHS 1								
RHS 1								
2. Secondary drive drum LHS 1								
RHS 1								
VII. Inspect the brakes for proper action and adjust, if necessary								
1. Primary drive drum LHS 1								
RHS 1								
2. Secondary drive drum LHS 1								
RHS 1								
VIII. Clean the breathers and check for leaks 4								
Zone B								
I. Grease the following points in self aligning return rollers (4 Nos.) 4								
(*a*) Centre bushing 4								
(*b*) Vertical guide rollers LHS 4								
RHS 4								
Zone C								
I. Grease the following points (Use Shell/Multi purpose Grease No. 3)								
Impact Idlers (24 Nos.) LHS 8×2								
RHS 8×2								
Middle 8×2								
Check and tighten bolts wherever necessary								
I. Drive station								
1. Primary drive drum pillow blocks 8								
2. Secondary drive drum pillow blocks 8								
3. Bend pulley I 8								
4. Bend pulley II 8								
5. Take-up wagon drum 8								
6. Mounting bolts for screw jacks LHS 8								
RHS 8								
7. Mounting bolts for rope sheave brackets 8								
8. Mounting bolts for winch in take-up tower 8								
II. Tail-end								
1. Return pulley pillow block mounting bolts 8								
2. Mounting bolts for impact idler supports								
LHS 8								
RHS 8								

APPENDIX 2 : JOB INSTRUCTION SHEET NO. 19
CONVEYOR MAINTENANCE DIVISION

JOB: Mounting of Fluid-Coupling (size-20) on G60 conveyors.

Procedure

A. Preparatory works
1. Prepare the ground well for easy approach of crane.
2. Inform Radio Control Room before starting the work.

B. Mounting
1. Measure the distance between motor face and the gear box flange face and keep it $15\frac{1}{2}$" (overall length to suit fluid coupling and multidisc unit).
2. Attach slings to the fluid coupling and insert the fluid coupling.
3. Fix the resilient plate bolts.
4. Insert the multidisc coupling and fix the bolts.
5. Screw a long stud into the tapped hole in the driving boss.
6. Clamp the dial gauge to this stud and set it to bear on the head of one of the driving bolts or a point on the driving plate near the edge.
7. Set the dial gauge at '0' at the top.
8. Turn the motor by hand and watch the dial gauge readings at bottom and sides over a complete revolution.
9. Adjust the position of the motor sideways, and the shims under the motor base until the dial gauge readings are constant within plus or minus 0.005" over a complete revolution.

Tools and Tackles Required
1. Ring and D.E. spanners of size 30–32, 24–27 and 19–22, 2 nos. each. Inch set spanner 1 set. Chisel 1 File 1, 2 lbs hammer 1 Sledge hammer 1 Crow bar 1 Pinch bar.
2. Different sizes shims. Tin cutter 1 Dial Gauge with stud 1 Screw driver 1 Manila rope. 24 box spanner.

Auxiliary Equipment and Special Gadgets Required
1. 'C' Crane for primary drum coupling and fowler crane for secondary drum.

Personnel Required

Supervisory: Foreman/JPE. 1
Labour: Mechanic Gr. II 1,
Fitters 2, Rigger 1, Helper 1.

Time Schedule

Preparatory works 1 hr.; Coupling Mounting $1\frac{1}{2}$ hrs.; Coupling alignment 4 hrs.; Final Checkup and Trial Run $\frac{1}{2}$ hr.; Total 7 hrs.

Safety Precautions

A. Special
1. The belt clear is to be obtained from the operation staff before starting the work.
2. Ensure that all isolators are put off from supply.
3. Emergencies should be applied wherever necessary.

B. General
1. Wire sling ropes and 'D' shackles should be checked before using on works.
2. Check the coupling oil level. The oil filling angle should be 50°.

Accuracy/Tolerance Permissible

The dial gauge readings must be constant within plus or minus 0.005" over a complete revolution

APPENDIX 3 : DELAY ANALYSIS CHART

	PLANNED STOPPAGES				PRODUCTION STOPPAGES															UNAVOIDABLE STOPPAGES								
EQUIPMENT NO.	Servicing	Trackshifting	Re-erection	Total	Overloading	Vulcanising	Electrical	Mechanical	Roller fixing	Slipping	Liningout	Chocking and cleaning	Boulder removal	Wiper adjustment	Belt piece cutting	Emergency acted	Transfer feeder and Table feeder	Buffer door changing	Total	Crawling	Blasting	Rain	Face preparation and servicing	Power failure	Others	Total	Loose soil handling and trimming	Rear loading
					1	2	3	4	5	6	7	8	9	10	11	12	13	14	15	16	17	18	19	20				
1144	40.30			40.30			0.20	12.10					0.25						12.55	0.40	0.25	1.50	1.00		1.05	5.00		6.30
B/89										0.15		0.10	0.25						0.50					2.25	0.05	2.30		
B1																												
B2																												
B3																									0.35	0.35		
B4						0.10						0.30	0.20						5.05						1.30	1.30		
B5												2.05							2.05									
B6					1.30							1.50							3.20						2.10	2.10		
B7							0.55					2.35		0.30					4.00						0.35	0.35		
B8					0.40			4.40	1.10			2.48							9.18						0.40	0.40		
Total	40.30			40.30	2.10	0.10	1.15	16.50	1.10	0.15		9.58	1.05	0.30					33.28	0.40	0.25	1.50	1.00	2.25	6.40	13.00		6.30

TOTAL STOPPAGES : 86 Hrs. 58 Mts. Machine No. ** 1144 OUTPUT (M3) 41744 B.W.E. B1.02 RATE (M3) 515

UNSOLVED QUESTIONS

1. What is maintenance ?
2. What is object of maintenance ?
3. Write about planned preventive maintenance.
4. Write about corrective maintenance.
5. Write about prediction maintenance.
6. Write about condition based monitoring.
7. Write about vibration monitoring.
8. Write about temperature monitoring.
9. Write about oil analysis.
10. Write about oil conditioning.
11. Write about crack monitoring.
12. Write about ultrasonic and dye penetrate test.
13. Write about sound monitoring.
14. Write about breakdown maintenance.
15. Write about maintenance of hoists cranes, conveyors, trucks, lift.

❑❑❑❑

17 Robots in Material Handling

In earlier chapters we had seen different material handling system which are operated by operator except few. Robot has wide application in material handling as they have certain characteristics such as accuracy, capability to do repetitive work, mobility, strength and most importantly controllable and programmable for a specific task. Robots has application in all fields such as mechanical, manufacturing space exploration, underwater application, assembly and maintenance, repair and material handling.

17.1 ROBOT BASICS

Most robots are designed to be a helping hand. They help people with tasks that would be difficult, unsafe, or boring for a real person to do alone. At its simplest, a robot is machine that can be programmed to perform a variety of jobs, which usually involve moving or handling objects. Robots can range from simple machines to highly complex, computer-controlled devices.

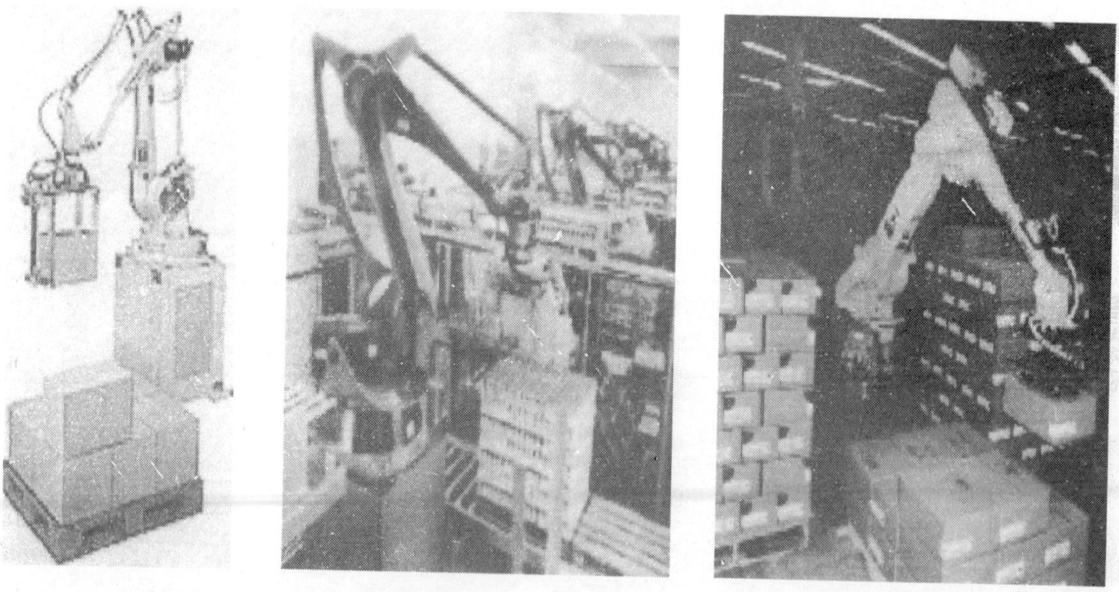

Figure 17.1 *Palletising Robots.*

Many of today's robots are robotic arms. In this exhibit, "flexible" kind of robot, which looks similar to a certain part of our body. It is called a jointed-arm robot is shown. The vast majority of robots do have several qualities in common.

(*i*) Almost all robots have a movable body.

(*ii*) Some only have motorized wheels, and others have dozens of movable segments, typically made of metal or plastic. Like the bones in our body, the individual segments are connected together with joints.

(*iii*) Robots spin wheels and pivot jointed segments with some sort of actuator. Some robots use electric motors and solenoids as actuators; some use a hydraulic system; and some use a pneumatic system (a system driven by compressed gases). Robots may use all these actuator types.

(*iv*) A robot needs a power source to drive these actuators. Most robots either have a battery or they plug into the wall. Hydraulic robots also need a pump to pressurize the hydraulic fluid, and pneumatic robots need an air compressor or compressed air tanks.

(*v*) The actuators are all wired to an electrical circuit. The circuit powers electrical motors and solenoids directly, and it activates the hydraulic system by manipulating electrical valves. The valves determine the pressurized fluid's path through the machine. To move a hydraulic leg, for example, the robot's controller would open the valve leading from the fluid pump to a piston cylinder attached to that leg. The pressurized fluid would extend the piston, swiveling the leg forward. Typically, in order to move their segments in two directions, robots use pistons that can push both ways.

(*vi*) The robot's computer controls everything attached to the circuit. To move the robot, the computer switches on all the necessary motors and valves. Most robots are reprogrammable—to change the robot's behaviour, you simply write a new program to its computer.

(*vii*) Not all robots have sensory systems, and few have the ability to see, hear, smell or taste. The most common robotic sense is the sense of movement—the robot's ability to monitor its own motion. A standard design uses slotted wheels attached to the robot's joints. An LED on one side of the wheel shines a beam of light through the slots to a light sensor on the other side of the wheel. When the robot moves a particular joint, the slotted wheel turns. The slots break the light beam as the wheel spins. The light sensor reads the pattern of the flashing light and transmits the data to the computer. The computer can tell exactly how far the joint has swiveled based on this pattern. This is the same basic system used in computer mice.

These are the basic nuts and bolts of robotics. Roboticists can combine these elements in an infinite number of ways to create robots of unlimited complexity.

17.2 MAIN PARTS OF ROBOT

For a machine to qualify as a robot, it usually needs these 5 parts:

(*i*) Controller

Every robot is connected to a computer, which keeps the pieces of the arm working together. This computer is known as the controller. The controller functions as the "brain" of the robot. The controller also allows the robot to be networked to other systems, so that it may work together with other machines, processes, or robots.

Robots today have controllers that are run by programs—sets of instructions written in code. Almost all robots of today are entirely pre-programmed by people; they can do only what they are programmed to do at the time, and nothing else. In the future, controllers with artificial intelligence, or AI could allow robots to think on their own, even program themselves. This could make robots more self-reliant and independent.

(*ii*) ARM

Robot arms come in all shapes and sizes. The arm is the part of the robot that positions the end-effector and sensors to do their pre-programmed business. Many (but not all) resemble human arms, and have shoulders, elbows, wrists, even fingers. This gives the robot a lot of ways to position itself in its environment. Each joint is said to give the robot 1 degree of freedom. So, a simple robot arm with 3 degrees of freedom could move in 3 ways : up and down, left and right, forward and backward. Most working robots today have 6 degrees of freedom.

(*iii*) DRIVE

The drive is the "engine" that drives the links (the sections between the joints) into their desired position. Without a drive, a robot would just sit there, which is not often helpful. Most drives are powered by air, water pressure, or electricity.

(*iv*) END-EFFECTOR

The end-effector is the "hand" connected to the robot's arm. It is often different from a human hand—it could be a tool such as a gripper, a vacuum pump, tweezers, scalpel, blowtorch—just about anything that helps it do its job. Some robots can change end-effectors, and be reprogrammed for a different set of tasks. If the robot has more than one arm, there can be more than one end-effector on the same robot, each suited for a specific task.

(*v*) SENSOR

Most robots of today are nearly deaf and blind. Sensors can provide some limited feedback to the robot so it can do its job. Compared to the senses and abilities of living things, robots have a very long way to go. The sensor sends information, in the form of electronic signals back to the controller. Sensors also give the robot controller information about its surroundings and lets it know the exact position of the arm, or the state of the world around it.

Sight, sound, touch, taste, and smell are the kinds of information we get from our world. Robots can be designed and programmed to get specific information that is beyond what our 5 senses can tell us. For instance, a robot sensor might "see" in the dark, detect tiny amounts of invisible radiation or measure movement that is too small or fast for the human eye to see.

17.3 TYPES OF ROBOTS

(*a*) MOBILE ROBOTS

Robotic arms are relatively easy to build and program because they only operate within a confined area. Things get a bit trickier when you send a robot out into the world. The first obstacle is to give the robot a working locomotion system. If the robot will only need to move over smooth ground, wheels or tracks are the best option. Wheels and tracks can also work on rougher terrain if they are big enough. But robot designers often look to legs, because they are more adaptable. Building legged robots also helps researchers understand natural locomotion—it's a useful exercise in biological research.

Typically, hydraulic or pneumatic pistons move robot legs back and forth. The pistons attach to different leg segments just like muscles attach to different bones. It's a real trick getting all these pistons to work together properly. As a baby, your brain had to figure out exactly the right combination of muscle contractions to walk upright without falling over. Similarly, a robot designer has to figure out the right combination of piston movements involved in walking and program this information into the robot's computer. Many mobile robots have a built-in balance system (a collection of gyroscopes, for example) that tells the computer when it needs to correct its movements.

Bipedal locomotion (walking on two legs) is inherently unstable, which makes it very difficult to implement in robots. To create more stable robot walkers, designers commonly look to the animal world, specifically insects. Six-legged insects have exceptionally good balance, and they adapt well to a wide variety of terrain.

Some mobile robots are controlled by remote—a human tells them what to do and when to do it. The remote control might communicate with the robot through an attached wire, or using radio or infrared signals. Remote robots, often called puppet robots, are useful for exploring dangerous or inaccessible environments, such as the deep sea or inside a volcano. Some robots are only partially controlled by remote. For example, the operator might direct the robot to go to a certain spot, but not steer it there—the robot would find its own way.

(*b*) Autonomous Mobility

Autonomous robots can act on their own, independent of any controller. The basic idea is to program the robot to respond a certain way to outside stimuli. The very simple bump-and-go robot is a good illustration of how this works.

This sort of robot has a bumper sensor to detect obstacles. When you turn the robot on, it zips along in a straight line. When it finally hits an obstacle, the impact pushes in its bumper sensor. The robot's programming tells it to back up, turn to the right and move forward again, in response to every bump. In this way, the robot changes direction any time it encounters an obstacle.

Advanced robots use more elaborate versions of this same idea. Roboticists create new programs and sensor systems to make robots smarter and more perceptive. Today, robots can effectively navigate a variety of environments. Simpler mobile robots use infrared or ultrasound sensors to see obstacles. These sensors work the same way as animal echolocation : The robot sends out a sound signal or a beam of infrared light and detects the signal's reflection. The robot locates the distance to obstacles based on how long it takes the signal to bounce back.

More advanced robots use stereo vision to see the world around them. Two cameras give these robots depth perception, and image-recognition software gives them the ability to locate and classify various objects. Robots might also use microphones and smell sensors to analyze the world around them. Some autonomous robots can only work in a familiar, constrained environment. Lawn-mowing robots, for example, depend on buried border markers to define the limits of their yard. An office-cleaning robot might need a map of the building in order to maneuver from point to point.

More advanced robots can analyze and adapt to unfamiliar environments, even to areas with rough terrain. These robots may associate certain terrain patterns with certain actions. A rover robot, for example, might construct a map of the land in front of it based on its visual sensors. If the map shows a very bumpy terrain pattern, the robot knows to travel another way. This sort of system is very useful for exploratory robots that operate on other planets.

An alternative robot design takes a less structured approach—randomness. When this type of robot gets stuck, it moves its appendages every which way until something works. Force sensors work very closely with the actuators, instead of the computer directing everything based on a program. This is something like an ant trying to get over an obstacle—it doesn't seem to make a decision when it needs to get over an obstacle, it just keeps trying things until it gets over it.

(c) Stationery Robots

These types of robots are fixed to a single site or point. They work in restricted area. Example of such kind of robot is pick and place robot. This kind of robot work similar to the mobile robot except they can't move bodily generally they are mounted on column or some kind of base. They have wide application in industries for material handling jobs. These kind of jobs are generally stereo type or repetitive in nature.

(d) Industrial Robots

These are the robots which are designed for specific tasks such as assembly, loading and unloading onto or from other handling equipments such as conveyors, storage systems, pick and place, etc. For details see Table 20.1.

(e) Special Purpose Robots

These are designed to perform special tasks. For details see Table 20.1.

17.4 APPLICATIONS OF ROBOTS IN MATERIAL HANDLING

Now days robots are used in almost all big industries, here are the mentioned few shown in tabular form.

Table 17.1 Applications of robots in Material Handling

Sr. no.	Type of Robots	Work or Operation	Features
1.	Industrial	1. Pick and place 2. Loading and unloading 3. Assembly 4. Inspection	1. Repetitive operations 2. Preprogrammed 3. Fixed 4. Rigid structured 5. Sensor guidance 6. In some cases mobile
2.	Special purpose	1. Nuclear material handling 2. Underwater handling and inspection 3. Space servicing, repair, assembly and exploration 4. Civil engineering	1. No Repetitive operations 2. Preprogrammed with multi arms 3. Mobile 4. Un-structured 5. Flexible behaviour 6. Multi Sensor guidance
3.	Autonomous	1. Transport and handling of tools and work pieces 2. Inspection 3. Exploration	1. No Repetitive operations 2. Preprogrammed with multi arms 3. Mobile 4. Un-structured 5. Flexible behaviour 6. Complex Sensor guidance

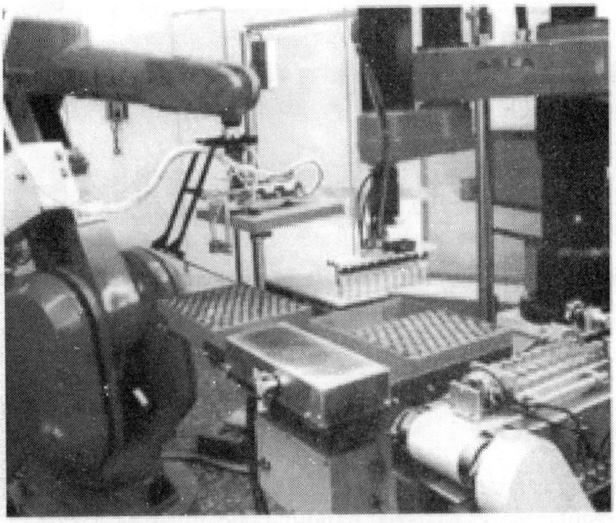

Figure 17.2 *Pick and Place Robot.*

UNSOLVED QUESTIONS

1. What are the important parts of any robots and functions of each part ?
2. Write down different types of robots.
3. What is the scope of robot in material handling equipments ?
4. What are the applications of robot material handling equipments ?

❑❑❑❑

18 Manual Handling

18.1 INTRODUCTION

Material handling is a task that almost every worker performs, either as a one-time or infrequent duty or as part of regular work. Material handling encompasses a wide range of work activities—from occasional movement of very large loads with cranes and powered industrial trucks to routine, repetitive lifting of relatively light objects and tasks that are incidental to a worker's regular, daily activities, such as an office move.

This chapter contains specific requirements and general guidance to protect workers from injuries when performing manual material-handling activities.

Manual handling is a major part of a worker's job in the Industries. Associated activities in these industries include stacking shelves, stacking cartons on pallets, moving trolleys, and packing stock, lifting and lowering load, pushing and pulling load.

Manual handling is not just about lifting heavy objects, but includes any activity requiring the use of force exerted by a person to lift, lower, push, pull, hold or restrain a person, animal or thing.

Manual handling is defined as unaided moving of the object combined with twisting and awkward postures, and contributing to musculoskeletal disorders (MSD).

18.2 HAZARDS

The hazards associated with improper material handling include being struck by a load, losing control of a load, physically overexerting oneself, and exceeding equipment capacities. Such accidents can lead to injuries (*e.g.*, abrasions, bruises, and broken bones) and even loss of life. Material-handling accidents account for a sizable percentage of all occupational injuries that occur in every work area throughout the Laboratory—not just in stockrooms or warehouses. The risk of injury is not confined to manual lifting of heavy objects. Some lost-time injuries have occurred in office situations where lifting is infrequent and involves only small items.

Manual handling injuries include :

1. Strains and sprains;
2. Neck and back injury;
3. Slips, falls and crush incidents;
4. Cuts, bruises and broken bones;

5. Hernia; and
6. Occupational overuse syndrome (OOS), once known as RSI.

People who suffer from a manual handling injury at work can spend the rest of their lives coping with pain and be unable to do a lot of things other people can do. Strain injuries occur through sudden overexertion or continuous overuse. Back injuries are one of the most serious types of strain. Excessive strain on your back can lead to long-term damage.

Manual material handling activity gives rise to problem of musculoskeletal disorders (MSD). Activities in the home, at the office, in industry and agriculture and in other areas almost inevitably bring with them the problem of musculoskeletal disorders (MSD). Since India has a largely labour-intensive economy, the MSD problem may in fact be acute, but insufficient awareness and a lack of proper documentation make it very difficult to quantify. With the globalization of India's capital-intensive market, a few major industries are being forced to follow international norms towards maintaining their work environments, but the majority do not comply with proper ergonomic standards.

The Indian workforce can be divided into three different categories : (1) the organized sector; (2) the unorganized sector; and (3) small-scale sector or cottage industries. Only a few top industrial organizations are trying to ensure work environments that meet the International Standards (ISO 9000–2000).

The MSD problem in an Indian context can be better illustrated by using examples from five different work environments : (1) heavy industry; (2) manual materials handling (MMH); (3) agricultural activities; (4) cottage or small-scale industries; and (5) use of imported tools or workstations.

In most cases, three basic factors—inadequate education, poverty and lack of awareness—hinder attempts to deal with the MSD problem. Poor working conditions and non-enforcement of statutory norms are additional factors.

18.3 HOW DO YOU KNOW IF A MANUAL HANDLING ACTIVITY IS A PROBLEM?

As a young worker you are more likely to be injured through manual handling work than an older worker. Your body is still developing and is less able to cope with strain. Persistent muscular aches or pains may indicate that there is excessive physical stress. There are a range of risk factors associated with manual handling activities. These risk factors increase the likelihood of the activity resulting in injury. Some of the main risk factors in the Industry are associated with—

(*a*) Frequent or prolonged handling involving :

Bending where the hands pass below mid-thigh level;

Reaching above the shoulder; or

Twisting the back.

(*b*) Occasional handling of objects which are difficult or awkward to move due to :

Shape, size or instability;

Heavy weight; or

Being placed or stored below mid-thigh level or above the shoulder.

18.4 CONTROLLING MANUAL HANDLING PROBLEMS

1. Eliminate double handling.

Example : Reduce the number of times an object is handled. Where possible move the stock directly from delivery to display.

2. Change the size or weight of packaging.

Example : Flour available in 25 kg bag for a bakery could also be available in smaller bags weighing 12.5 kg. Smaller bags can be lifted and handled more easily.

3. Reduce push/pull forces.

Example : Employers could limit the number of shopping trolleys to be collected at one time by employees.

4. Work at the correct height.

Example : Use stable, sturdy, step stools to enable easy access to stock that are out of reach.

5. Training and Education.

You should be informed and trained in :

1. Safe manual handling methods;
2. Specific manual handling hazards;
3. Safe work procedures;
4. Using manual handling aids; and
5. The right to ask for help.

Most manual handling injuries to young people could have been prevented by instruction, training and supervision. Safe work procedures should be prepared by employers with the help of employees to care for the special needs of young and inexperienced workers.

Even though you may be young and new to the job, as a young employee you have a responsibility to speak up if you feel your job is too heavy, too difficult, too tiring or puts you at risk of injury. It is the employer's responsibility to provide you with safe work procedures, and with instruction, training and supervision for manual handling tasks.

Ways to reduce the risk :

(*a*) Lighten loads (break loads into smaller quantities);

(b) Reduce bending, twisting, reaching movements; and

(c) Prevent muscle strain and fatigue. This includes warming up before working, allocating time for rest breaks, and allowing time to gradually get used to a new job.

18.5 PRECAUTIONS FOR SAFELY LIFTING AND HANDLING MATERIALS

When lifting or handling materials manually, use only methods that ensure your safety and that of the material.

1. Never attempt to lift objects that are too heavy or bulky to handle safely. Never overestimate your ability to perform a task.
2. Whenever possible, push rather than pull loads : Pushing uses the strong leg muscles, whereas pulling uses the easily strained back muscles.

3. When occasional lifts of compact loads (<70 kg) are required, observe the following precautions :
   (*a*) Loads should be handled no more than 18 cm in front of the body as measured from the ankles. The heavier a load, the more closely to the body the load should be held.
   (*b*) Very low lifts [25 cm or less from the floor] are not desirable because of the difficulty of maintaining balance when squatting to lift.
   (*c*) Medium lifts [75–135 cm from the floor] are more desirable because more strength is available in the lower part of this height range.
   (*d*) High lifts [135–188 cm. from the floor] are not recommended, except when the item to lift weighs less than 5 kg.
4. Muscle fatigue can potentially occur where lifting is required more than once every 2–5 min. Therefore, the weight for repetitive lifts should be less than that for occasional lifts. If the lifting rate exceeds six lifts per minute, lifting should be limited to 20 min or less to allow for muscle recovery.
5. Moving objects by sliding, rather than lifting, is recommended for repetitive handling tasks. For ease of manipulation during subsequent moves, heavy items should be stored on racks, shelves, or the like at elbow level. Lighter items may be stored at either higher or lower levels.

18.6 SAFE LIFTING PRACTICES

The following are recommended safe practices for lifting and moving heavy objects.

(*a*) Provide good handholds on an object to be carried.

(*b*) Firmly grip an object to be carried.

(*c*) Use carts and handling aids to support an object's weight.

(*d*) Try to minimize the distance that an object is moved.

(*e*) Carry a load close to your body and at a proper height from the floor. When standing, avoid lifting an item to a height greater than 50 cm or lowering an item to a height less than 20 cm.

(*f*) Push or pull objects whenever possible, rather than lifting or lowering objects.

(*g*) Position storage materials on racks or shelves in ways that make handling easier (*e.g.*, place the heaviest objects at elbow height).

(*h*) Keep your torso in a neutral, upright position whenever possible when lifting a heavy load. The neutral position is 0–15°. An angle greater than 15° is considered high risk.

(*i*) Do not twist or bend while lifting or handling a heavy load.

(*j*) Do not using jerking motions to move a load.

(*k*) Provide elbow support for long reaches.

(*l*) Warm up before starting a strenuous task after a less-strenuous task (*e.g.*, changing from sitting to performing a rapid, dynamic task.)

(*m*) Avoid lowering materials that need to be later lifted.

(*n*) Plan movements of material in direct paths (*i.e.*, do not zigzag or backtrack).

(*p*) Use an adjustable/tilt table or a reach extender to reduce forward bending or extended reaches.

(*q*) Use a step stool or low platform to reduce back hyperextension.

(*r*) Use roller bearings or roller conveyors to reduce twisting. Use sheet supporters, hand trucks, and small overhead hoists to improve posture.

18.7 MANUAL MATERIAL HANDLING ASSISTED WITH MECHANIZED EQUIPMENTS WITH SOME GUIDELINES

MANUAL MATERIALS HANDLING - Vacuum Systems

DESCRIPTION :

Vacuum systems for lifting and transport of materials.

WHEN TO USE :

- Vacuum systems can be used for lifting and transporting poultry parts, boxes of product, spice bags, ice and other materials. Applications of vacuum systems include :
  - ★ Lifting of individual boxes and placement onto racks or pallets for storage or transport.
  - ★ Vacuum systems connected to chutes or transport tubes can transport individual poultry parts, or collect poultry parts in a container.

POINTS TO REMEMBER :

- Vacuum entry points can be placed at individual work areas to gather product for transport to chilers or other holding areas for further processing or packaging.
- Vacuum systems can be designed specifically to handle internal transport of hearts, livers, gizzards, and necks from harvesting area to giblet handling area.

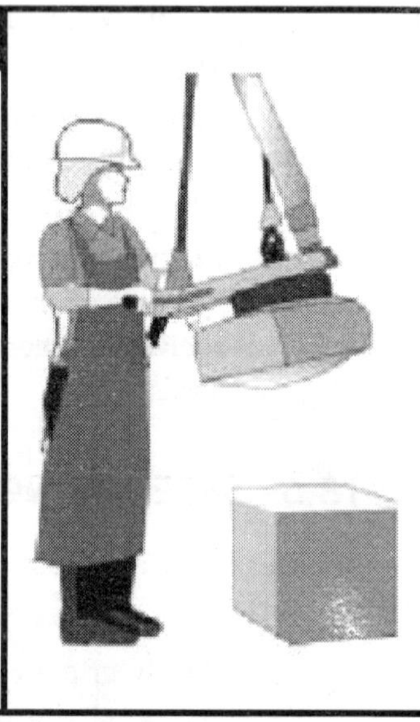

Figure 18.1

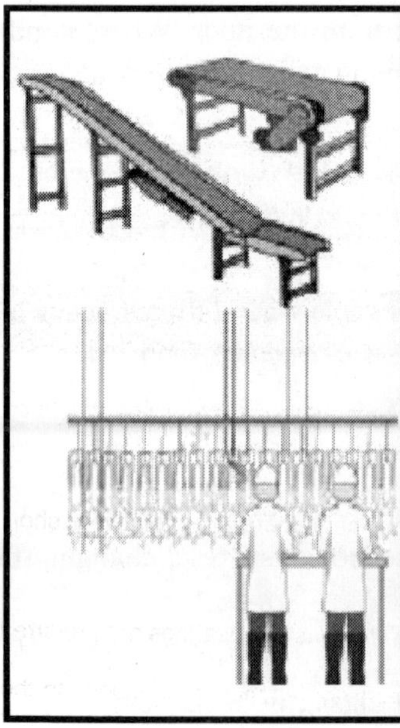

MANUAL MATERIALS HANDLING - Belt and Overhead Conveyors

DESCRIPTION :

Mechanical systems such as belt-driven sheets of textured fabric or suspended shackles/hooks that continually move product.

WHEN TO USE :

For processing and cutting tasks, placing bags and smaller packages in boxes, and loading or unloading trucks. Eliminates lifting and carrying when transporting whole birds and parts through and between processing areas.

POINTS TO REMEMBER :

- Ensure that requried forward reach does not extend beyond arm's length in front of the body.
- Overhead conveyors should be designed so that the employee can grasp birds without reaching higher than mid-chest level. This can be accomplished by lowering the line or providing access for employees at their work area.
- Belt conveyors should be installed so that the employee can grasp birds or bird parts while keeping their elbows close to the body. The conveyor speed should allow sufficient time for the task being performed.
- Conveyors can be designed to reorient birds or bird parts.

Figure 18.2

MANUAL MATERIALS HANDLING - Racks and Shelves

DESCRIPTION :
Rack and shelf design to optimize manual access.

WHEN TO USE :
Shelf systems can be configured to minimize excessive lifting, carrying and awkward postures associated with storage of any item used or produced at a workstation.

POINTS TO REMEMBER :
- For items that will be manually lifted, shelf height range should be from approximately knee to shoulder height with an optimal range between hip and shoulder height. Store heavy and frequency used items in this optimal range.
- Bulk packages of supplies that are transported by mechanical devices can be stacked and stored as needed. However, once packages are opened and removed manually, placement on shelves should follow the recommendations listed above.
- Avoid closely spaced shelves as they limit visual and hand access.

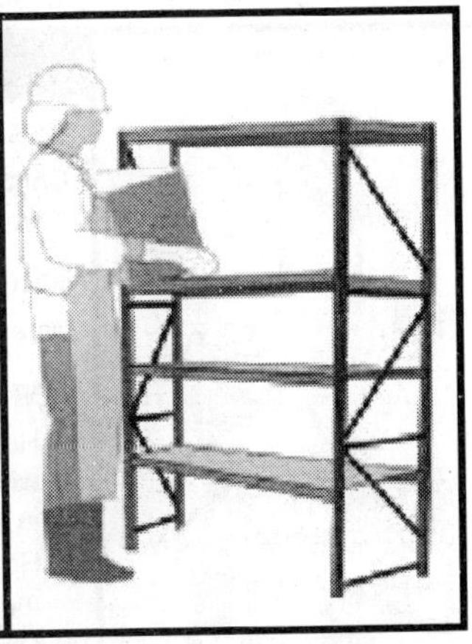

Figure 18.3

MANUAL MATERIALS HANDLING - Carts and Hand Trucks

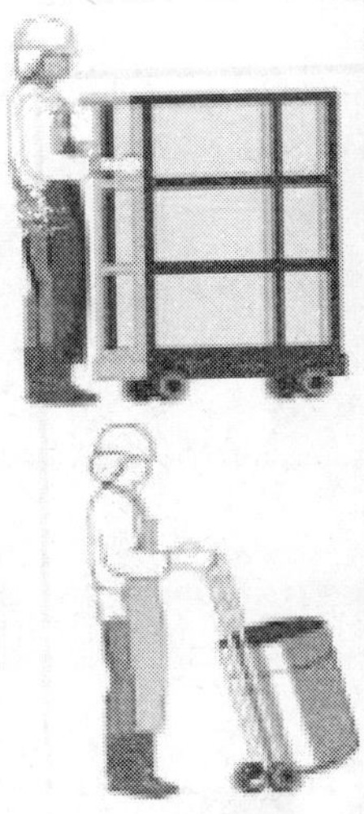

DESCRIPTION :
Wheeled devices designed to transport materials.

WHEN TO USE :
For transporting containers of whole birds, bird parts, processed meats, waste products or supplies between work areas.

POINTS TO REMEMBER :
- A small lip around the edges is helpful to prevent items from slipping off.
- Ensure that the top stack height does not obstruct vision.
- Optimal shelf height range should be between hip and shoulder height.
- Pushing is preferred to pulling.
- Balance loads and keep loads under manufacturer's weight restrictions.
- Handles that are vertical, with some horizontal adjustability, will allow employees to adjust how they push carts. A swing-out design may be useful to improve access.
- Manually pushed carts and hand trucks should have full bearing wheels, made of a material designed for the floor surface in the facility as well as brakes that are easy to operate. Rear swivel and front fixed wheel design improves ease of pushing, especially if the cart is long. In some designs, a third set of non-swivel wheels is placed centrally to add stability and improve ease of turning.
- Carts may be designed for general use or for specialized applications. Specially carts that have contours or clamps to hold specific supplies (*e.g.*, rolls of plastic) may be designed to aid in supply transfer directly to the machine or work surface that uses that product.
- Large, wheeled containers should have an angled front to allow for easy removal of contents.
- Pallet jacks may be manual or motorized. The motorized version is preferred for frequent or long distance travel.

MANUAL MATERIALS HANDLING - Roller Tables

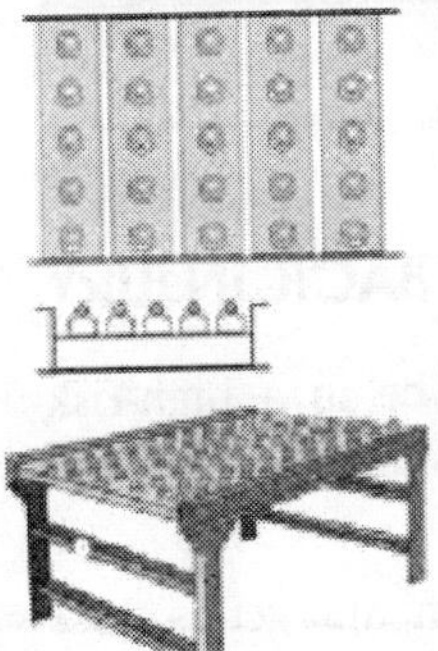

DESCRIPTION :
Table top or work surface embedded with rollers or ball bearings to reduce friction and force when sliding items.

WHEN TO USE :
Recommended for transporting boxes, bins, and other containers over relatively short distances.

POINTS TO REMEMBER :
- Rollers are preferred for transport in a linear direction whereas ball bearings are preferred when change of direction is required.
- Pushing or sliding containers eliminates lifting and carrying that can be strenuous to the hands, arms, and back.
- Appropriate for product in containers, but not loose parts due to sanitation issues.

Figure 18.4

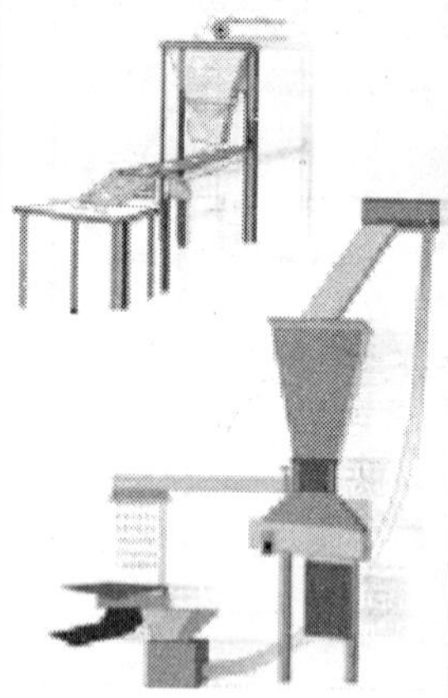

MANUAL MATERIALS HANDLING - Hoppers and Augers

DESCRIPTION :

Container used to hold and dispense contents into a machine, a new container or, onto a workstation through an open gate or using a screw-type mechanism.

WHEN TO USE :

When storage of product or other items is required at a workstation.

POINTS TO REMEMBER :

- Hoppers are generally preferred for dispensing larger objects such as poultry parts whereas augers are preferred for smaller product such as ice, spices and tenderizers in predetermined quantity to match recipe or packaging units.
- Hoppers are loaded from the top of the unit whereas augers have screw-type mechanisms that lift smaller particles from a container.
- Operation can be automatic or employee initiated with activation controls located to avoid reaching and bending.
- May replace the need for shoveling, especially for ice.

Figure 18.5

PERSONAL PROTECTIVE EQUIPMENT - Selection Considerations

DESCRIPTION :

Clothing and other work accessories designed to create a barrier against workplace hazards that take ergonomics considerations into account.

WHEN TO USE :

When selecting personal protective equipment for processing operations.

POINTS TO REMEMBER :

- Protective clothing should be selected for the specific task performed.
- Employees should be provided with a variety of options for rubber gloves worn under mesh gloves for optimal fit, dexterity and flexibility.
- Mesh gloves can include small barbs on the palm surface to reduce finger force when pinching and pulling meat away from bone. A mesh or cut resistant glove and forearm sleeve should be worn on the hand that does not hold the knife to protect against accidental cuts.
- Gloves should minimize the loss of tactile sensitivity.
- Gloves should provide increased friction between the hand and the material being handled.
- Use non-slip footwear with well-cushioned insteps and insoles.

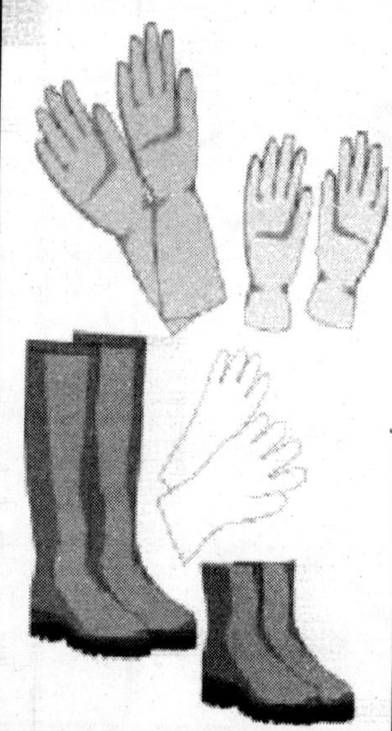

Figure 18.6

18.8 MANUAL MATERIAL HANDLING METHODS TO PREVENT BACK INJURY

If your job involves heavy labour or manual material handling, your may be in a high-risk category. Manual material handling entails lifting, but also includes climbing, pushing, pulling and pivoting.

Lifting from the floor is hard on your lumbar spine. Use a diagonal foot position, and get as close to the load as possible. Keep the load as close to your body as you can when standing up (See Fig. 18.7 Lifting from the floor).

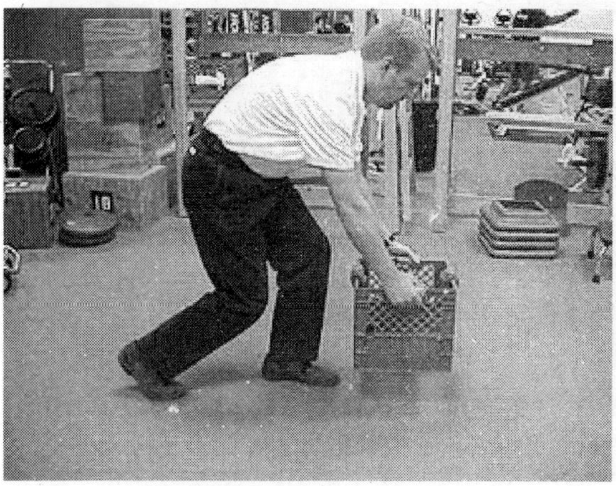

Figure 18.7 *Lifting from the floor.*

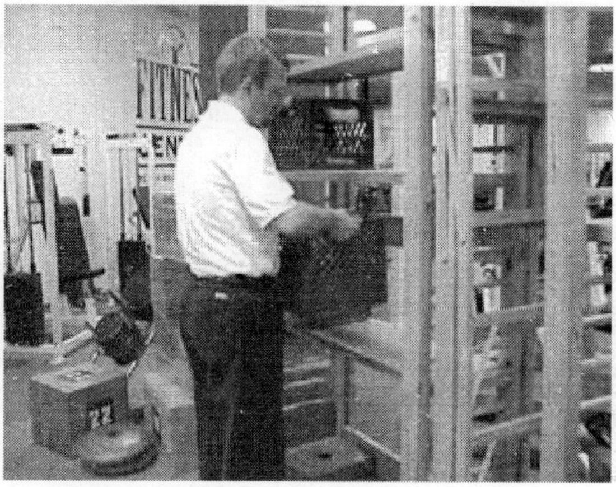

Figure 18.8 *Stacking pallets or loads.*

It is easier to move loads that are waist high or off the floor. Stacking pallets to raise the height of the load is one ergonomic solution. A scissors lift will mechanically raise the load to comfortable lifting level. Repetitive lifting form the floor is particularly risky, so try to get the material off the floor (See. Fig. 18.8 Stacking pallets).

Keep all loads as close to your center of gravity as possible. Carrying loads on your shoulder is safer for long and narrow material. This would include construction material or rolls of carpet. (See Fig. 18.9 Carrying loads on your shoulder).

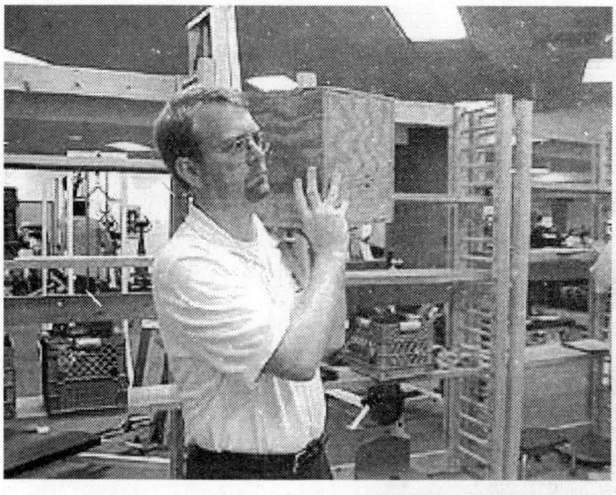

Figure 18.9 *Carrying loads on shoulder.*

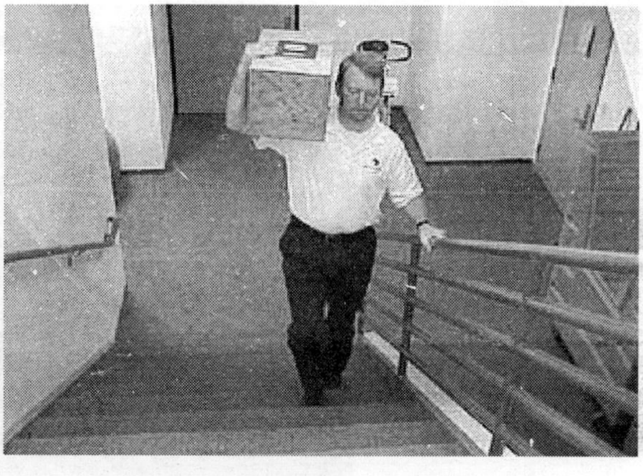

Figure 18.10 *Climbing with a load.*

When lifting anything with a handle, place a hand on your knee to get additional leverage and use a diagonal foot position. Carrying two objects of the same weight will balance the load as long as the weight of the load is reasonable.

When climbing with a load, "three-point" contact is important for safety. This means two hands and a foot or both feet and a hand must be in contact with the ladder or stairs at all times. If the load is bulky, get another person or a mechanical device to assist you (See Fig. 18.10 Climbing with a load).

Manual material handling may require pushing or pulling. Pushing is generally easier on your back than pulling. Use your arms and legs to provide the leverage to start the push. (See Fig. 18.11)

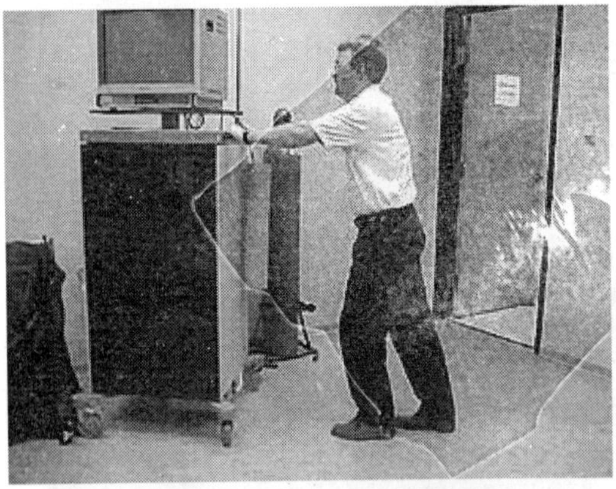

Figure 18.11 *Pushing and pulling.*

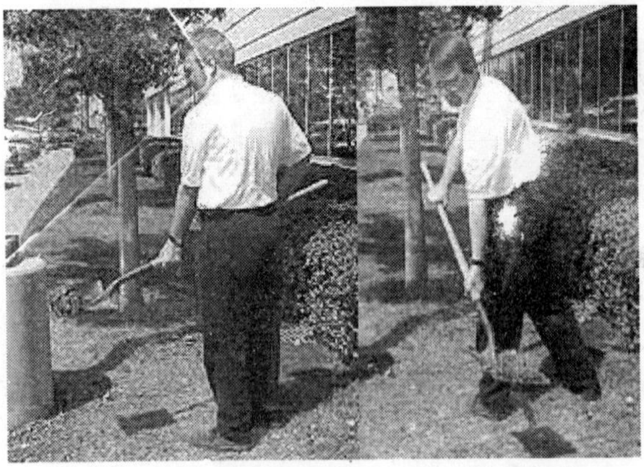

Figure 18.12 and 18.13 *Twisting and Pivoting.*

A handle would ideally be waist high for ease of pushing. If you have to pull, keep the cart at your side to avoid twisting your lower back.

Sometimes, for very large loads, turning around and using your back to push against an object allows you to get the best effort from your legs while protecting your back.

The opposite of twisting is pivoting. Pivoting means moving your shoulders, hips and feet with the load in front of you at all times. The lower back is not designed to torque or twist repetitively. Whether you are using a shovel or moving material or products, avoid twisting your back. (See. Fig. 18.12 and Fig. 18.13 twisting and pivoting). Only you can take care of your back. Practice these techniques, both at work and at home, and soon they will become a habit.

18.9 A CASE STUDY ON MANUAL MATERIALS HANDLING (MMH) : AN INDIAN SCENARIO

Out of total Indian population of around 850 million, 290 million (34%) are manual labourers and constitute the main workforce of the country (Census 1991). According to the Indian Factory Act, 1948, the limit for weights to be lifted is 30 kg in the case of adult males and females.

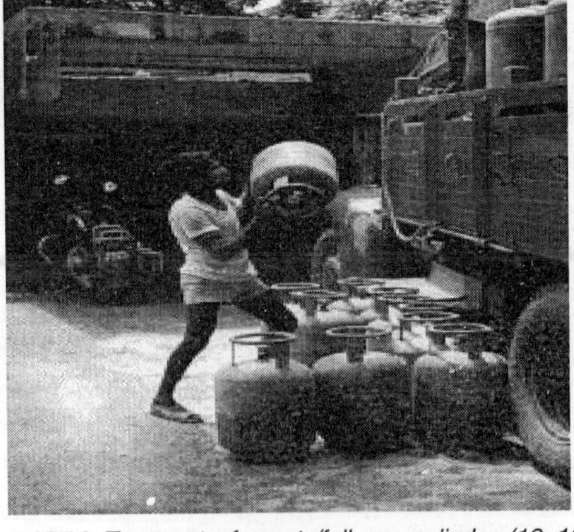

Figure 18.14 *Transport of empty/full gas cylinder (13–16 kg).*

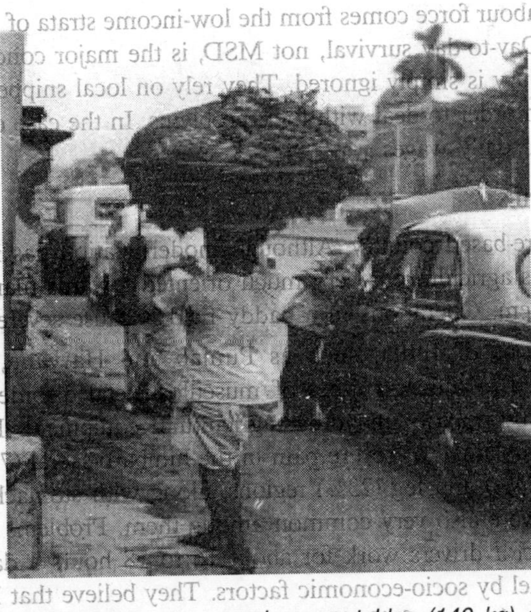

Figure 18.15 *Male porter carrying vegetables (140 kg) on head.*

But no proper scientific documentation is available to justify this weight limit. Figures 18.14–18.17 show some situations involving manual materials handling by both male and female individuals. In the case of males the weight may be as much as 140–145 kg, and for females 50 kg, when loads are carried on the head.

The average body weight of Indian male porters is 52 kg, while that of females is around 42 kg. In a survey conducted in Mumbai (Bombay), 6.5% of women, employed as professional water carriers reported miscarriages. Of those women, 64.5% complained of whole-body fatigue at the end of work. About 75% of women, employed as labourers in the construction industries reported problems related to the lumbar and calf regions : these were followed by problems in the cervical region (62.5%). 67% of women engaged in construction activities complained of aches affecting the whole body.

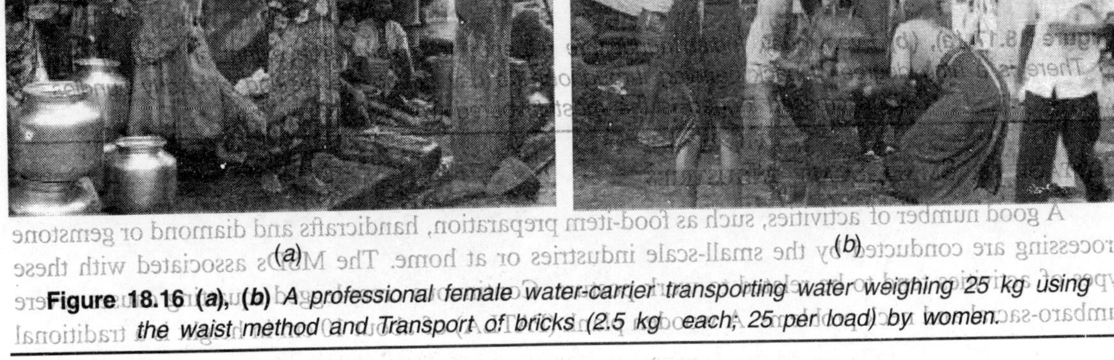

(*a*) (*b*)

Figure 18.16 (*a*), (*b*) *A professional female water-carrier transporting water weighing 25 kg using the waist method and Transport of bricks (2.5 kg each; 25 per load) by women.*

Most of the Indian labour force comes from the low-income strata of society and has a negligible educational background. Day-to-day survival, not MSD, is the major concern of these workers. Any kind of small muscular injury is simply ignored. They rely on local snippets of information and tend to consult quack doctors in order to deal with their problems. In the case of very severe injuries, they take some rest, losing their daily wages.

Agricultural Activities

India is an agriculture-based country. Although modernization has reached some parts of the subcontinent, in most areas agriculture is very much oriented towards manual labour because of the feudal land-ownership system. Operations in a paddy field is cause severe MSDs.

Some northern regions of India, such as Punjab and Haryana, have begun agricultural mechanization. This has led to different types of musculoskeletal disorders resulting from the high levels of vibration caused by tractors and other agricultural equipment. In one study, most of the tractor drivers (N = 13) complained of severe pain in the lumbaro-sacral (77%), neck (54%), shoulder (39%), chest (23%), arm (23%) and leg (23%) regions, along with stomach pain and a high level of acidity (23%). Headaches were also very common among them. Problems become very acute during the peak season, when tractor drivers work for about 16 to 18 hours a day but are prevented from consulting medical personnel by socio-economic factors. They believe that liquor will take care of all these aches and pains. After the age of 40–45 years, it becomes difficult for them to work on tractors. They are unemployed and still suffer from unbearable pains. They use medicines obtained from the quack doctor, as they have no money for proper medical treatment.

Nowadays manually or electrically operated threshing machines are available throughout the country. Though these have improved productivity, they have also caused an enormous increase in the rate of accidents, especially hand injuries, because of their poor and unsafe design.

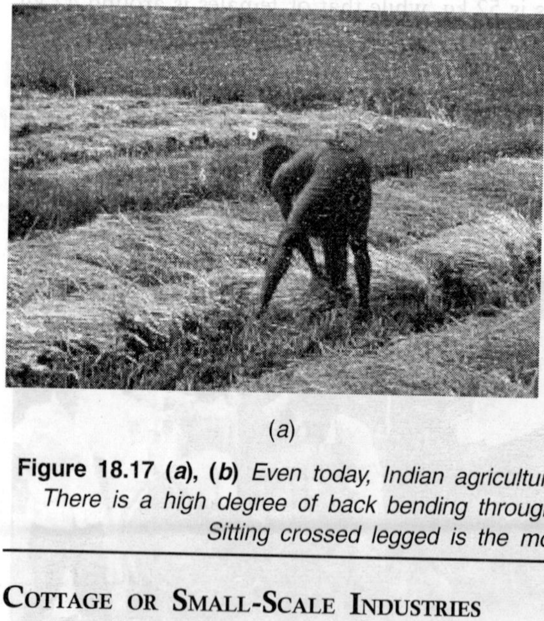

(a)

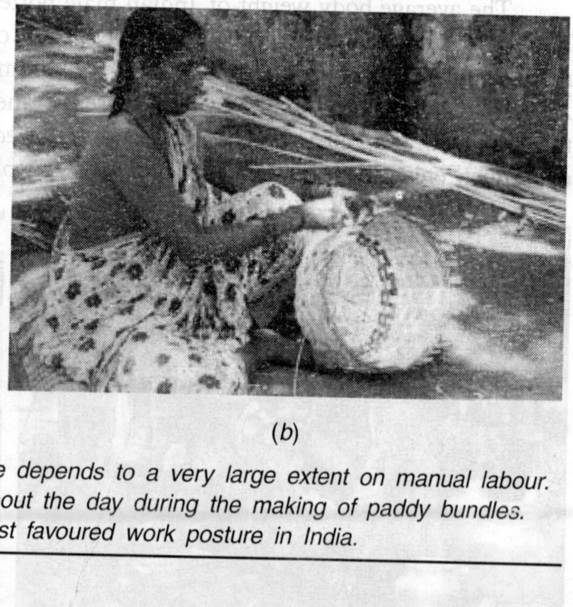

(b)

Figure 18.17 (*a*), (*b*) *Even today, Indian agriculture depends to a very large extent on manual labour. There is a high degree of back bending throughout the day during the making of paddy bundles. Sitting crossed legged is the most favoured work posture in India.*

Cottage or Small-Scale Industries

A good number of activities, such as food-item preparation, handicrafts and diamond or gemstone processing are conducted by the small-scale industries or at home. The MSDs associated with these types of activities tend to be related to work posture. Continuous cross-legged squatting causes severe lumbaro-sacral and neck problems. A wooden plank (PATLA) of about 10 cm in height is a traditional low seat used by the Indians. Sitting crossed legged on the floor might have some benefit in comparison to sitting on a chair when relative assessments are made with regard to the chances of developing varicose vein problems in the lower limbs. But the drawbacks of the PATLA are that it causes kyphotic back and pain in the hip, knee and ankle joints during prolonged sitting. At older ages, people have difficulty in squatting crosslegged. They also face problems when sitting down and getting up. A modified PATLA with a forward downward slope might solve some of these problems.

MSD IN OTHER SECTORS

In the household, office and other activities spheres, such as transport and hospital work, people are relatively aware about their health. Although they cannot relate MSDs to work habits or the poorly designed products, they do visit the doctor once problems appear.

Dealing with the Situation

Indians are fortunate in having an abundant labour force, but this does not necessarily mean that they can destroy the labour resource at their own will. Poverty and lack of education are the fundamental problems of India, and they must be tackled by the Government. But to make people conscious of the MSD problem and its impact on productivity, we can certainly launch some awareness programmes aimed at both employees and employees. This can be done by using various audiovisual training programmes suitably designed for different sectors of society.

Basic research in the area of ergonomics and occupational health should be strengthened if we are to understand cause-and-effect relationships. The findings of such research should be widely disseminated in order to enable mass education through proper media, and to assist the statutory bodies in amending and enforcing of statutes and laws. The statutory bodies should also encourage the implementation of the statutes and laws, and should periodically evaluate their effectiveness.

UNSOLVED QUESTIONS

1. What is meant by Manual material handling ? What is a standard definition ?
2. What is meant by MSD ? How to minimize them ?
3. Write down about Indian scenario with respect to Manual material handling.
4. What corrective measures are necessary to minimize Manual material handling ?
5. What are the correct and safe lifting practices ?
6. What is an approach to control the risk which is involve din Manual material handling ?
7. Explain the standard methods to minimize back injury in Manual material handling.
8. What are the different types of mechanical aids which can be used for Manual material handling ?

□□□□

19 Safety in Handling

19.1 INTRODUCTION

Although all the industries are going through automation and modernization, safety is always gained importance in every field of engineering. Safety is important in material handling also.

(*i*) Because it will damage the life as well as property if properly not taken care off.

(*ii*) In every industries material handling equipments are used and mostly they are having operator handling it. Hence, safety in handling mostly concerned with the operator.

The hazards associated with improper material handling (*i*) being struck by a load, (*ii*) losing control of a load, (*iii*) physically overexerting oneself, (*iv*) and exceeding equipment capacities. Such accidents can lead to injuries (*e.g.*, abrasions, bruises, and broken bones) and even loss of life. The risk of injury is not confined to manual lifting of heavy objects. Some lost-time injuries have occurred in office situations where lifting is infrequent and involves only small items.

19.2 MATERIAL HANDLING EQUIPMENT SAFETY

It can be achieved by either designing the equipment with necessary norms and specification for safety or operating the equipment safely. Second part of this deals with the operator who actually operates the equipments. In industries most of the accidents occur due to the operators fault. Hence, operator must be trained and motivated for safety training whether it is manual handling or mechanized handling.

19.3 OPERATOR TRAINING

During the training session, reviews of specific procedures that are part of the everyday work, performed by the attendees are done.

The inspection and operating procedures that are taught are focused on all aspects of proper operation, including rigging and slinging. It is advised the implementation of proper overhead material handling procedures that will increase the life of equipment and reduce maintenance costs.

19.3.1 Training Sessions Include

Classroom Training including : Video, overhead transparencies, handout materials and inspection of defective samples.

Question and answer period. Hands on, shop floor, check-out of daily safety inspection requirements. (If available) Sign out sheet and certificate presentation.

Topics Covered in Training Sessions :

- Proper material handling equipments terminology.
- Listing of codes pertaining to material handling equipments.
- Daily material handling equipments operator inspection and proper inspection methods.
- Proper handling practice and use of safety devices.
- Basic maintenance procedures and requirements.

19.3.2 Course Length

Typically 8 hours, depending on course content, questions and number of attendees. At the end of this course, the operator will be able to answer 10 important questions. If the operators in your company cannot answer the following questions, perhaps it is time for a review of procedures.

1. What does one cubic metre ot of steel weigh and how can one determine the correct loading for equipment ?
2. What is the defined purpose of the equipment ?
3. What are the accepted and standard methods of operating ?
4. What is the minimum and maximum load handling capacity as a standard practice ?
5. How does the angle affect the capacity of equipment ?
6. What is resultant sling angle and why is it a killer ?
7. How does an operator gain control of an equipment ?
8. What is the difference between a cable and a wire rope ?
9. Different attachment which can be attached to standard equipment ?
10. What required safety precautions must be taken before working on an equipment ?

19.3.3 Training Aids

It includes on line and off line training. (*i*) On line training includes training the operators to work safely on equipment and use of standard practice. (*ii*) Offline training includes graphs, figures, postures and video films to demonstrate graphically the correct procedures and results when they are not applied. These programs are designed to determine the safest and most productive methods to accomplish the specific tasks that are required by the company.

19.4 SAFETY NORMS AND STANDARDS

All material handling equipments are designed by adhering to standards and norms which are predefined by BIS ISO. For material handling equipments also there are norms and standards which relates with design of material handling equipments, operating norms of material handling equipments, safety norms of material handling equipments etc. For details see ISI Standards which are given in appendix B.

19.5 CRANES SAFETY

Improper use of cranes, hoists, and rigging devices can cause equipment to fail or loads to drop from the lifting system, resulting in personnel injury, death, significant property loss, or damage to the environment. Following should be considered while working on crane and operator should have following Worksite Considerations, Operational Considerations, Technical knowledge, Manufacturers Load Charts. Safe operation of overhead cranes requires operators to have the knowledge and competence to employ safe rigging practices. The rigger must rig the load to ensure its stability when lifted. The following points highlight safety tips for overhead crane operations :

1. Before use, ensure the crane is suitable for the planned hoisting task(s). Confirm it has appropriate travel, lift and capacity.
2. Visually and physically inspect the crane before use. Check for damage, wear and proper operation of all functions.
3. Confirm the load weight. Check the capacity of all equipment including the hardware, rope and slings. Do not exceed these capacities.
4. Select the right sling for each lift. Inspect slings and other rigging hardware before use for wear, stretch, or other damage. Do not use damaged or defective slings. Use softeners around sharp corners. Do not splice broken slings.
5. When communicating with a crane operator, use clear agreed-upon signals. Except for the stop signal, the crane operator should follow instructions from only one person—a designated signaler. Where a wired or remote controller is used, the operator should become familiar with all of its functions before lifting the load.
6. Warn all people in the load lift area before starting the lift. Ensure that the path of the load is clear of persons and obstructions. Do not lift loads over anyone. Centre the crane hoist over the load before hoisting to prevent swinging of the load.
7. Slide the sling fully onto the hoisting hook and ensure the safety catch is closed. Do not load the hook tip or hammer a sling into place.
8. Secure unused sling legs. Do not drag slings or leave loose materials on a load being hoisted.
9. Keep hands and fingers from being trapped when slack is taken out of a sling. Step away before the lift is made. Move the load and controls smoothly. Minimize load swing.
10. Walk ahead of the load during travel and warn people to keep clear. Use a tag line to prevent rotation or other uncontrolled motion. Raise the load only as high as necessary to clear objects. Do not ride on hook or load.
11. Set loads down on blocking, never directly on a sling. Do not pull or push loads out from under the hoist.
12. Do not leave the load (or the crane) unattended while the load is suspended.
13. Where crane operation by other personnel must be restricted, employ lockout and tagging procedures. Store slings off the floor in a clean, dry location on hooks or racks. Do not leave slings, accessories, or blocking lying on the floor.

Ensure that the load is free to move. If a load is stuck and the crane begins or continues to lift, it may reach its full capacity quickly. There may be little or no warning of this condition and rigging components may fail.

19.6 CONVEYOR SAFETY

In a recent year period, accidents related to miners were increased considerably mostly died in conveyor-belt accidents. The leading cause of these accidents was trying to remove material sticking to a roller or pulley while the belt was running. Consider the following :

- In these fatal accidents, belt speeds were typically in the range of 90 to 150 m per minute (rmpm). This means the belt was moving from 1.5 to over 2.4 m per second.
- For the average person, the time needed to react to an unexpected sensation is about one second.

Putting these facts together shows why working near a running conveyor belt is so dangerous. Even at 100 mpm, in that one second that the average person needs to react, the belt moves over the pulleys and rollers a distance of 1.5 m. That's far enough to draw your tool, your loose clothing, gloved or bare hand and arm, into a pinch point before you can react.

Safety tips to work with conveyor :

1. Don't take chances around a conveyor belt.
2. Don't perform work that will expose you to a pinch point, like clearing caked material from a roller, while a belt is in operation.

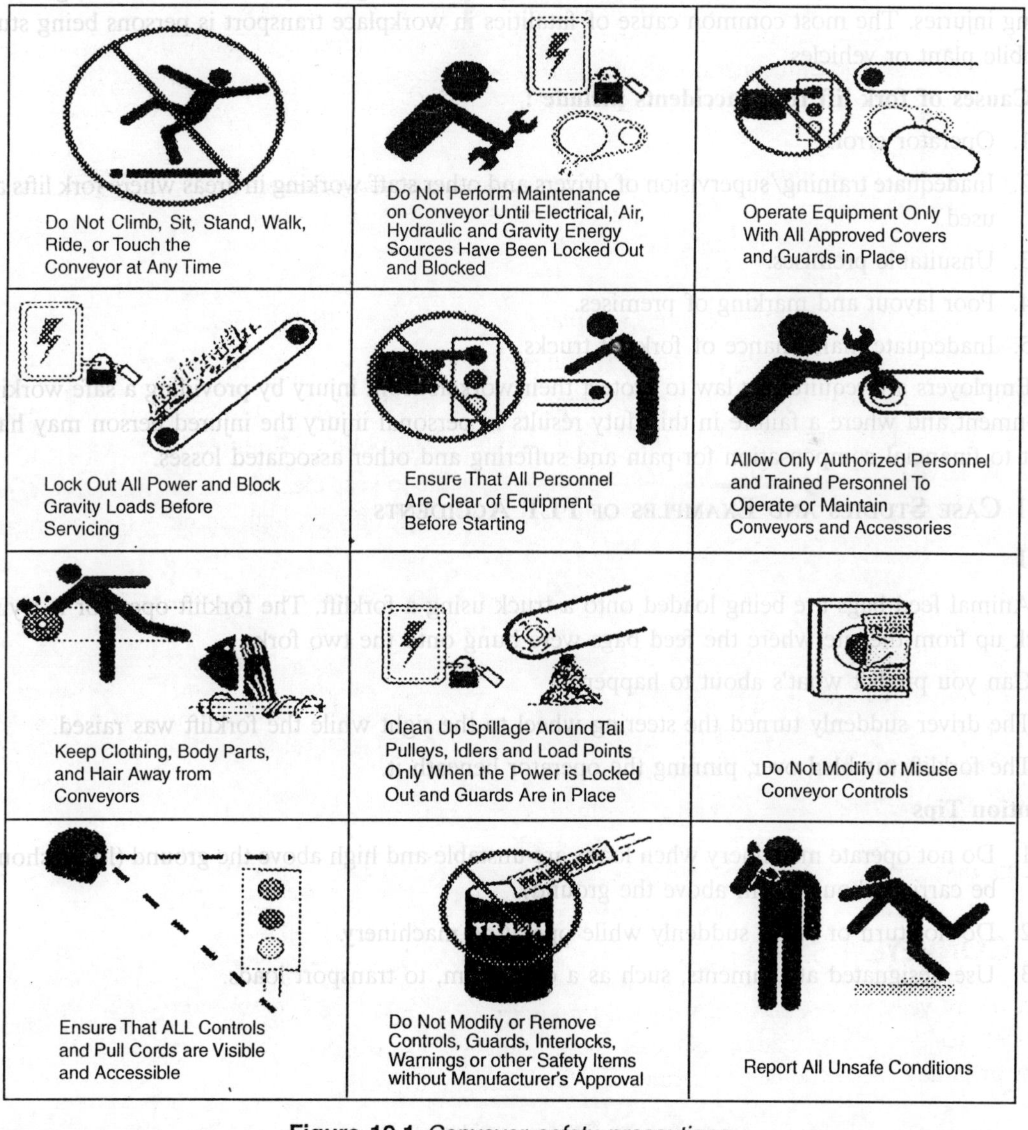

Figure 19.1 *Conveyor safety precautions.*

3. Only work on a conveyor that is locked and tagged out.
4. If you are working near a stopped belt, and a start-up warning is given, stop what you're doing and get clear !
5. Never cross a moving belt.
6. Always cross at a bridge or designated crossing point.

19.7 FORK LIFT TRUCK SAFETY

Fork lift trucks are potentially dangerous items of equipment if not used correctly and are the cause of several thousand industrial accidents each year.

Common types of accident are loads being dropped or pushed on to employees, collisions with static items knocking them on to other persons, employees being hit by trucks, overturning trucks, trapping injuries. The most common cause of fatalities in workplace transport is persons being stuck by mobile plant or vehicles.

Causes of fork lift truck accidents include :

1. Operator error.
2. Inadequate training/supervision of drivers and other staff working in areas where fork lifts are used.
3. Unsuitable premises.
4. Poor layout and marking of premises.
5. Inadequate maintenance of fork lift trucks.

Employers are required by law to protect their workers from injury by providing a safe working environment and where a failure in this duty results in personal injury the injured person may have a right to financial compensation for pain and suffering and other associated losses.

19.7.1 Case Studies and Examples of FLT Accidents

Case 1

Animal feed bags are being loaded onto a truck using a forklift. The forklift operator is trying to back up from the site where the feed bags were hung onto the two forks.

Can you predict what's about to happen ?

The driver suddenly turned the steering wheel to the right while the forklift was raised.

The forklift tumbled over, pinning the operator beneath it.

Prevention Tips

1. Do not operate machinery when loads are unstable and high above the ground (loads should be carried about 15 cm above the ground).
2. Do not turn or brake suddenly while operating machinery.
3. Use designated attachments, such as a crane arm, to transport loads.

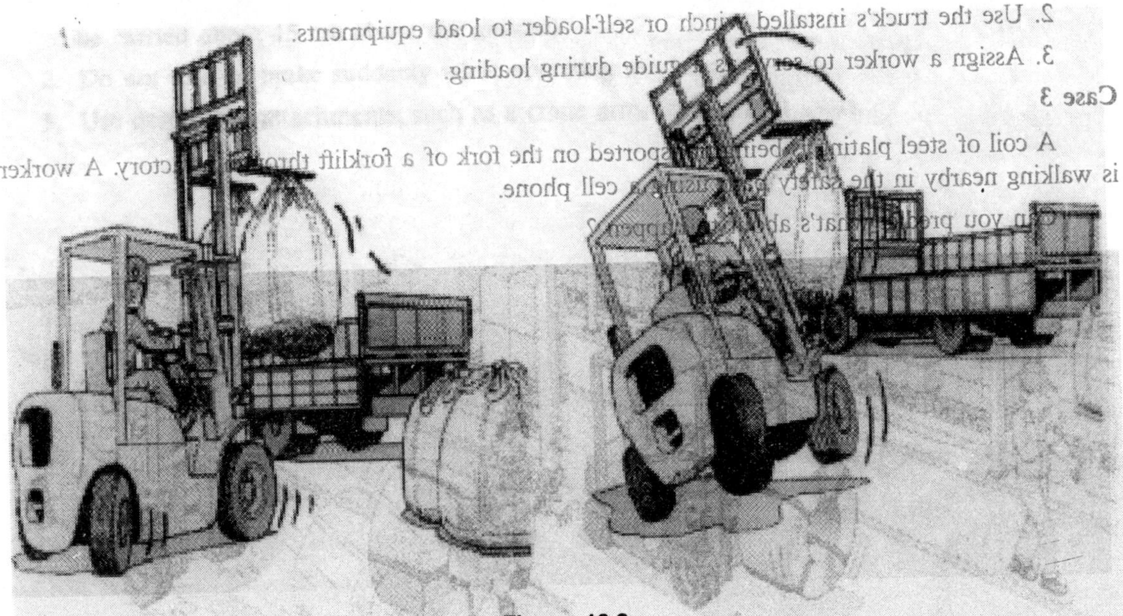

Figure 19.2

Case 2

A forklift operator is using two aluminium gangplanks to load a forklift onto the bed of a truck at a materials depot.

Can you predict what's about to happen ?

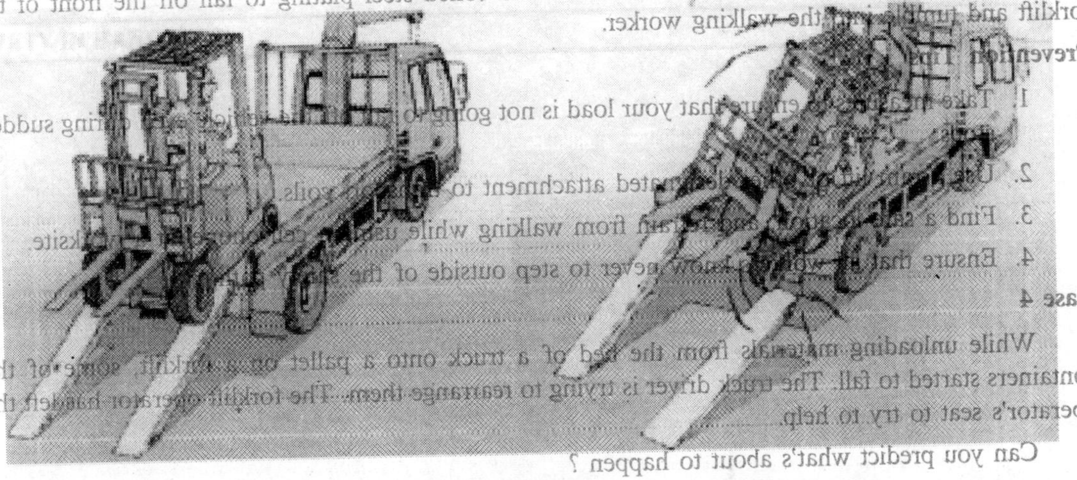

Figure 19.3

When the front wheels of the forklift were loaded onto the bed of the truck, one of the gangplanks slipped and the forklift suddenly tilted over. When the operator tried to get out, he tumbled over the side of the truck with the forklift and was pinned beneath it.

Prevention Tips

1. Gangplanks should be placed securely at an appropriate incline.

2. Use the truck's installed winch or self-loader to load equipments.
3. Assign a worker to serve as a guide during loading.

Case 3

A coil of steel plating is being transported on the fork of a forklift through a factory. A worker is walking nearby in the safety path using a cell phone.

Can you predict what's about to happen ?

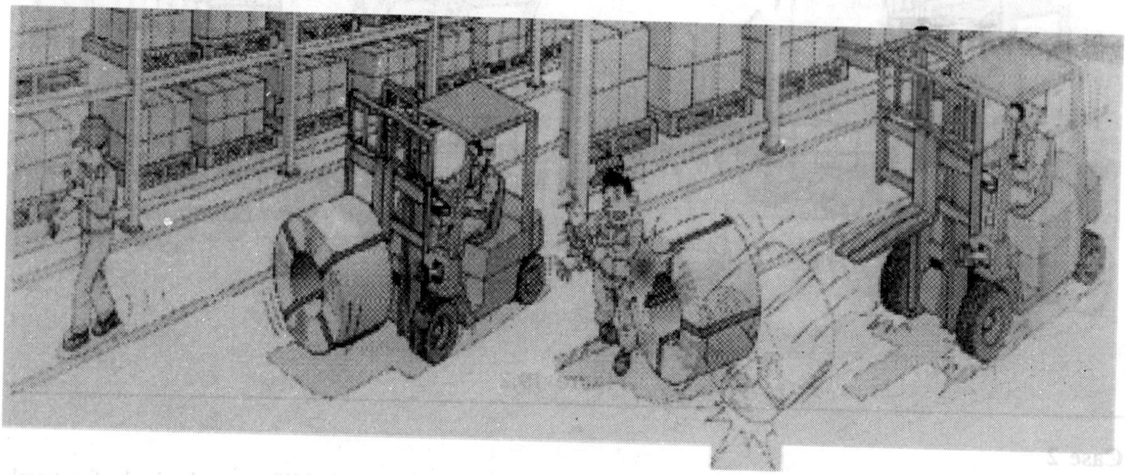

Figure 19.4

The forklift operator panicked and slammed on the brakes when the worker suddenly stepped outside the lines of the safety path. This caused the coiled steel plating to fall off the front of the forklift and tumble into the walking worker.

Prevention Tips

1. Take measures to ensure that your load is not going to fall off the vehicle even during sudden stops.
2. Use a ram lift or other designated attachment to transport coils.
3. Find a safe location, and refrain from walking while using a cell phone, at a worksite.
4. Ensure that all workers know never to step outside of the safety path.

Case 4

While unloading materials from the bed of a truck onto a pallet on a forklift, some of the containers started to fall. The truck driver is trying to rearrange them. The forklift operator has left the operator's seat to try to help.

Can you predict what's about to happen ?

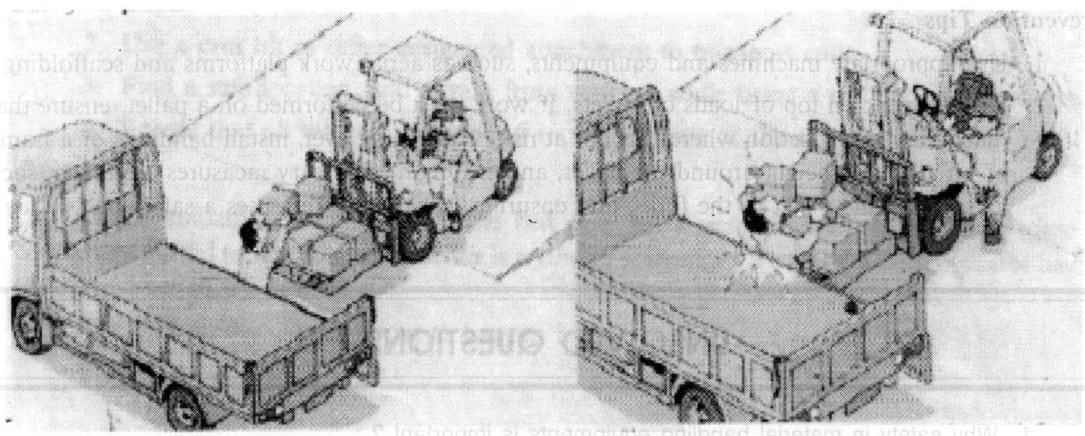

Figure 19.5

When the forklift operator left his seat to help adjust the load, the forklift moved forward, crushing the truck driver between the pallet and the truck.

Prevention Tips

1. Check the status of a load before loading and unloading.
2. When trying to adjust a load on a forklift, lower the load to the ground first.
3. When leaving the operator's seat, set the side brake, turn off the engine, and remove the key.
4. Do not stand in front of or behind a forklift.

Case 5

To replace a light bulb suspended from a warehouse ceiling, pallets have been piled onto the forks of a forklift and are being raised with a worker on top of them toward the light bulb.

Can you predict what's about to happen ? When the forklift operator moved the forklift slightly to adjust its position, the pallets became unstable and fell over, causing the worker to fall to the warehouse floor.

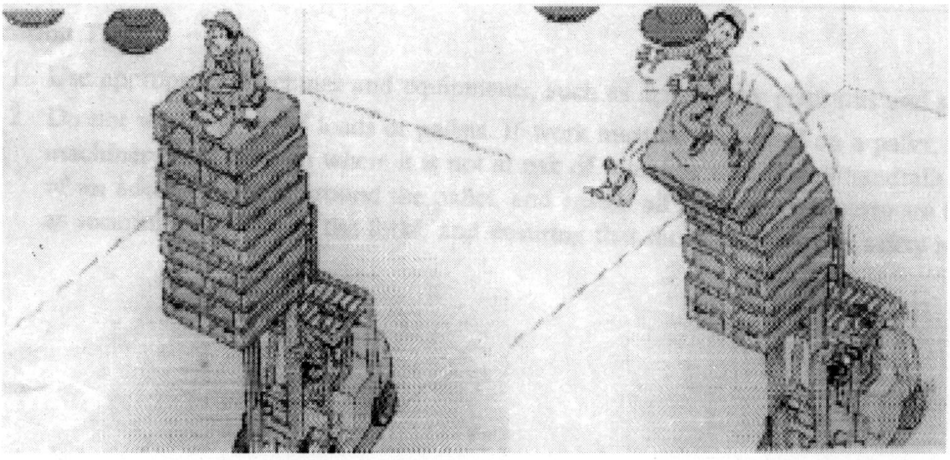

Figure 19.6

Prevention Tips

1. Use appropriate machines and equipments, such as aerial work platforms and scaffolding.
2. Do not work on top of loads or pallets. If work must be performed on a pallet, ensure that machinery in a location where it is not at risk of tumbling over, install handrails or a frame of an adequate height around the pallet, and ensure all necessary measures are taken, such as securing the pallet to the forks, and ensuring that the workers uses a safety belt.

UNSOLVED QUESTIONS

1. Why safety in material handling equipments is important ?
2. Explain how safety can be implemented in material handling equipments ?
3. Explain how operator can be trained for safety ?
4. List out safety norms for conveyor and cranes.
5. Explain reason or causes of accidents in case of cranes.
6. Explain reason or causes of accidents in case of conveyors.
7. Explain reason or causes of accidents in case of FLT.
8. Give few types of examples of crane failures.
9. Give few types of examples of conveyor failures.
10. Give few types of examples of truck failures.

❑❑❑❑

20 Design of Material Handling Equipments

20.1 INTRODUCTION

Generally, all the material handling equipments were designed keeping in the view material to be handled and application where is to be put in use.

20.2 DESIGN

Design always starts from the need which arises from the problems faced in day today life. In case of material handling it might be the problem of moving very heavy load, wet powder, handling of garbage or handling of ash in power plants and so many examples of given. The problem of speed, efficiency and economy are the other aspects as well. Taking into consideration problems of some kind design is made. In case of material handling the important factor which is considered in design are in general listed below :

1. Material to be handled (weight, geometry, shape and size, toxicity, physical state, flowability, hardness, grain size, etc.).
2. Speed of handling (slow, medium, fast).
3. Working Environment (temperature, humidity, soil, wind speed, dust, etc.).
4. Reliability and safety.
5. Plant layout.
6. Work area (stationery, mobile, portable, overhead, etc.)

20.3 DESIGN OF CRANES

Cranes are designed for lifting and lowering the load and moving with the same (mobile crane) or moving it (stationery cranes) with great reliability and high speed. Cranes capacity is generally given in terms of tones depending on maximum load which can be lifted. According the application this is calculated for a crane taking into consideration load characteristics. Different parts of it designed by different design considerations, *e.g.,* crane boom and girders, steel structures are designed for bending while crane hook is designed for combined bending and shear; wire ropes are designed for tensile strength, while gears and drives are designed for fatigue and compression. Design of cranes changes from work areas and environment, *e.g.,* tower crane design for urban areas and for rural areas will different as wind speed will be different as well as soil conditions will be different. Mobile crane working on hard surface and loose surface will have different design.

In cranes different attachments were used will depends on material characteristics. Different type of loading is considered such as live load, static load, wind load, fluctuating load, etc.

20.4 DESIGN OF CONVEYORS

Conveyors are designed for transporting (belt, slat, chain, etc.), lifting and lowering the load and moving with the same (elevators, inclined conveyors, mobile and portable conveyors) or moving it (stationery conveyors) with great reliability and high speed. Conveyors capacity is generally given in terms of tonnes/hr depending on load which can be handled per unit time. According the application this is calculated for a conveyors taking into consideration load characteristics. Different parts of it designed by different design considerations, *e.g.,* conveying body (actual load carrier) designed for bending while frames are designed for combined bending and shears; chains and wires are designed for tensile strength, while gears, sprockets and drives are designed for fatigue and compression. Design of conveyors changes from work areas and environment, *e.g.,* belt conveyors design for power plant and for cement plant will different as material characteristics and dust, toxicity will be different as well as layout and distance by which material is to be conveyed will be different.

In conveyors different attachments (buckets or slats) were used will depends on material characteristics. Different type of loading is considered such as live load, static load, wind load, fluctuating load, etc.

20.5 DESIGN OF TRUCKS

Trucks are designed for transporting, lifting and lowering, storing the load and moving with the same (all types of trucks, dumpers) with great reliability and high speed. Trucks capacity is generally given in terms of tonnes/hr. depending on load which can be handled per unit time. According the application this is calculated for a Trucks taking into consideration load characteristics. Different parts of it designed by different design considerations, *e.g.,* trucks body, frame designed for bending and compression while forks and attachments are designed for combined bending and shears; chains and wires are designed for tensile strength, while gears, sprockets and drives are designed for fatigue and compression. Design of Trucks changes from work areas and environment, *e.g.,* fork lift truck design for power plant and for cement plant will different as material characteristics and dust, toxicity will be different as well as layout and distance by which material is to be transported will be different.

Distinguishing characteristics of truck design is operator's cabin design for front and rear vision. Also aisle width requirement and work area design. In case of AGV's special flooring is designed and circuits were installed. In case of stacker and order picker they are designed for maximum reach.

In Trucks different attachments were used will depends on material characteristics. Different type of loading is considered such as live load, static load, wind load, fluctuating load, etc.

20.6 DESIGN STANDARDS

In design of material handling equipments certain standards are followed that are listed in appendix 'B'. These norms basically deal with approaches to be followed in design and the dimensions of individual parts. It also deals with safety requirements and work place design.

UNSOLVED QUESTIONS

1. Explain the design methodology followed for different material handling equipments.
2. Explain in brief design of cranes.
3. Explain in brief design of conveyors.
4. Explain in brief design of trucks.

❑❑❑❑

APPENDIX 1

Indian Standards for Cranes and Hoists

Document Number	Standard Title	Status
IS 807 : 1976	Code of practice for design, manufacture, erection and testing (structural portion) of cranes and hoists	Active
IS 3177 : 1999	Code of Practice for Electric Overhead Travelling Cranes and Gantry Cranes other than Steel Work Cranes	Active
IS 4137 : 1985	Code of practice for heavy duty electric overhead travelling cranes including special service machines for use in steel work	Active
IS 4410 : Part 23 : 1999	Glossary of Terms Relating to River Valley Projects —Part 23 : Hoists, Cranes and Other Related Terms	Active
IS 4573 : 1982	Specification for Power Driven Mobile Cranes	Active
IS 4594 : 1968	Code of practice for design of portal and semiportal wharf cranes (electrical)	Active
IS 6521 : Part 1 : 1972	Code of practice for design of tower cranes : Part 1 Static and rail mounted	Active
IS 13367 : Part 1 : 1992	Safe use of cranes—Code of Practice, Part 1 General	Active
IS 13473 : Part 1 : 1992	Cranes—Vocabulary, Part 1 General	Active
IS 13473 : Part 2 : 1992	Cranes—Vocabulary, Part 2 Mobile Cranes	Active
IS 13473 : Part 3 : 1993	Cranes—Vocabulary, Part 3 Tower Cranes	Active
IS 13558 : Part 1 : 1992	Cranes—Controls—Layout and Characteristics : Part 1 General Principles	Active
IS 13558 : Part 2 : 1995	Cranes—Controls—Layout and Characteristics : Part 2 Mobile Cranes	Active
IS 13558 : Part 3 : 1995	Cranes—Controls—Layout and Characteristics : Part 3 Tower Cranes	Active

IS 13558 : Part 4 : 1993	Cranes—Controls—Layout and Characteristics : Part 4 Jib Cranes	Active
IS 13558 : Part 5 : 1993	Cranes—Controls—Layout and Characteristics : Part 5 Overhead travelling cranes and Postal bridge cranes	Active
IS 13559 : Part 1 : 1993	Cranes—Information to be provided, Part 1 General	Active
IS 13559 : Part 4 : 1992	Cranes—Information to be provided, Part 4 Jib Cranes	Active
IS 13572 : Part 1 : 1992	Cranes—Crane driving manual : Part 1 General	Active
IS 13583 : Part 1 : 1993	Cranes—Training of drivers : Part 1 General	Active
Is 13834 : Part 1 : 1994	Cranes—Classification : Part 1 General	Active
IS 13834 : Part 2 : 1993	Cranes—Classification : Part 2 Mobile Cranes	Active
IS 13834 : Part 3 : 1995	Cranes—Classification : Part 3 Tower Cranes	Active
IS 13834 : Part 4 : 1993	Cranes—Classification : Part 4 Jib Cranes	Active
IS 13834 : Part 5 : 1993	Cranes—Classification : Part 5 Overhead travelling and portal bridge cranes	Active
IS 13870 : Part 1 : 1993	Cranes and lifting appliances—Selection of wire ropes : Part 1 General	Active
IS 13870 : Part 2 : 1993	Cranes and lifting appliances—Selection of wire ropes : Part 2 Mobile cranes—Coefficient of utilization	Active
IS 14467 : 1997	Cranes—Wind load assessment	Active
IS 14468 : 1997	Cranes other than mobile and floating cranes—General requirements for stability	Active
IS 14469 : 1997	Mobile cranes—Determination of stability	Active
IS 14470 : 1997	Cranes—Test code and procedures	Active
IS 14471 : 1997	Cranes and lifting appliances—Technical characteristics and acceptance documents	Active
IS 14472 : Part 5 : 1997	Cranes—Information to be provided : Part 5 travelling cranes and portal bridge	Active
IS 14473 : Part 1 : 1997	Cranes—Inspections : Part 1 General	Active
IS 14474 : Part 1 : 1997	Mobile Cranes—Experimental Determination of Crane Performance : Part 1 Tipping Loads and Radii	Active

IS 14475 : Part 1 ; 1997	Cranes—Condition monitoring : Part 1 General	Active
IS 15419 : 2004	Jib Cranes—Code of practice	Active
SP 38 (S and T) : 1987	Handbook of Typified Designs for Structures with Steel Roof Trusses (with and without Cranes) (Based on is Codes)	Active
SP 43 (S and T) : 1987	Handbook on Structures with Reinforced Concrete Portal Frames (Without Cranes)	Active
SP 47 (S and T) : 1988	Handbook on Structures with Steel Lattice Portal Frames (Without Cranes)	Active
IS/ISO 8686-1 : 1989	Cranes—Design Principles for Loads and Load Combination	Active
IS/ISO 8686-5 : 1992	Cranes—Design Principles for Loads and Load Combinations—Part 5 : Overhead Travelling and Protal Bridge Cranes	Active
IS/ISO 9373 : 1989	Cranes and Related Equipment—Accuracy Requirements for Measuring Parameters During Testing	Active
IS 8730 : 1997	Classification and codification of bulk materials for continuous material handling equipments	Active

APPENDIX 2

Indian Standards

CODE OF RECOMMENDED PRACTICE FOR CONVEYOR SAFETY

PART I : GENERAL INFORMATION

1. **Scope.** Covers the general information regarding the recommended practice to be adopted in the safe use of conveyors and conveying machinery used for transportation of bulk materials or unit loads.

   1.1. The conveyors covered in this standard may be of temporary or permanent nature.

   1.2. This code does not apply to industrial trucks, tractors or trailers, tiering machines, cranes, hoists, skip hoists, monorails, power and hand shovels, power scoops, bucket drag lines, platform elevators designed to carry passengers or the elevator operator, moving stairways and highway or rail vehicles, or conveyors specifically designed for the conveyors of passengers.

2. **Terminology**

   2.1. For the purpose of this code, the following definition, in addition to those given in IS : 42–40 shall apply.

   2.2. **Conveyor System.** An installation of two or more conveyors which transport or transfer loads from one conveyor to another and whose control is integrated.

   2.3. **Normal Working Conditions.** Specified conditions as agreed by the interested parties for the use of continuous mechanical handling equipment.

   2.3.1 These normal working conditions will either be defined by the user and agreed with the manufacturer at the time of passing the order, or defined for standard equipment in the manufacturer's catalogue when it is issued.

   2.4. **Personnel.** Persons essential to the use (the operation) of the continuous mechanical handling equipment itself and persons legitimately in the vicinity.

3. **Materials**

   3.1. Unless otherwise specified all materials used in the construction and operation of a conveyor and its supporting structure shall comply with the requirements of relevant Indian Standards where available. A list of such Indian Standards for guidance is given in Appendix A. Materials equivalent to or superior to those specified may be used subject to agreement between the manufacturer and the user.

4. **General Information**

   4.1. **Stages.** The safety requirements of the conveyors of various types may be divided into three stages as follows :

   (*a*) Construction stage (design and manufacture);

   (*b*) Installation stage (layout, erection and entry into service); and

   (*c*) Utilization stage (operation and maintenance).

4.1.1. The safety requirements during construction stage shall be those to be taken care of during the design and manufacture of the conveyor system equipments and machinery.

4.1.2. The safety requirements during installation stage shall be those to be taken care of during layout, erection and entry into service of the conveyor system, equipments and machinery.

4.1.3. The safety requirements during utilization stage shall be those to be taken care of during operation and maintenance of the conveyor systems, equipments and machinery.

4.2. **Lighting.** Suitable lighting of conveyor systems, equipments and machinery shall be provided. For guidance recommendations of IS : 3646 'Code of practice for interior illumination' in following parts may be referred :

IS : 3646 (Part 1)	Principles of good lighting and aspects of design.
IS : 3646 (Part 2)	Schedule for values of illumination and glare index.
IS : 3646 (Part 3)	Calculation of coefficients of utilization by the BZ method.

4.2.1. Permanent lighting shall be installed in tunnels and pits with a switch located convenient to the entrance(s).

4.2.2. Electric outlets for handlamps shall be provided at convenient locations.

4.2.3. The lamps/lights shall be cleaned regularly.

4.3. **Electrical Equipment.** Electrical equipment shall be installed and maintained to meet the statutory requirements. For guidance following Indian Standards may be referred :

(*a*) IS : 4051	Code of practice for installation and maintenance of electrical equipment in mines.
(*b*) IS : 732	Code of practice for electrical wiring installations.
IS : 732 (Part 1)	Definitions and general requirements *(second revision).*
IS : 732 (Part 2)	Design and construction *(Second revision).*
IS : 732 (Part 3)	Inspection and testing of installations *(second revision).*
(*c*) IS : 5571	Guide for selection of electrical equipment for hazardous areas *(first revision).*
(*d*) IS : 5572 (Part 1)	Classification of hazardous areas (other than mines) for electrical installations: Part 1 : Areas having flammable gases *(first revision).*
(*e*) IS : 8239	Classification of maximum surface temperatures of electrical equipment for use in explosive atmospheres.

4.4. **Fire Protection.** Adequate fire protection facilities shall be provided and shall be to the approval of the statutory authority. For guidance following Indian Standards may be referred :

(*a*) IS : 1642	Code of practice for fire safety of buildings (general) : Materials and details of construction.
(*b*) IS : 1641	Code of practice for fire safety of buildings (general) : General principles and fire grading.
(*c*) IS : 1643	Code of practice for fire safety of buildings (general) : Exposure hazard.

(*d*) IS : 1644 Code of practice for fire safety of buildings (general) : Personal hazard.

(*e*) IS : 3034 Code of practice for fire safety of industrial buildings : Electrical generating and distributing stations *(first revision).*

4.5. **Dust and Fumes.** Where dust or fumes may become a hazard, means shall be taken to prevent injury to persons. The design of the system shall be such as to prevent or control the emission of dust and fumes.

4.6. **Static Charge/Electricity.** It is known that continuous rotation and friction of non-metallic components in a conveyor system produces a lot of static electricity. For normal run-off mine handling plants for coal, iron, limestone, etc., the accumulated static charge may not pose much of a problem. However, for special application of conveyors, which carry explosive dusts and compounds, limiting the accumulated static charge/electricity in a conveying system is extremely important.

4.6.1. Where accumulated static charge/electricity is considered important, the limit of such a charge shall be as agreed to between the user and the manufacturer.

4.7. List of Indian Standards on conveyors, vertical hoists, bulk elevators and bulk handling equipment is given in Appendix B for information.

APPENDIX A
(Clause 3.1)

INDIAN STANDARD SPECIFICATION ON MATERIALS

A–1. Structural Steel

IS : 226	Structural steel (standard quality) *(fifth revision).*
IS : 808	Rolled steel beam, channel and angle sections *(revised).*
IS : 811	Cold formed light guage structural steel section *(revised).*
IS : 961	Structural steel (high tensile) *(second revision).*
IS : 1161	Steel tubes for structural purposes *(third revision).*
IS : 2239 (Part 1)	Mild steel tubes *(fourth revision).*
IS : 1239 (Part 2)	Mild steel tubulars and other wrought steel pipe fittings *(third revision).*
IS : 1252	Rolled steel sections bulb angle.
IS : 2062	Structural steel (fusion welding quality) *(second revision).*
IS : 3039	Structural steels for construction of hulls of ships *(first revision).*
IS : 3443	Crane rail sections *(first revision).*
IS : 3601	Steel tubes for mechanical and general engineering purposes.
IS : 3954	Hot rolled steel channel sections for general engineering purposes.
IS : 4923	Hollow steel sections for structural use.

A–2. Forgings and Castings

IS : 210	Grey iron castings *(third revision).*
IS : 1030	Carbon steel castings for general engineering purposes *(third revision).*

IS : 1865	Iron castings with spheroidal or nodular graphite *(second revision).*
IS : 2004	Carbon steel forgings for general engineering purposes *(second revision).*
IS : 2107	Whiteheart malleable iron castings *(first revision).*
IS : 4367	Alloy and tools steel forgings for general industrial use.
IS : 4771	Abrasion-resistant iron castings *(first revision).*
IS : 4896	One-percent chromium steel castings for resistance to abrasion *(first revision).*

A–3. Rivets

IS : 1929	Hot forged steel rivets for hot closing (12 to 36 mm diameter) *(first revision).*
IS : 2998	Cold forged steel rivets for cold closing (1 to 16 mm diameter) *(first revision).*

A–4. Bolts

IS : 1363	Hexagon head bolts, screws and nuts of product grade C.
IS : 1364	Hexagon head bolts, screws and nuts of product grade A and B.
IS : 2585	Black square bolts and nuts (diameter range 6 to 39 mm) and black square screws (diameter range 6 to 24 mm) *(first revision).*
IS : 3138	Hexagonal bolts and nuts.

A–5. Conveyor Belts

IS : 1891 (Part I)	General purpose belting *(second revision).*
IS : 1891 (Part II)	Heat-resistant belting *(first revision).*
IS : 1891 (Part III)	Oil-resistant belting *(first revision).*
IS : 1891 (Part IV)	Hydienic belting.
IS : 3181	Fire resistant conveyor belting for underground use in coal mines *(first revision).*

A–6. Chain

IS : 2403	Transmission steel roller chains and chain wheels *(first revision).*
IS : 3542	Extended pitch transmission precision roller chains and chain wheels *(first revision).*
IS : 3560	Short pitch transmission precision bush chains and chain wheels *(first revision).*
IS : 3948	Calibrated high tensile steel chain (round link) for chain conveyors and coal ploughs used in mines.
IS : 6834 (Part I)	Chains.

A–7. Gears

IS : 2535	Basic rack and modules of cylindrical gears for general engineering and heavy engineering *(second revision).*
IS : 3734	Dimensions for worm gearing *(first revision).*
Is : 5037	Basic rack and modules of straight bevel gears for general engineering and heavy engineering.

A–8. Wire Ropes

IS : 1856	Steel wire ropes for haulage purposes *(second revision).*
IS : 2266	Steel wire ropes for general engineering purposes *(second revision).*

A–9. Concrete

IS : 456	Code of practice for plain and reinforced concrete *(third revision).*

NOTE : Updated information shall be available in the latest available ISI Handbook.

APPENDIX B

(Clause 4.7)

LIST OF INDIAN STANDARDS ON CONVEYORS, HOISTS AND RELATED MATERIAL HANDLING EQUIPMENT

1. IS : 3938	Electric wire rope hoists *(second revision).*
2. IS : 3948	Calibrated high tensile steel chain (round link) for chain conveyors and coal plough used in mine.
3. IS : 4240	Glossary of conveyor terms and definitions.
4. IS : 4776 (Part 1)	Troughed belt conveyors : Part 1 : Surface installations *(first revision).*
5. IS : 4776 (Part 2)	Troughed belt conveyors : Part 2 : Troughed belt conveyors for underground installation.
6. IS : 5228	Code of practice for the construction of continuous movement monocable ropeways with automotive grips intended for transportation of passengers.
7. IS : 5229	Code of practice for the construction of continuous movement monocable ropeways transportation of passengers.
8. IS : 5230	Code of practice for the construction of continuous to-and-fro movement bicable ropeways intended for the transportation of passengers.
9. IS : 5563	Screw conveyors for industrial use.
10. IS : 5895	Steel roller conveyors.
11. IS : 5972	Glossary of pneumatic and hydraulic conveyor terms and definitions.
12. IS : 6832	Fixing screws and washers for buckets for bucket elevators.
13. IS : 6833	Buckets for bucket elevators.
14. IS : 6834 (Part 1)	Conveyor chain, chain wheels and attachment : Part 1 Chains.
15. IS : 6834 (Part 2)	Conveyor chain, chain wheels and attachment : Part 2 : Chainwheels.
16. IS : 6834 (Part 3)	Conveyor chain, chain wheels and attachment : Part 3 : Attachments.
17. IS : 6930	Dimensions for fixing arrangements of buckets for bucket elevators.
18. IS : 7054	Casing for bucket elevators.

19. IS : 7155	Code of recommended practice for conveyor safety.
20. IS : 7167	Code for selection and use of bucket elevators.
21. IS : 7423	Dimensions for apron conveyors.
22. IS : 7439	Dimensions for aeroslides.
23. IS : 7465	Portable and mobile troughed belt conveyors.
24. IS : 7649	Glossary of terms used in connection with aerial ropeways.
25. IS : 8531	Pulleys for belt conveyors.
26. IS : 8597	Flat belt conveyors.
27. IS : 8598	Idlers and idler sets for belt conveyors.
28. IS : 8599	Recommendations for selection of apron conveyors.
29. IS : 8646	Scrapper (flight) conveyors.
30. IS : 8647	Design criteria for pneumatic conveying system.
31. IS : 8723	Dimensions for vibrating conveyors and feeders with rectangular or trapezodial trough.
32. IS : 8730	Classification of bulk materials handled by continuous mechanical handling equipment.
33. IS : 9047 (Part 1)	Data sheet for aerial ropeways : Part 1 : Data to be supplied by intending purchaser for aerial system for transportation of goods.
34. IS : 9047 (Part 2)	Data sheet for aerial ropeways and cableways : Part 2 : Data to be supplied by intending purchaser for aerial system for transportation of passengers, surface.
35. IS : 9047 (Part 3)	Data sheet for aerial ropeways and cableways : Part 3 : Data to be supplied by intending purchaser for aerial system for transportation of passenger, under ground.
36. IS : 9047 (Part 4)	Data sheet for aerial ropeways and cableways : Part 4 : Data to be supplied by intending purchaser for aerial system for transportation of goods, under ground.
37. IS : 9047 (Part 6)	Data sheet for aerial ropeways and cableways : Part 6 : Data to be supplied by intending purchaser for aerial system for transportation of forestry products.
38. IS : 9228	Mono seat chairs for chairlifts.
39. IS : 9413	Rope guide rollers for haulage ropes for bicable aerial ropeways for transportation of goods.
40. IS : 9454	Bicable load carrier for aerial ropeways.
41. IS : 9706	Code of practice for the construction of aerial ropeways for the transportation of material.
42. IS : 9776	Data sheet for selection of weighing-cum-tipping/bag tilling machines for bulk handling equipment.
43. IS : 9777	Data sheet for weigh bridges for bulk handling equipment.
44. IS : 10095	Code of practice for design, construction, installation, operation and maintenance of tipplers for railways wagons.
45. IS : 10280	Guidelines for format and contents of manuals for earth moving equipment.

46. IS : 10303 Code of practice for preservation and storage of earth moving equipment.
47. IS : 10443 Data sheet for selection of marine loading arms.
48. IS : 10463 (Part 6) Glossary of terms for bulk handling equipment : Part 6 : Cyclic loads bulk handling equipment (non-stationary).
49. IS : 10464 Data sheet for selection of wagon puller/pusher.
50. IS : 10678 Data sheet for selection of vacuum producer.
51. IS : 10689 Minimum dimensions for access opening on earth moving machinery.
52. IS : 10704 Rules for design of access system for earth moving machinery.
53. IS : 10856 Volumetric ratings of elevating scrappers.
54. IS : 10859 Volumetric ratings for dumper bodies.

NOTE : (1) ISO 1819—Continuous Mechanical Handling Equipment–Safety Code.
(2) ISO 7149—Continuous Handling Euipment–Safety Code.

In preparation of above standards help of these two publications were taken when revision has been carried out.

LIST OF INDIAN STANDARDS

- **Belt Conveyor :**
  - Portable and mobile IS 7465
  - Safety code IS 7155
  - Selection and design IS 11592
  - Belt Fastener
  - Conveyor belting IS 12091
  - Belt Feeders selection practice IS 12215
- **Bins :**
  - Steel, for storage of bulk material criteria for design IS 9178
  - Reclaimer, Data sheet for collection IS 12663
  - Terminology is 10463
  - Wagon marshalling Equipment
- **Conveyors :**
  - Dimensions for IS 7423
  - Selection of IS 8599
  - Belt troughed, portable and mobile IS 7465
  - Belting fire resistant, for coal mines IS 3181
  - Rubber belting IS 1891
- **Conveyor Belting :**

  Selection Storage installation and maintenance IS 6687
- **Bulk Materials Classification :** IS 8730

TABLE : MATERIAL CLASSIFICATION CODE CHART

MAJOR CLASS	MATERIAL CHARACTERISTIC INCLUDED	CODE DESIGNATION
Density	Very Fine	A*
Size	Fine (–) 3 mm & Under	B
	Granular (+) 12 mm Under	C
	Granular (+) 12 mm But (–) 75 Mm	D
	Lumpy (+) 75 mm	DX
	Irregular, Stringy, Interlocking, etc.	E
	Very Free Flowing, Angle of Repose < 20°	1
Flowability	Free Flowing > 20° < 30°	2
	Average Flowing 30° < 45°	3
	Sluggish > 45°	4
Hardness	Non Abrasive	5
	Mildly Abrasive	6
	Abrasive	7
	Very Abrasive or Sharp Edges	8
	Builds Up And Hardens	F
	Generates Static Electricity	G
	Decomposes-Deteriorates in Storage	H
	Flammability	J
	Becomes Plastic or Tends to Soften	K
	Very Dusty	L
	Aerates and Develops Fluid Characteristics	M
Miscellaneous Characteristics Or Hazards	Contains Explosive Dust	N
	Stickiness-Adhesion	O
	Contaminable, Affecting Use	P
	Degradable, Affecting Use/Sale	Q
	Gives of Harmful or Toxic Gas/Fume	R
	Highly Corrosive	S

APPENDIX 3

UNIVERSITY PRACTICE PAPERS

(MITS, Gwalior)

B.E. EXAMINATION, Dec. 2004

(Seventh Semester)

MECHANICAL ENGINEERING

715

MATERIAL HANDLING

Time : Three Hours *Maximum Marks : 100* *Minimum Pass Marks : 35*

Note : *Attempt any five questions.*

1. (*a*) Explain how good plant layout minimizes handling problems. 8

(*b*) Explain different cost factors considered while selecting any material handling equipment. 8

(*c*) Define material handling. State its importance. 4

2. (*a*) Classify unrestricted equipments. Explain tractors and trolley. 8

(*b*) What is a function of side loader ? Explain. 4

(*c*) Explain with neat sketches counter balanced trucks. 8

3. (*a*) Classify conveyors. Write in detail about screw conveyor. 10

(*b*) Differentiate with sketches between belt and roller conveyor. 10

4. (*a*) Why is there a need of sorting installation ? What are different methods of sorting in industries ? 10

(*b*) Explain automatically guided vehicles in detail. 10

5. (*a*) Explain with neat sketch any two types of mobile cranes after classifying them. 10

(*b*) Write about gantry crane and its types, applications, advantages and disadvantages. 10

6. (*a*) Explain good reception equipment minimizes material handling. 5

(*b*) Write about slat conveyors. 5

(*c*) Explain in detail lifts. 10

7. (*a*) Explain various types of maintenance policies adopted for maintenance of material handling equipments. 10

(*b*) Why is there a need of pallets and containers ? Explain different types of pallets. 10

8. Write short notes on any four of the following : $4 \times 5 = 20$

(*a*) Plant layouts (*b*) Carts and trolleys

(*c*) Chutes (*d*) Jib crane

(*e*) Order pickers (*f*) Warehouse.

B.E. (Fourth Year) EXAMINATION, Dec. 2004

MATERIAL HANDLING

715

Time : Three Hours *Maximum Marks : 100* *Minimum Pass Marks : 35*

***Note :** Attempt any five questions. All questions carry equal marks.*

1. (*a*) Explain factors affecting selection of material handling equipments. 8

(*b*) What are the principles of material handling ? Explain. 8

(*c*) Define :

Unit load, Bulk material, Batched goods and material handling. 4

2. (*a*) Explain what is meant by counter-balancing of trucks ? Explain its types. 10

(*b*) Explain with neat sketch function of straddle carrier. 5

(*c*) Explain concept of air cushion handling frames. 5

3. (*a*) Explain in detail about belt conveyors. 10

(*b*) Write about slat conveyor. 5

(*c*) Explain with neat sketch roller conveyors. 5

4. (*a*) What is meant by AGV ? Explain in detail. 10

(*b*) Write about trolley and tractor. 5

(c) Explain sorting installations. 5

5. (*a*) Explain with neat sketch different types of important cranes. 10

(*b*) Explain concept of palletisation and containerisation. 5

(*c*) Explain one example of "Goods Reception equipment". 5

6. (*a*) Explain economics of material handling. 10

(b) What is meant by "Side Loader" ? Give applications. 5

(c) What is the use of chain conveyors ? Explain. 5

7. (*a*) Explain different types of maintenance policies adopted for maintenance of material handling equipments. 10

(*b*) Explain principle of working of lifts, classify lifts. 5

(*c*) Write about carts and trolleys. 5

8. Write short notes on any four of the following : $4 \times 5 = 20$

(*a*) Order pickers

(*b*) Warehouse layouts

(*c*) Elevators

(*d*) Relation between layout and material handling.

(*e*) Importance of material handling.

(*f*) Mobile lifting frames.

APPENDIX 4

List of Manufactures in India

1. Belts

1. **ABRAZO INDUSTRIES**
Abdulla Mansions, 32 Mirza St.,
Mumbai–400003.

2. **DARSH ENGINEERING COMPANY**
5-148, Lal Temple Road, Ranigunj,
Secunderabad–500003.

3. **MIK RUBBER PRODUCTS PVT. LTD.**
Ke Road, Bhandup, Mumbai–400078.

4. **ANDREW YULE & CO. LTD.**
Belting Division, 8, Dr. Rajendra Prasad,
Saranl, Kolkata–700001.

5. **ASHA CASTINGS**
Plot O.28, IDA Nacharam,
Hyderabad–501507.

6. **ASHA LATEX & ALLIED INDS. PVT. LTD**
LBS Marg, Nandawan Indl. Est,
Mulund (W), Mumbai–400080.

7. **ASHUTOSH TEXMACH PVT. LTD.**
707, Hamkool Complex, Opp. Capital
Comm. Centre, Ashram Road,
Ahmedabad–380009.

8. **ASIAN MARKETING & MFG CO.**
97 B Way, Chennai–600018.

9. **ASIAN RUBBER & PLASTIC INDS.**
603, Viman Darshan Sahar Road,
Andheri (E), Mumbai–400069.

10. **AVALANI ENTERPRISES**
23, Strand Road, Kolkata–700001.

11. **IF BELTINGS PVT. LTD.**
34, LLT Nagar,
Hyderabad–500004.

12. **M.P. BELTINGS LTD.**
32 Phase IV, IDA Patancheru,
Hyderabad–502319.

13. **BAJAJ AUTO LTD.**
Akurdi, Pune–411045.

14. **BALAJI ENGG. EQUIPMENTS**
33, New Avadi Road, Chennai–600010.

15. **BELTING ENTERPRISES**
A/10, Chand Society, Juhu Curch Road,
Juhu, Mumbai–400079.

16. **BENGAL BELTING COMPANY**
83/85, N.S. Road, Shop No. ¾,
Kolkata–700001.

17. **BOMABY CO. PVT. LTD.**
34, A to Z Idl Est, Lower Parel,
Mumbai–400013.

18. **BOMBAY CO. PVT. LTD.**
Wallace St., Bombay–400001.

2. Conveyors

19. **A.G. ENGINEERS**
33A Ajmeri Udayog Nagar, O/S, Shahpur
Gate, Ahmedabad–380001.

20. **A.G. ENGINEERS**
104/3 Shri Ram Market, D Pura,
G.T. Road, Delhi–110001.

21. **A.G. ENGINEERING SERVICES**
Pune.

22. A.C.C. VICKERS BABCOCK LTD.
Shahbad, Karnataka.

23. CIKRWAL ENGINEERING CO.
Stephen House, B.B.D. Bagh.

24. AJAY ENGINEERING PVT. LTD.
148/25 Indl. Suburb, Ypt,
Bangalore–560022.

25. ALFA ENGINEERING CO. LTD.
Station Road, Bhandup, Mumbai–400078.

26. ALUMINIUM INDUSTRIES LTD.
Material Handling Division 'Alind House',
5.9.92, Chapel Road, Hyderabad–500001.

27. AMALGAMATED ENGINEERING PVT. LTD.
004, 2ND Block 21st cross,
Rajaji Nagar, Bangalore–560010.

28. BOMBAY STAR ENGINEERING
Errabalu Chetty Street,
Chennai–600001.

29. CHAMPION INDUSTRIES
A-40, Giriraj Est, Mahakali Road,
Andheri (E), Mumbai–400093.

30. CHELLAM ENGINEERING WORKS
39-B, South Baug Road, Chennai–600017.

31. CHEVI ENTERPRISES
A-211, B.G. Tower, O/s Delhi Gate,
Shahi Baug Road, Ahmedabad–380004.

32. COLOSEUM PRODUCTS PVT. LTD.
53 Appolo Indl. Est, Andheri (East),
Mumbai–400093.

33. COMMERCIAL ENGINEERING CORPN.
5-3-85, Sara Iron Market, Ranigunj,
Secunderabad–500003.

34. DECON SYSTEMS
10, Sargam, Paud Road, Pune 411038
Fac : W-254, J. Block, MIDC, Bhosari,
Pune–411026.

35. DHAWAN ASSOCIATES
660, Kapasia Bazart, Ahmedabad–380002.

3. Feeders

36. TANCE DYNAMICS
36, Kirti Nagar, New Delhi–110015.

37. EGIS ENGG. COMPANY PVT. LTD.
0.7, G.I.D.C. Ind., Estate, Vatva.

38. AUTO MEASUREMATIC LTD. PEBCO (I) LTD.
402–405, Swastik Chambers, Umarashi,
Bappa Chopwk, Chember,
Mumbai–400071.

39. BASP (INDIA)
71, Canning Street, Room A/512,
5th Floor, Kolkata–700001.

40. BEACON ELEVATOR CO.
52 Vandana, 11 Tolstoy Marg,
New Delhi–110001

41. BEEKEY ENGINEERING CORPORATION
45/47, Industrial Estate, Bhilali–490026

42. BHI SUNDER DAS & SONS CO. PVT. LTD.
21, Mile Stone, Faridabad–121007.

43. BUCKAU WOLF INDIA LTD.
Pimpri, Pune–411018.

4. Stackers and Reclaimers

44. BEEKEY ENGINEERING CORPORATION
45/47, Industrial Estate, Bhilai–490026,
(M.P.).

45. BUCKAU WOLF INDIA LTD.
Pimpri, Pune–411018.

46. CMC WORKS
P.B. No. 2001, 19/1 Paltform Road,
Seshadripuram, Bangalore–560020.

47. COFITO ENGG. CO.
J-315, S Block, Bhosari Indl. Est,
Pune–411026.

S.NO.	Industry's Name	Address	M.H. Equipments
48.	**Action Construction Equipment (P) Ltd.** **ACE** **ISO9001**	Marketing H.Q. Plot No. 7, 2nd Floor, Pocket-H Market, Sarita Vihar, N. Delhi-110044.	1. Hydraulic Mobile Pick-n-Move Crane. 2. Wheeled Loaders. 3. Tower Cranes. 4. Road Making Machinery.
49.	**Ingersoll-Rand (India) Ltd.** **Construction & mining**	Peenya Ind. Area, Peenya, Bangalore-560058 *Mfg. Plants :* Peenya (Bangalore), Naroda, Ahmedabad.	1. Crawlair drills, Blast Hole Drills for Mining. 2. Truck/Trailer mounted Water well drills, DHD Hammers and Bits. 3. Pile Drills, Anchor Drills. 4. Battery operated load haul Dumpers. 5. Compressors for Coal Mining.
50.	***Indef*** **Authorized Distributor : Mach Mill Engineers**	2-E/3, Jhandewalan Extension New Delhi-55.	1. Chain electric Hoists. 2. Chain Pulley Blocks. 3. Wire Rope Hoists. 4. EOT/HOT cranes.
51.	**Mahindra Stiller Auto Trucks Ltd.**	B-72, Sector 27-A, Mathura Road, Faridabad, Haryana.	1. Battery Operated Stacker. 2. Hydraulic Pallet Truck. 3. Mechanical Pallet Truck. 4. Hydrulic Elevating Table. 5. Golf Cart.
52.	**Automech Industries (P) Ltd**	E-2, Indl.Area, Ghaziabad–201001, (India).	1. EOT/HOT/JIB Cranes. 2. Electric Hoists. 3. Electric Winches.
53.	**Pilco Storage Systems (P) Ltd.**	60/19 1ST Floor, Old Rajinder Nagar, N. Delhi-110060.	1. Storage Racks. 2. Heavy duty Pallet Racks. 3. Slotted angle Racks. 4. Cable Trays. 5. Mezzanine Floor. 6. 2/3 Tier Structures.
54.	**Kelson Steel Products Private Ltd.**	22, Radhey Puri, Ext.-2, Near Main Jagat Puri Chowk, Delhi-51.	1. Slotted angle storage sys. 2. Palletise Racking system. 3. Cable Trays. 4. Mezzanine floor. 5. 2/3 Tier Structures. 6. Industrial Shed.

55.	**Greeves Limited**	Hd.Off. Power Transmission, Unit-1, Chinchwad, Pune.	1. Planetary Gear Boxes. 2. Geared Motors. 3. Worm Gear Boxes. 4. Bevel Helical Gear Box. 5. Light weight Motorized Worm gear Boxes.
56.	**Indital Construction Machinery Ltd. ICM**	10-B, Peenya Indl. Area, Bangalore-560058.	1. Dumpher. 2. Pick-n-carry Cranes. 3. Winches. 4. Forklifts. 5. Stackers.
57.	**Maini Materials Movement (P) Ltd.**	122 D & E Bommasandra Indusrial Estate, Bommasandra, Bangalore-562158.	Stackers, Forklifts, Narrow Aisles, Dock Levelers, Battery Operated Trucks.
58.	**Hydrolines**	497 C-4 Phase Peenya Indl. Area, Bangalore 560058.	Giant Manufacturers of Earth Moving Equipments
59.	**L & T Ltd.**	Laxminarayan Complex, Palace Road, Bangalore–560052	1. Hydraulic excavator 2. Wheel Loaders 3. Vibratory compactors
60.	**BEML-Bharat Earth Movers Ltd.**	BEML, SR Nagar, Bangalore-560027.	1. Hydraulic excavator 2. Wheel Loaders

❑❑❑❑

Index